Biogenic Amines on Food Safety

Biogenic Amines on Food Safety

Special Issue Editors

Claudia Ruiz-Capillas
Ana M. Herrero

MDPI • Basel • Beijing • Wuhan • Barcelona • Belgrade

MDPI

Special Issue Editors

Claudia Ruiz-Capillas
Department of Products (DPRD), Institute
of Food Science, Technology and Nutrition
(ICTAN) of the Spanish Science Research
Council (CSIC)
Spain

Ana M. Herrero
Department of Products (DPRD), Institute
of Food Science, Technology and Nutrition
(ICTAN) of the Spanish Science Research
Council (CSIC)
Spain

Editorial Office
MDPI
St. Alban-Anlage 66
4052 Basel, Switzerland

This is a reprint of articles from the Special Issue published online in the open access journal *Foods* (ISSN 2304-8158) from 2018 to 2019 (available at: https://www.mdpi.com/journal/foods/special_issues/Biogenic_Amines_Foods)

For citation purposes, cite each article independently as indicated on the article page online and as indicated below:

LastName, A.A.; LastName, B.B.; LastName, C.C. Article Title. *Journal Name* **Year**, *Article Number*, Page Range.

ISBN 978-3-03921-054-1 (Pbk)
ISBN 978-3-03921-055-8 (PDF)

Contents

About the Special Issue Editors

Claudia Ruiz-Capillas earned her Ph.D. degree in Veterinary Science from the Complutense University of Madrid in 1997. She is currently a senior research scientist at the Products Department of the Meat and Meat Products Science and Technology laboratory in the Institute of Food Science Technology and Nutrition (ICTAN-CSIC) of the Spanish Science Research Council (CSIC). Dr. Ruiz-Capillas has been working for several years on the production of biogenic amines and free amino acids in fish and meat products obtained under different manufacturing and storage conditions. In 2001 she was granted a Ramón y Cajal Contract (MCYT/01) to study "Actuating mechanisms of biogenic decarboxylase amine-producing enzymes in meat and meat products". She is currently conducting research on animal fat replacers in order to develop healthier meat products and enhance their quality. She has participated in and led a number of national and international research projects. She has more than 120 SCI scientific contributions in Food Science and Technology to her credit. She has co-authored various chapters in different books; she has also been a guest co-editor of several Special Issues of journals and the co-editor of a book.

Ana M. Herrero received her degree in chemistry from the Complutense University (UCM) in Spain and finished her Ph.D. in chemistry in 2004 at the same university. She is currently a scientific researcher at the Products Department of the Institute of Food Science, Technology, and Nutrition (ICTAN) of the Spanish National Research Council (CSIC). Her research focuses on both basic and applied science as well as technological applications designed to improve the quality of fish and meat and their products through studies of the implications of chemical, physical, and structural properties using vibrational spectroscopy techniques and applying traditional and emerging technologies to these foods. In recent years, her research has focused on developing meat product reformulation processes with the aim of obtaining healthier products and particularly on how these processes impact biogenic amine formation. Throughout her scientific career, she has participated in nationally and internationally funded projects and she is also involved in technology transfer through projects with various industrial sectors. Dr. Herrero has published over 70 peer-reviewed scientific papers and various book chapters in the area of meat science. She has also supervised Ph.D. theses and given lectures at seminars, courses, etc.

Preface to "Biogenic Amines on Food Safety"

The study of biogenic amines (BA) in food is important for safety and quality reasons. The consumption of foods with high concentrations of biogenic amines has been associated with health issues. Biogenic amines also play an important role as indicators of food quality and/or acceptability. This is especially important because biogenic amines are present in varying concentrations in a wide range of foods and their formation is influenced by different factors. The book consists of 11 chapters of original contributions and reviews aimed at gaining a better understanding of biogenic amines and their impact on food quality and safety. It addresses the presence of BA in different foods (fresh meat, milk, cheese, plants, fermented soybeans, etc.) and provides a deeper understanding of the factors affecting the formation of these compounds in relation to the raw material, amino acid levels, physicochemical characteristics, microorganisms, lactic acid bacteria, animal feed, etc. It reviews the analytical techniques used to determine biogenic amines (HPLC, GC, MS-MS, FIA, etc.) and the screening methods to evaluate amino acid decarboxylase activity. It also describes different tools that can be used to reduce biogenic amines (biocontrols, starters, etc.) and the impact of different processing (high hydrostatic pressure, etc.) and storage conditions (chilled, packaging, MAP, etc.). Moreover, this book reviews the biogenic amines derived from basic and aromatic amino acids involved in the pathophysiology of the gastrointestinal tract. Furthermore, it outlines legal limits and legislation concerning biogenic amines in food and beverages and the difficulties encountered in establishing BA toxicity ranges given the dual importance of these compounds (quality and health implications). Lastly, it reports on the problems of low-histamine diets of plant origin in relation to histamine intolerance (food histaminosis or food histamine sensitivity) and the difference of these issues in comparison to the more established histamine intoxication.

Claudia Ruiz-Capillas, Ana M. Herrero
Special Issue Editors

foods

MDPI

Review

Impact of Biogenic Amines on Food Quality and Safety

Claudia Ruiz-Capillas * and Ana M. Herrero

Department of Products, Institute of Food Science, Technology and Nutrition, ICTAN-CSIC, Ciudad Universitaria, 28040 Madrid, Spain; ana.herrero@ictan.csic.es
* Correspondence: claudia@ictan.csic.es; Tel.: +34-5492300; Fax: +34-91-549-36-27

Received: 20 December 2018; Accepted: 5 February 2019; Published: 8 February 2019

Abstract: Today, food safety and quality are some of the main concerns of consumer and health agencies around the world. Our current lifestyle and market globalization have led to an increase in the number of people affected by food poisoning. Foodborne illness and food poisoning have different origins (bacteria, virus, parasites, mold, contaminants, etc.), and some cases of food poisoning can be traced back to chemical and natural toxins. One of the toxins targeted by the Food and Drug Administration (FDA) and European Food Safety Authority (EFSA) is the biogenic amine histamine. Biogenic amines (BAs) in food constitute a potential public health concern due to their physiological and toxicological effects. The consumption of foods containing high concentrations of biogenic amines has been associated with health hazards. In recent years there has been an increase in the number of food poisoning cases associated with BAs in food, mainly in relation to histamines in fish. We need to gain a better understanding of the origin of foodborne disease and how to control it if we expect to keep people from getting ill. Biogenic amines are found in varying concentrations in a wide range of foods (fish, cheese, meat, wine, beer, vegetables, etc.), and BA formation is influenced by different factors associated with the raw material making up food products, microorganisms, processing, and conservation conditions. Moreover, BAs are thermostable. Biogenic amines also play an important role as indicators of food quality and/or acceptability. Hence, BAs need to be controlled in order to ensure high levels of food quality and safety. All of these aspects will be addressed in this review.

Keywords: biogenic amines; food products; food quality; food safety; quality control; quality indexes; public health; legislation–regulation; analytical determination

1. Biogenic Amines and Food Safety

Food safety is one of the main concerns of consumer and health agencies around the globe (European Food Safety Authority (EFSA), Food and Drug Administration (FDA), Food Safety Commission of Japan (FSCJ), World Health Organization (WHO), etc.). According to the WHO, more than 200 diseases are transmitted by food and the vast majority of the population will contract a foodborne disease at some point in their lifetime. For example, in the U.S. 48 million people (one in six) suffer a foodborne disease each year. Of these, 128,000 are hospitalized and 3000 die from such diseases [1,2]. Moreover, the real numbers are higher as many cases of foodborne disease go undetected and are not recorded as such, due to the difficulty in establishing a causal relationship between food contamination and illness or death. This highlights the importance of making sure that the food we consume is not contaminated with potentially harmful elements at any point along the food chain. Because food can become contaminated at any point along the global supply chain during production, distribution, and preparation–consumption, each individual along this chain, from producer to consumer, has a role to play in ensuring that the food we eat does not cause disease.

Furthermore, if we are to prevent such disease, we must gain a deeper understanding of the origin of foodborne illness and the way to control it.

The origin of foodborne illness could be bacteria, virus, parasite, mold, contaminants, metals, allergens, pesticides, natural toxins, etc., that can contaminate food and cause disease. In general, most food poisoning is caused by bacteria, viruses, and parasites as opposed to toxic substances. Nonetheless, there are cases of food poisoning that can be linked to chemical or natural toxins. From among these toxins, the FDA and EFSA pay particular attention to aflatoxins, mycotoxins, histamine, etc. Of these, it is worth noting that histamine, a biogenic amine, is present in most foods but in greater abundance in fish and fishery products. This biogenic amine is the main component in "scombrid poisoning" or "histamine poisoning" since these intoxications are related to the consumption of fish of the *Scombridae* and *Scomberesocidae* families (tuna, mackerel, bonito, bluefish, etc.) containing high levels of histamine. These species contain high levels of the free amino acid histidine in their muscle tissue, which is decarboxylated to histamine. However, other non scombroid species also contain high levels of free histamine in their muscle tissue [3,4], which is why this illness came to be known as "histamine poisoning". There have been recent cases involving vacuum-packed salmon. The most common symptoms of histamine poisoning are due to the effects it has on different systems (cardiovascular, gastrointestinal, respiratory, etc.) producing low blood pressure, skin irritation, headaches, edemas, and rashes typical of allergic reactions [5,6]. Furthermore, histamine plays a role in the health problem known as histaminosis or histamine intolerance associated with the increase of histamine in plasma [4]. It is also important to point out that histamine is a mediator of allergic disorders. Biogenic amines are released by mast cell degranulation (in response to an allergic reaction) and the consumption of foods containing histamine can have the same effect. Since food allergy symptoms are similar to those of histamine poisoning (food intolerance), physicians occasionally make a faulty diagnosis. For all these reasons, histamine is the biogenic amine (BA) causing major concerns in clinical and food chemistry. However, we would note that apparently histamine is not the only agent causing scombroid poisoning [7–12]. Other amines, such as putrescine and cadaverine, are also associated with this illness, although both seem to have much lower pharmacological activity on their own but enhance the toxicity of histamine and decrease the catabolism of this amine when they interact with amine oxidases, thus favoring intestinal absorption and hindering histamine detoxification [13,14].

Another important biogenic amine related to food poisoning is tyramine. In this case, intoxication is known as the "cheese reaction" as it is associated with the consumption of foods with high concentrations of tyramine, mainly associated with the consumption of cheese [10,14–17]. However, high levels of tyramine have also been observed in meat and meat products [14,18–21]. As in the case of histamine, this illness came to be known as "tyramine reaction" because of the main compound involved. Typical symptoms of tyramine poisoning are migraines, headaches, and increased blood pressure, since tyramine sparks the release of noradrenaline from the sympathetic nervous system [5,6,10].

Other Bas, such as spermidine or spermine, have also been associated with food allergies [6,22,23]. Tyramine and β-phenylethylamine are suspected of triggering hypertensive crises in certain patients and of producing dietary-induced migraines. Although tryptamine has toxic effects on humans (causing blood pressure to increase, thus leading to hypertension), the maximum amount of tryptamine permitted in sausages is not regulated in some countries [23]. It is worth noting an additional toxicological risk associated with BAs, mainly secondary BAs (putrescine and cadaverine), which are involved in other kinds of food poisoning, such as the formation of nitrosamines, that are believed to be cancer causing compounds [24,25]. This risk is greatest in meat products with high biogenic amine levels and which contain nitrite and nitrate salts used as curing agents, and also with heat treated products, as these factors favor interaction between BAs and nitrites to form nitrosamines [25,26]. However, under normal circumstances, the human body possesses detoxification systems to take care of these BAs, mainly in the intestine through the action of monoamine oxidase (MAO; CE 1.4.3.4), diamine oxidase (DAO; CE 1.4.3.6), and polyamine oxidase (PAO; CE 1.5.3.11). However, in certain cases this mechanism can be hindered by a variety of factors or circumstances, and BAs could accumulate in the body and cause serious toxicological problems and a high risk of poisoning [4,14]. Factors that could alter the detoxification mechanism include the consumption of amine oxidase inhibitors (mono and diamine oxidase inhibitors (MAOI/DAOI)), alcohol, immune deficiency of the consumer, gastrointestinal disorders, large amounts of BA, for example in the case of spoiled or fermented foods, etc. [5,13,27]. When calculating BA intake, one must consider that foods are not typically consumed in isolation but rather in the context of a meal where several foods are eaten simultaneously (meat, fish, cheese, wine, vegetables, etc.). Therefore, the aggregate amount of BA consumed would be the sum of all the amines from the different foods rather than one food considered individually. The potential toxicological effect would be the sum of the amines in all of the different foods, the synergies between them, and the other personal factors mentioned above. The role of various substances that enhance the toxicity of BA and the existence of synergic effects have been demonstrated. For example, in Europe approximately 20% of the population regularly takes MAOI and/or DAOI antidepressant drugs. In such circumstances, not even low amounts of biogenic amines can be metabolized efficiently, the result being increased sensitivity to BAs [14]. Some authors [28,29] have suggested that ripened meat products ("chorizo", "salchichón", "salami", etc.) contain enough tyramine to poison people taking MAOI even with low levels of tyramine (in the 6–9 mg/kg range). The consumption of 100 g of any of these products would interact with MAOI, while in the absence of MAOI none of these processed meats would be toxic if ingested in normal amounts, always depending of course on individual susceptibility. A new generation of MAOI has been developed that diminishes this sensitivity. The ingestion of even small amounts of tyramine has been known to cause severe migraines with intracranial hemorrhaging in patients treated with classic MAOIs, while tyramine between 50 and 150 mg is better tolerated by patients treated with a new generation of MAOIs, i.e., the so-called RIMA (reversible MAO-A inhibitor) [4,30]. The market is currently offering pharmaceutical preparations based on the DAO enzyme for the treatment of migraines whose fundamental function is to mitigate deficiencies of this enzyme (DAO), thus favoring the metabolism of histamine. It is very difficult to establish toxicity parameters for BAs considering the number of factors that affect their toxicity.

2. Biogenic Amines and Quality Control of Food Products

It is important to control and monitor biogenic amines not only for toxicological and health reasons as mentioned above, but also because they may play an important role as quality and/or acceptability indicators in some foods, and managing this quality is also a way to guarantee and ensure food safety. Food quality refers to main characteristics having to do with safety, nutrition, availability, convenience, integrity, and freshness [31].

BAs have been frequently employed as quality indexes in various foods (meat, fish, wines, etc.) to signal their degree of freshness and/or deterioration and also to control the processing and development of food and beverages. Individual BAs, such as histamine, tyramine, cadaverine, or a combination of various amines (putrescine–cadaverine, spermidine–spermine, etc.), have likewise been used as a quality index [19,26,32–40]. Also, different BA-based quality indexes have been proposed, such as the traditional one developed by Miet and Karmas [32] used as an indicator of the decomposition of fish. This index is based on the increase in putrescine, cadaverine, and histamine levels and the decrease in spermidine and spermine levels throughout the fish storage process. Scores of 0 and 1 are indicative of good quality fish, between 1 and 10 are tolerable, and a score of over 10 indicates decomposition of the product. However, in the case of other foods, such as cheese, meat, and meat products, this index has not yielded good results mainly because it does not include levels of tyramine, the main biogenic amine in these products. An alternative biogenic amine index (BAI) has been proposed for meat that consists of the sum of putrescine, cadaverine, histamine, and tyramine [40,41]. Hernández-Jover et al. [41] also suggested quality ranges for the index: BAI <5 mg/kg indicating good quality fresh meat, between 5 and 20 mg/kg for acceptable meat but with signs of initial spoilage, between 20 and 50 mg/kg for low quality meat, and >50 mg/kg for spoiled meat. However, the usefulness of BAs as a quality index depends on many factors, mainly concerning the nature of the product (fresh, canned, modified atmosphere, fermented, etc.). For example, BA indexes have proven to be more satisfactory in fresh meat and meat products and heat-treated products than in fermented products [40]. This is at least partly because biogenic amine concentrations vary much more in fermented products than in fresh and cooked meat products owing to the number of different factors involved in their processing (ripening, maturation, starter, additives, etc.) [13,14,20,21,42–44]. Therefore, establishing a biogenic amine index that reliably predicts product quality is no simple matter. It is important to note that sometimes foods with toxic levels of BAs, such as histamine or tyramine often appear organoleptically "normal". This could be the case of tuna, salmon, or fermented chorizo where unacceptable and toxic levels of histamine are undetectable prior to consumption and therefore consumers are unable to reject products based on sensorial parameters. This is another important reason to control these compounds.

3. Biogenic Amines in Food

Biogenic amines are compounds that are commonly found in food and beverages such as meat, fish, cheese, vegetables, wine, etc. The most important BAs found in food are histamine, tyramine, putrescine, cadaverine, β-phenylethylamine, agmatine, tryptamine, serotonin (SRT), spermidine, and spermine. These dietary amines are classified according to their chemical structure as aromatic amines (histamine, tyramine, serotonin, phenylethylamine, and tryptamine), aliphatic diamines (putrescine and cadaverine), and aliphatic polyamines (agmatine, spermidine, and spermine) [33,45]. In terms of origin or synthesis, they are classified as polyamines when they are endogenous and formed naturally by animals, plants, and microorganisms, which play an important role in physiological functions (neurotransmitter, psychoactive, vasoactive, regulating gene expression, cell growth and differentiation, gastric secretions, immune response, inflammatory processes, etc.), and biogenic amines, when formed mainly by the decarboxylation of free amino acids (FAAs) from the action of decarboxylase enzymes, which are mainly of microbial origin (Figure 1).

BA formation is influenced by numerous factors (Figure 1) that can be divided into three groups: raw materials (composition, pH, ion strength, etc.), microorganisms (decarboxylase activity is attributed chiefly to *Enterobacteriaceae*, *Pseudomonadaceae*, *Micrococcaceae*, lactic acid bacteria, etc.), and processing and storage conditions (fresh, cured, fermented, refrigerated, modified atmosphere, etc.) [9,14,17,43,46–48]. These factors do not act in isolation but rather have combined effects that determine the final concentration of BAs in food. Therefore, to ensure food quality from the perspective of BAs, it is vital to use suitable raw materials to limit the presence of BAs in the end product and hence assure better quality. It should be noted that these BAs are thermostable. In other words, once these biogenic amines are produced they are very difficult to destroy by subsequent processing (pasteurization, cooking, etc.) meaning that if they are present in the raw material or product, they will still be present in the final product.

In the case of factors such as microorganisms, it is necessary to control not only the microbial load in the product but also the type of microbiota constituting that load (bacterial species and strain), that in turn depends on factors associated with the raw material and processing and storage conditions [40]. These conditions directly or indirectly affect substrate and enzyme concentrations and determine the presence of other compounds or conditions that modulate (favor or not) decarboxylase activity (pH, temperature, co-factors, etc.). Therefore, there are many factors to be considered, especially in connection with the technology applied (thermal treatments, additives, fermentation, refrigeration, packaging, etc.). Hence, suitable raw materials are not enough to limit BA formation. Processing conditions must also be optimized as they are responsible for the specific profile of the biogenic amine in the different products. For example, fermentation generally promotes BAs, and in fact, this is the group of meat products with the greatest amount and diversity of these compounds. This has to do with several factors, such as the raw material, temperature of the medium (assuring conditions favorable to starter growth), the presence and concentration of additives (sugar, salt, antimicrobial agents, etc.), the microorganisms present, etc. The large quantities of microorganisms in these products, accompanied by proteolysis, gives rise to high concentrations of the amino acids constituting the nutrients required by the bacteria and the substrate on which decarboxylase enzymes work. In some cases, the presence of BAs in fermented products has been attributed to the poor quality of raw materials and defective processing.

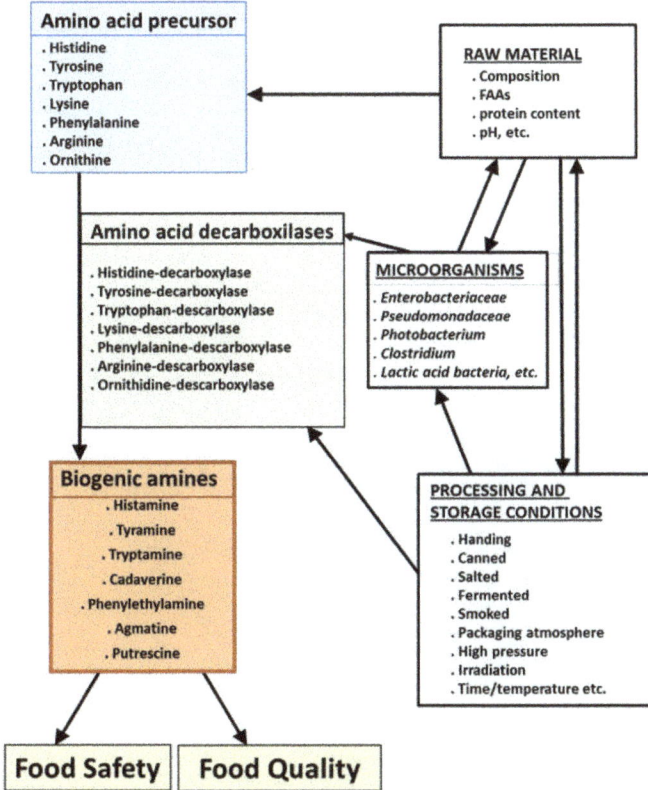

Figure 1. Formation of biogenic amines and factors influencing their formation. FAAs are free amino acids.

The storage temperature of final products is also one of the critical factors in the formation of BAs. Freezing temperatures inhibit microbial growth and therefore the production of biogenic amines. In contrast, higher chilled storage temperatures (>5 °C in fresh meat or fish) or poor temperature control foster the growth of microorganisms in products, which results in an increase in proteolysis in muscle tissue and an increase in decarboxylase enzymes and activity. Hence, low storage temperatures can make for improved quality and longer shelf-life of products. However, an increase in BAs is also related to processing and packaging conditions (modified atmosphere, vacuum, high hydrostatic pressure, irradiation, cooking products, etc.) that have an important influence on microbial flora. Controlling all of these factors improves the quality and shelf-life of food [14,34,48–50]. Today's lifestyle and global markets have led to the massive consumption of food and with this the development of new production and conservation systems and a complex food chain, that in many cases requires a deeper knowledge of how these foods are handled and forces us to face new challenges and problems in supplying safe foods.

4. Legislation Concerning Biogenic Amines in Food and Beverages

While it is very difficult to establish BA toxicity ranges owing to the many factors involved as described in the foregoing, given the dual importance of BAs (quality and health implications), efforts are being made to control BAs in food products and all countries have enacted legislation in this respect [4]. However, specific legislation only covers histamine in fishery products and no criteria have been established for other BAs or other food products, such as meat, dairy, or other products, despite the presence of important levels of BA in all types of food and the potential health risk in certain sectors of society where these products are consumed. However, in general the same legislation applicable to fish is applied to these products [19,22,40,44,51,52]. European Commission Regulations (2073/2005, 144/2007, 365/2010) set food safety criteria for histamine in fish. This legislation applies to particular fish species within the *Scombridae, Clupeidae, Eugraulidae, Coryphenidae, Pomatomidae,* and *Scomberesocidae* families throughout their shelf life with a sampling plan comprising nine units, two of which may be between 100–200 mg/kg of histamine and none above the limit of 200 mg/kg. This legislation also covers histamine levels in the processing (brine, enzyme maturation, curing, etc.) of these species with a sampling plan comprising nine units, two of which may be between 200–400 mg/kg of histamine and none above the limit of 400 mg/kg. The Australian and New Zealand standard codex feature similar levels between 100 mg/kg and none may exceed the limit of 200 mg/kg. In the U.S. the Food and Drug Administration [3] has set histamine limits in food in general at 50 mg/kg. This legislation is more advanced than its counterpart in the EU insofar as it applies to all food products.

Notwithstanding the difficulties and limitations in determining the real risk of toxicity for consumers posed by BAs in food, we should be aware that this legislation has its limitations. It is designed for one single biogenic amine (histamine) that, while admittedly one of the most important amines from a toxicological point of view, is not the only cause of toxicity. Limits should also be established for other amines, particularly tyramine, that have toxic effects, while also bearing in mind the other factors contributing to toxicity such as toxicity enhancers (individual susceptibility, consumption of MAOI, synergies resulting from the consumption of different foods during the same meal, etc.), with a view to establishing more restrictive legislation in certain cases. Although these aspects are truly difficult to address, they should be studied and included in future regulations to guarantee food safety and consumer health.

5. Analytical Determination of Biogenic Amines

As noted above, from the point of view of food safety and to assess the potential toxic effect of BAs, it is important to control and determine which BAs should be addressed. A number of swift and accurate analytical methods have been developed to determine BA levels in different foods and they were collected in various reviews [36,53–58]. These methods range from the more traditional colorimetric and fluorometric methods focused mainly on determining histamine individually, as is also the case with fast commercial kits based on the Elisa enzyme immunoassay to detect histamine in fish, to methods allowing for the simultaneous determination of several BAs (preferable) using chromatography methods such as: gas chromatography (CG) and gas chromatographic–mass spectrometry, high-performance liquid chromatography (HPLC), HPLC-tandem mass spectrometry, flow injection analysis (FIA), capillary electrophoresis, etc. (Table 1). Of all of these methods, HPLC is the most popular and frequently reported for the separation and quantification of BAs. This is the specific analytical method in European Commission (EC) [4]. This procedure offers high resolution, sensitivity, and versatility, and sample treatments are generally simple. Moreover, it offers the advantage of analyzing several BAs simultaneously. The HPLC method involves the first phase of BA extraction from the products and a second phase of determination. The extraction of BA is conducted using different solvents, such as hydrochloric acid, trichloroacetic acid, perchloric acid, methanol, etc., for the extraction procedure depending on the type of matrix (Table 1). The complexity of these matrices is a critical consideration for the adequate recovery of all BAs and to prevent interference with other compounds in the samples. This phase is also necessary in many other methods (Table 1). Chromatographic determination by HPLC is generally used pre- and post-column with reverse phase or ion exchange columns. Depending on the type of column employed, different derivate reagents are used to increase the sensitivity of the determination since BAs have low volatility and lack chromophores. The reagents commonly used in the literature are: dansyl chloride, ortho-ophtaldehyde (OPA), benzoyl chloride, p-phenyldiazonium sulfonate, 3-(4-fluorobenzoyl)-2-quinolinecarboxaldehyde, methanesulfonic acid, etc. Of these, OPA and dansyl chloride are the ones most widely used. The type of derivatization reagent used has implications for detection systems: UV/Vis, diode array, and fluorescence detector (Table 1). Important advances in analytical methods have paved the way for the use of more routine methods, such as flow injection analysis (FIA), which has been successfully used to determine BAs. This methodology offers a number of advantages, such as easy control of the chemical reaction, rapid reaction in the system, all reagent additions are performed automatically, etc. Moreover, FIA methods have been extensively used in combination with mass spectroscopy and with immobilized enzymes and electrodes or reactors using several different enzymes (amine oxidase, peroxidase, histaminase, etc.) to determine BAs in various elements by means of amperiometry or chemiluminescence. This has marked a major step forward in biosensor-assisted FIA determination of BAs [53].

Table 1. Methods and conditions for the determination of biogenic amines in food samples.

Analyte	Method/ Equipment	Sample	Extraction Solvents	Separation Technique	Derivatization Reagents	Detection System	Time of Analysis (min)	LOD	Ref
HIS	Fluorometric	Seafood, meats, cheeses, sauerkraut, etc.	MeOH	—	OPA	PF	—	0.02 mg/100 g	[59]
HIS	Fluorometric	Fish (fresh, dry, salted, frozen, brine, etc.)	MeOH	Ion exchange resin	OPA	PF	—	-	[60]
HIS	Colorimetric	Fish (tuna, mackerel),	NaCl solution	—	P-phenyldiazonium sulfonate	UV/vis	—	1 mg/100 g	[61]
HIS	ELISA immunoassay	Fish, wine		—	—	UV/vis	10	-	[62]
HIS, Tyr, Cad, Put, Phe, Trp, Spd, Spm	TLC	Cod, squid, MRS, TSB	TCA	—	Dansyl chloride	PF	—	5 ng–10 ng (1 mg/L per 10 µL spotted)	[63, 64]
HIS, Cad, Put, Spm	IEC	Tuna fish	MSA, HCL, PCA, PB	IonPac CS17	MSA	EDC	20	0.15–0.50 mg/kg	[65]
HIS, Tyr, Cad, Put, Spd	HPLC	Milk (cow, goat)	PCA	ODS2-C18	Benzoyl Chloride	UV/vis	13	0.03–1.30 mg/L	[66]
His, Tyr, Phe, Try, Cad, Put Spm Spd	HPLC	Sausages, cheese	PCA, HCL	Eclipse XDB-C18	Dansyl chloride, Fluorenylmethoxy-carbonyl chloride, Benzoyl chloride, Dansyl chloride	UV/vis	25–50	0.03–0.38 mg/kg,	[67]
HIS, Tyr, Cad, Put, Phe, Trp. Spd, Spm	HPLC or UHPLC	Meat, beer, wine, rice, mushroom, sausage, juice, oil, peanut butter, fish, shrimp sauce, etc.	PCA TCA HCL	ODS2-C18, Nova-Pak C18, Zorbax XDB C18	Dansyl chloride	UV/vis	6–30	0.01–0.10 mg/kg 4.43–6.96 µg/L	[68–71]
HIS, Tyr, Cad, Put, Phe, Spd, Spm, Ser, Met, Etm	HPLC or UHPLC	Wines, meat, beverages, coffee	TCA	Phenomenex Luna 5u RP-18 Kromasil	Dansyl chloride	DAD	35	0.5 mg/kg	[72, 73]
Trp, Phe, Put, Cad, HIS, Tyr, Spm, Spd	HPLC	Chicken carcasses	PCA	C18	Dansyl chloride	FLD	32	0.05–25 µg/mL	[74]
HIS, Met, Etm, Tyr, Phe, Put, Cad	HPLC	Wine	—	Nova-Pak C18	OPA	FLD	42	0.006–0.057 mg/L	[62]
HIS, Tyr, Cad, Put, Phe, Agm, Trp, Spd, Spm	HPLC or UHPLC	Meat, fish, squid, prawn	TCA PCA	Cation exchange-Capcell Pak MG-C18	OPA	FLD	25-55	0.05–0.2 mg/L 0.2–2.0 µg/L	[75–77]

Table 1. *Cont.*

Analyte	Method/ Equipment	Sample	Extraction Solvents	Separation Technique	Derivatization Reagents	Detection System	Time of Analysis (min)	LOD	Ref
Met, Etm, HIS, Tym, Trp, Phe, Put, Cad	HPLC	Canned tuna fish	TCA	Inertsil ODS-3	Naphthalene-2,3-dicarboxaldehyde	FLD	50	2.5–330 mg/kg	[78]
HIS, Tyr, Phe, Ser, Trp, Oct, Dopa, Cad, Put, Agm, Spd, Spm	HPLC	Wine, cider, spinach hazelnut, banana, potato, milk, chocolate, meat	PCA	Nova-Pak C18	OPA	FLD	55–60	0.03–0.06 mg/L 0.07–0.2 mg/L ≤1.5 mg/kg	[41, 79, 80]
HIS, Tyr, Phe, Ser, Trp, Oct, Dopa, Cad, Put, Agm, Spd, Spm	UHPLC	Wine, fish, cheese, sausage	PCA	Acquity BEH C18	OPA	FLD	7	0.2–0.3 mg/L	[81]
Put, HIS, Cad, Phe, Tyr, Spd, Spm	HPLC-	Beer, cheese, fish, sausage, shrimp	TCA	Hypersil BDS C18	EAC	FLD	6	0.27–0.69 ng/mL	[82]
HIS, Tyr, Cad, Put, Phe, Agm, Trp, Spd, Spm, Ser, Oct, Dopa	HPLC	Wines	—	A Zorbax C18	NQS	DAD	45	0.2–3 mg/L,	[83]
Met, HIS, Put, Cad, Tyr, Spm, Spd, Trip, Phe, Etm	HPLC	Wine, beer	PVP	Inertsil ODS-3 column	Dansyl chloride	DAD–APCI-MS	35	0.008–40.0 mg/L	[84]
His, Tyr, Spd, Spm, Cad, Put, Agm	HPLC	Fish, cheese, meat, vegetable	HCL, TCA, MeOH	LiChrospher RP 18,	OPA	FLD–DAD	20	0.5–8.5 mg/kg	[85]
HIS, Cad, Agm, Tyr, Put, Phe	HPLC	Beer, wines	BB	Gemini C-18	p-toluenesulfonyl chloride	MS	22	0.023–12 µg/dm³	[86]
HIS, Tyr, Phe	HPLC	Cheese	HCL	Luna C18	—	MS	11	0.05–0.25 mg/kg	[87]
Cad, Put, HIS	GC	Cheese, fish	—	OV-225	Perfluoropropionyl derivatives	ECD	20	<1.1 µg/g	[88]
Etm, HIS, Put, Spm, Trp, Tyr, Phe, Met, Prp	GC	Wine	MeOH, CHCl3 (DLLME)	ZB-5MS capillary column	IBCF; PCF	MS-MS	25	<4.1 µg/L.	[89]
Cad, Put HIS	GC	Apple juice	DLLME	CC-DB-5	—	MS	8	0.06–2.20 µg/L	[90]
Put, Cad, HIS, Phe, Tyr	GC	Alcoholic beverages	Toluene	CC-HP-5MS	Isobutyl chloroformate	MS	12	1–10 µg/L	[91]
HIS	FIA	Mackerel, mahi-mahi	MeOH	—	OPA	FLD	—	0.8–6 mg/kg	[92, 93]
HIS	FIA	Cider, wine		Anion exchange mini-column	OPA	FLD	—	30–101 µg/L	[94]
HIS, Tyr, Put, Cad, Agm, Spm	FIA	Tuna	Water	Electrode-Biosensor (AO, HmDH)	OPA	APMD	—	100 pmol	[95, 96]

Table 1. *Cont.*

Analyte	Method/Equipment	Sample	Extraction Solvents	Separation Technique	Derivatization Reagents	Detection System	Time of Analysis (min)	LOD	Ref
HIS, Phe	CE-FIA	Standard solutions	Water	—	—	MS	22	0.018–0.09 µg/mL	[97]
Put, Cad, Spm, Spd, Trp, Tyr, HIS	CE	Sauerkraut	PCA	Silica capillary	Benzoyl chloride	UV/vis	35	0.2–0.7 mg/L	[98]
HIS, Tyr, Phe, Put, Cad, Spm, Spd	CE	Soy sauce, fish, wine	TCA	Silica capillary	FBQCA	LIFD	14	0.4–10 nM	[99]
Put, Cad, Spd, Spm	CE	Fresh milk	PCA	Ag/AgCl electrode	—	APMD	27	100–400 nM	[100]
Put, HIS, Try, Phe, Spd	CE	Oyster	PCA	capillary column-Ag/AgCl	—	ECHL	30	9.2×10^{-4}–9.6×10^{-2} µg/mL	[101]
HIS, Tyr	CE	Meat, cheese, fish, vegetable	HCL, MeOH, TCA		—	DAD	9	2–6 mg/kg	[85]
Spm, Spd, Put, Cad, HIS, Phe, Trp, Tyr	CE	Beer, wine	—	Electrophoretic separation	—	MS	10	1–2 µg/L	[102]

Agm: agmatine, AO: amine oxidase, APMD: amperometric detection, BB: borate-buffer, Cad: cadaverine, CC: capillary column, CE: capillary electrophoresis, CHCl3: chloroform, CHMD: chemiluminescence detector, DAD: diode-array detector, DAD–APCI-MS: diode array detection–atmospheric pressure chemical ionization mass spectrometry system, DLLME: dispersive liquid microextraction, DOPA: dopamine, EAC: ethyl-acridine-sulfonyl chloride, EB: electrode-biosensor, ECD: electron capture detector, ECHL: electrochemiluminescence, EDC: electrochemical detector-conductivity, Etm: ethylamine, FBQCA: 3-(4-fluorobenzoyl)-2-quinolinecarboxaldehyde, FLD: fluorescence detection, HCL: chloridric acid, HIS: histamine, HmDH: histamine dehydrogenase, HPLC: high-performance liquid chromatography or high pressure liquid chromatography, HS-SPME: head space solid phase microextraction; IBCF: isobutyl choloroformate; IBUT: isobutylamine, IEC: ion-exchange chromatography, LIFD: laser-induced fluorescence detection, LLE: liquid-liquid extraction; LOD: limits of the detection, MAS: methanesulfonic acid, MeOH: methanol, Met: methylamine, MRS: Man, Rogosa and Sharpe Broth, MSA: methanesulphonic acid, MS: mass spectrometry, NQS: 1,2-naphthoquinone-4-sulfonate, Oct: octopamine, OPA: *o*-phthaldialdehyde, PB: phosphate buffer, PCA: percloric acid, PCF: propyl chloroformate, PF: photofluorometer, Phe: β-phenylethylamine, Put: putrescine, PVP: polyvinylpyrrolidone, Ser: serotonin, Spd: spermidine, Spm: spermine, TCA: tricloroacetic acid, TLC: thin layer chromatography, Trp: tryptamine, TSB: tryptic soy broth, Tyr: tyramine, UHPLC: ultra-high performance liquid chromatography, UV: ultraviolet.

6. Conclusions

There are many reasons to prevent the accumulation of biogenic amines in food products, mainly related to their utility as food quality indicators and their potential implications for consumer health. Controlling these compounds implies a deep understanding of the formation, monitoring, and reduction of biogenic amines during the processing and storage of food, and even of the effects of biogenic amines in consumers after the digestion of foods containing different levels of these compounds. Moreover, it is important to have quick, reliable, and precise analytical techniques to determine not only histamine and tyramine levels individually, but also to analyze other biogenic amines (putrescine, cadaverine, β-phenylethylamine, etc.) with implications for health and metabolic processes.

Such control of biogenic amines would benefit public authorities, industry, and consumers as it would help put higher quality products with fewer health implications on the market. However, guaranteeing the quality and safety of food requires a commitment not only from public institutions but also from production sectors, commercial processors, and ultimately from consumers who must play an important and active role in achieving food safety.

7. Future Trends and Perspectives

There are many lines of research looking into BAs in food and there are also many possibilities to be explored with regard to this subject from the technological, microbiological, analytical, and toxicological points of view.

Work should focus on determining the real risk of toxicity for consumers posed by BAs in food and should not be limited to a single amine or food product but should rather cover all the amines involved and in all foods consumed. Attention should also be given to the other factors contributing to toxicity, such as toxicity enhancers (individual susceptibility, consumption of MAOI, synergies resulting from the consumption of foods, etc.). Although these aspects are truly difficult to address, they should be studied and included in future regulations to guarantee food safety and consumer health.

Another important reason to control these compounds is the fact that often foods with toxic levels of BAs, such as histamine or tyramine, appear organoleptically 'normal' and consumers are unable to reject products based on sensorial parameters.

Moreover, today's market is trending towards the development of new products with new ingredients and new processing technologies, which create new conditions that could either favor or reduce the formation of biogenic amines. This is the case, for example, of the effect of decarboxylase enzymes responsible for their formation and the factors that modulate this activity. Therefore the implications of these new factors must be taken into account in new projects.

Important research efforts should continue in the field of analysis and determination of these BAs, always focused on the simultaneous determination of all of them, and on the different matrices, in order to solve the problems of extraction and interference of complex matrices. Also, advances need to be made in the search for more accurate, swift, simple, and unified determination methods that can easily be transferred to laboratories, industry, and the public administration.

Consequently, all research efforts should focus on the overarching goal of food safety and on providing the authorities with the tools they need to conduct swift checks of these compounds to reduce risk to consumers.

Author Contributions: C.R.-C. and A.M.H. designed and wrote the paper.

Funding: This research was supported by the Intramural Project 201470E073 of CSIC financial and the project MEDGAN-CM S2013/ABI2913 of CAM and ESI financial.

Acknowledgments: The authors wish to express heartfelt thanks to Francisco Jimenez-Colmenero for his welcome and generous support during his long, illustrious, and productive scientific career. Also thanks to Mehdi Triki for work performed during his stay in our lab.

Conflicts of Interest: The authors declare no conflict of interest.

References

1. FDA (Food and Drug Administration). Food Safety Modernization Act (FSMA). Available online: https://www.fda.gov/food/guidanceregulation/fsma/ (accessed on 18 September 2018).

2. CDCP (Centers for Disease Control and Prevention). Available online: https://www.cdc.gov/foodsafety/index.html (accessed on 25 September 2018).

3. FDA (Food and Drug Administration). Fish and Fishery Products Hazards and Controls Guidance - Fourth Edition. Available online: https://www.fda.gov/Food/GuidanceRegulation/GuidanceDocumentsRegulatoryInformation/Seafood/ucm2018426.htm (accessed on 25 October 2018).

4. EFSA. Scientific Opinion on risk based control of biogenic amine formation in fermented foods. *EFSA J.* **2011**, *9*, 2393. [CrossRef]

5. Bardócz, S. Polyamines in food and their consequences for food quality and human health. *Trends Food Sci. Technol.* **1995**, *6*, 341–346. [CrossRef]

6. Kalač, P. Health effects and occurrence of dietary polyamines: A review for the period 2005-mid 2013. *Food Chem.* **2014**, *161*, 27–39. [CrossRef] [PubMed]

7. Taylor, S.L.; Eitenmiller, R.R. Histamine food poisoning: Toxicology and clinical aspects. *Crit. Rev. Toxicol.* **1986**, *17*, 91–128. [CrossRef] [PubMed]

8. Lehane, L.; Olley, J. Histamine fish poisoning revisited. *Int. J. Food Microbiol.* **2000**, *58*, 1–37. [CrossRef]

9. Kim, M.K.; Mah, J.H.; Hwang, H.J. Biogenic amine formation and bacterial contribution in fish, squid and shellfish. *Food Chem.* **2009**, *116*, 87–95. [CrossRef]

10. Pegg, A.E. Toxicity of polyamines and their metabolic products. *Chem. Res. Toxicol.* **2013**, *26*, 1782–1800. [CrossRef]

11. Kovacova-Hanuskova, E.; Buday, T.; Gavliakova, S.; Plevkova, J. Histamine, histamine intoxication and intolerance. *Allergologia et Immunopathologia* **2015**, *43*, 498–506. [CrossRef]

12. Prester, L. Biogenic amines in fish, fish products and shellfish: A review. *Food Addit. Contam. Part A Chem. Anal. Control Expo Risk Assess.* **2011**, *28*, 1547–1560. [CrossRef]

13. Halász, A.; Baráth, Á.; Simon-Sarkadi, L.; Holzapfel, W. Biogenic amines and their production by microorganisms in food. *Trends Food Sci. Technol.* **1994**, *5*, 42–49. [CrossRef]

14. Ruiz-Capillas, C.; Jiménez-Colmenero, F. Biogenic amines in meat and meat products. *Crit. Rev. Food Sci. Nutr.* **2004**, *44*, 489–499. [CrossRef] [PubMed]

15. Karovičová, J.; Kohajdová, Z. Biogenic amines in food. *Chem. Papers* **2005**, *59*, 70–79.

16. Linares, D.M.; Martín, M.C.; Ladero, V.; Alvarez, M.A.; Fernández, M. Biogenic amines in dairy products. *Crit. Rev. Food Sci. Nutr.* **2011**, *51*, 691–703. [CrossRef] [PubMed]

17. Benkerroum, N. Biogenic Amines in Dairy Products: Origin, Incidence, and Control Means. *Compr. Rev. Food Sci. Food Saf.* **2016**, *15*, 801–826. [CrossRef]

18. Rice, S.L.; Eitenmiller, R.R.; Koehler, P.E. Biologically active amines in food: A review. *J. Milk Food Technol.* **1976**, *39*, 353–358. [CrossRef]

19. Hernández-Jover, T.; Izquierdo-Pulido, M.; Veciana-Nogués, M.T.; Vidal-Carou, M.C. Biogenic Amine Sources in Cooked Cured Shoulder Pork. *J. Agric. Food Chem.* **1996**, *44*, 3097–3101. [CrossRef]

20. Suzzi, G.; Gardini, F. Biogenic amines in dry fermented sausages: A review. *Int. J. Food Microbiol.* **2003**, *88*, 41–54. [CrossRef]

21. Stadnik, J.; Dolatowski, Z.J. Biogenic amines in meat and fermented meat products. *Acta Sci. Pol. Technol. Aliment* **2010**, *9*, 251–263.

22. Kalač, P.; Krausová, P. A review of dietary polyamines: Formation, implications for growth and health and occurrence in foods. *Food Chem.* **2005**, *90*, 219–230. [CrossRef]

23. Shalaby, A.R. Significance of biogenic amines to food safety and human health. *Food Res. Int.* **1996**, *29*, 675–690. [CrossRef]

24. Al Bulushi, I.; Poole, S.; Deeth, H.C.; Dykes, G.A. Biogenic amines in fish: Roles in intoxication, spoilage, and nitrosamine formation-A review. *Crit. Rev. Food Sci. Nutr.* **2009**, *49*, 369–377. [CrossRef] [PubMed]

25. De Mey, E.; De Klerck, K.; De Maere, H.; Dewulf, L.; Derdelinckx, G.; Peeters, M.C.; Fraeye, I.; Vander Heyden, Y.; Paelinck, H. The occurrence of N-nitrosamines, residual nitrite and biogenic amines in commercial dry fermented sausages and evaluation of their occasional relation. *Meat Sci.* **2014**, *96*, 821–828. [CrossRef] [PubMed]

26. Ruiz-Capillas, C.; Carballo, J.; Jiménez Colmenero, F. Biogenic amines in pressurized vacuum-packaged cooked sliced ham under different chilled storage conditions. *Meat Sci.* **2007**, *75*, 397–405. [CrossRef] [PubMed]

27. Alvarez, M.A.; Moreno-Arribas, M.V. The problem of biogenic amines in fermented foods and the use of potential biogenic amine-degrading microorganisms as a solution. *Trends Food Sci. Technol.* **2014**, *39*, 146–155. [CrossRef]

28. Vidal-Carou, M.C.; Izquierdo-Pulido, M.L.; Martín-Morro, M.C.; Mariné, F. Histamine and tyramine in meat products: Relationship with meat spoilage. *Food Chem.* **1990**, *37*, 239–249. [CrossRef]

29. Santos, C.; Jalón, M.; Marine, A. Contenido de tiramina en alimentos de origen animal. I. Carne, derivados cárnicos y productos relacionados. *Rev. Agroquim Technol. Aliment.* **1985**, *25*, 362–368.

30. McCabe-Sellers, B.J.; Staggs, C.G.; Bogle, M.L. Tyramine in foods and monoamine oxidase inhibitor drugs: A crossroad where medicine, nutrition, pharmacy, and food industry converge. *J. Food Composit. Anal.* **2006**, *19*, S58–S65. [CrossRef]

31. Herrero, A.M. Raman spectroscopy a promising technique for quality assessment of meat and fish: A review. *Food Chem.* **2008**, *107*, 1642–1651. [CrossRef]

32. Mietz, J.L.; Karmas, E. Polyamine and histamine content of rockfish, salmon, lobster, and shrimp as an indicator of decomposition. *J. Assoc. Off. Anal. Chem. (USA)* **1978**, *61*, 139–145.

33. Smith, T.A. Amines in food. *Food Chem.* **1980**, *6*, 169–200. [CrossRef]

34. Ruiz-Capillas, C.; Moral, A. Production of biogenic amines and their potential use as quality control indices for hake (*Merluccius merluccius*, L.) stored in ice. *J. Food Sci.* **2001**, *66*, 1030–1032. [CrossRef]

35. Rokka, M.; Eerola, S.; Smolander, M.; Alakomi, H.-L.; Ahvenainen, R. Monitoring of the quality of modified atmosphere packaged broiler chicken cuts stored in different temperature conditions: B. Biogenic amines as quality-indicating metabolites. *Food Control* **2004**, *15*, 601–607. [CrossRef]

36. Ruiz-Capillas, C.; Jiménez-Colmenero, F. Biogenic amines in seafood products. In *Handbook of Seafood and Seafood Products Analysis*; Leo, M.L., Nollet, F.T., Eds.; CRC Press Taylor & Francis Group: Boca Raton, FL, USA, 2009; pp. 833–850.

37. Galgano, F.; Favati, F.; Bonadio, M.; Lorusso, V.; Romano, P. Role of biogenic amines as index of freshness in beef meat packed with different biopolymeric materials. *Food Res. Int.* **2009**, *42*, 1147–1152. [CrossRef]

38. Vinci, G.; Antonelli, M.L. Biogenic amines: Quality index of freshness in red and white meat. *Food Control* **2002**, *13*, 519–524. [CrossRef]

39. Kalač, P.; Křížek, M. A review of biogenic amines and polyamines in beer. *J. Inst. Brewing* **2003**, *109*, 123–128. [CrossRef]

40. Triki, M.; Herrero, A.M.; Jiménez-Colmenero, F.; Ruiz-Capillas, C. Quality Assessment of Fresh Meat from Several Species Based on Free Amino Acid and Biogenic Amine Contents during Chilled Storage. *Foods* **2018**, *7*, 132–148. [CrossRef] [PubMed]

41. Hernández-Jover, T.; Izquierdo-Pulido, M.; Veciana-Nogués, M.T.; Vidal-Carou, M.C. Ion-Pair High-Performance Liquid Chromatographic Determination of Biogenic Amines in Meat and Meat Products. *J. Agric. Food Chem.* **1996**, *44*, 2710–2715. [CrossRef]

42. Latorre-Moratalla, M.L.; Veciana-Nogués, T.; Bover-Cid, S.; Garriga, M.; Aymerich, T.; Zanardi, E.; Ianieri, A.; Fraqueza, M.J.; Patarata, L.; Drosinos, E.H.; et al. Biogenic amines in traditional fermented sausages produced in selected European countries. *Food Chem.* **2008**, *107*, 912–921. [CrossRef]

43. Gardini, F.; Özogul, Y.; Suzzi, G.; Tabanelli, G.; Özogul, F. Technological factors affecting biogenic amine content in foods: A review. *Frontiers in Microbiology* **2016**, *7*. [CrossRef]

44. Eerola, H.S.; Roig Sagués, A.X.; Hirvi, T.K. Biogenic amines in Finnish dry sausages. *J. Food Saf.* **1998**, *18*, 127–138. [CrossRef]

45. Silla Santos, M.H. Biogenic amines: Their importance in foods. *Int. J. Food Microbiol.* **1996**, *29*, 213–231. [CrossRef]

46. Bodmer, S.; Imark, C.; Kneubühl, M. Biogenic amines in foods: Histamine and food processing. *Inflamm. Res.* **1999**, *48*, 296–300. [CrossRef]

47. Komprda, T.; Smělá, D.; Pechová, P.; Kalhotka, L.; Štencl, J.; Klejdus, B. Effect of starter culture, spice mix and storage time and temperature on biogenic amine content of dry fermented sausages. *Meat Sci.* **2004**, *67*, 607–616. [CrossRef] [PubMed]

48. Roig-Roig-Sagués, A.X.; Ruiz-Capillas, C.; Espinosa, D.; Hernández, M. The decarboxylating bacteria present in foodstuffs and the effect of emerging technologies on their formation. In *Biological Aspects of Biogenic Amines, Polyamines and Conjugates*; Dandrifosse, G., Ed.; Transworld Research Network: Kerala, India, 2009.

49. Naila, A.; Flint, S.; Fletcher, G.; Bremer, P.; Meerdink, G. Control of biogenic amines in food - existing and emerging approaches. *J. Food Sci.* **2010**, *75*, R139–R150. [CrossRef] [PubMed]

50. Kim, J.H.; Ahn, H.J.; Lee, J.W.; Park, H.J.; Ryu, G.H.; Kang, I.J.; Byun, M.W. Effects of gamma irradiation on the biogenic amines in pepperoni with different packaging conditions. *Food Chem.* **2005**, *89*, 199–205. [CrossRef]

51. Ten Brink, B.; Damink, C.; Joosten, H.M.; Huis in 't Veld, J.H. Occurrence and formation of biologically active amines in foods. *Int. J. Food Microbiol.* **1990**, *11*, 73–84. [CrossRef]

52. Bover-Cid, S.; Miguélez-Arrizado, M.J.; Vidal-Carou, M.C. Biogenic amine accumulation in ripened sausages affected by the addition of sodium sulphite. *Meat Sci.* **2001**, *59*, 391–396. [CrossRef]

53. Ruiz-Capillas, C.; Herrero, A.M.; Jiménez-Colmenero, F. Determination of biogenic amines. In *Flow Injection Analysis of Food Additives*; Ruiz-Capillas, C., Nollet, L.M.L., Eds.; CRC Press Taylor & Francis Group: Boca Raton, FL, USA, 2015; pp. 675–690.

54. Rivoira, L.; Zorz, M.; Martelanc, M.; Budal, S.; Carena, D.; Franko, M.; Bruzzoniti, M.C. Novel approaches for the determination of biogenic amines in food samples. *Stud. u. Babes-Bol. Chem.* **2017**, *62*, 103–122. [CrossRef]

55. Önal, A. A review: Current analytical methods for the determination of biogenic amines in foods. *Food Chem.* **2007**, *103*, 1475–1486. [CrossRef]

56. Mohammed, G.I.; Bashammakh, A.S.; Alsibaai, A.A.; Alwael, H.; El-Shahawi, M.S. A critical overview on the chemistry, clean-up and recent advances in analysis of biogenic amines in foodstuffs. *Trends Anal. Chem.* **2016**, *78*, 84–94. [CrossRef]

57. Ordóñez, J.L.; Troncoso, A.M.; García-Parrilla, M.D.C.; Callejón, R.M. Recent trends in the determination of biogenic amines in fermented beverages–A review. *Anal. Chim. Acta* **2016**, *939*, 10–25. [CrossRef] [PubMed]

58. Papageorgiou, M.; Lambropoulou, D.; Morrison, C.; Kłodzińska, E.; Namieśnik, J.; Płotka-Wasylka, J. Literature update of analytical methods for biogenic amines determination in food and beverages. *Trends Anal. Chem.* **2018**, *98*, 128–142. [CrossRef]

59. Taylor, S.L.; Lieber, E.R.; Leatherwood, M. A simplified method for histamine analysis of foods. *J. Food Sci.* **1978**, *43*, 247–250. [CrossRef]

60. AOAC. Histamine in seafood: Fluorometric method Sec. 35.1.32, Method 977.13. In *Official Methods of Analysis of AOAC International*; Cunniff, P.A., Ed.; AOAC International: Gaithersburg, MD, USA, 1995; pp. 6–17.

61. Patange, S.B.; Mukundan, M.K.; Kumar, K.A. A simple and rapid method for colorimetric determination of histamine in fish flesh. *Food Control* **2005**, *16*, 465–472. [CrossRef]

62. Marcobal, A.; Polo, M.C.; Martín-Álvarez, P.J.; Moreno-Arribas, M.V. Biogenic amine content of red Spanish wines: Comparison of a direct ELISA and an HPLC method for the determination of histamine in wines. *Food Res. Int.* **2005**, *38*, 387–394. [CrossRef]

63. Lapa-Guimarães, J.; Pickova, J. New solvent systems for thin-layer chromatographic determination of nine biogenic amines in fish and squid. *J. Chromatogr.* **2004**, *1045*, (1–2). [CrossRef]

64. Latorre-Moratalla, M.L.; Bover-Cid, S.; Veciana-Nogués, T.; Vidal-Carou, M.C. Thin-layer chromatography for the identification and semi-quantification of biogenic amines produced by bacteria. *J. Chromatogr.* **2009**, *1216*, 4128–4132. [CrossRef] [PubMed]

65. Cinquina, A.L.; Calì, A.; Longo, F.; De Santis, L.; Severoni, A.; Abballe, F. Determination of biogenic amines in fish tissues by ion-exchange chromatography with conductivity detection. *J. Chromatogr.* **2004**, *1032*, 73–77. [CrossRef]

66. Costa, M.P.; Balthazar, C.F.; Rodrigues, B.L.; Lazaro, C.A.; Silva, A.C.O.; Cruz, A.G.; Conte Junior, C.A. Determination of biogenic amines by high-performance liquid chromatography (HPLC-DAD) in probiotic cow's and goat's fermented milks and acceptance. *Food Sci. Nutr.* **2015**, *3*, 172–178. [CrossRef] [PubMed]

67. Liu, S.J.; Xu, J.J.; Ma, C.L.; Guo, C.F. A comparative analysis of derivatization strategies for the determination of biogenic amines in sausage and cheese by HPLC. *Food Chem.* **2018**, *266*, 275–283. [CrossRef]

68. Eerola, S.; Hinkkanen, R.; Lindfors, E.; Hirvi, T. Liquid chromatographic determination of biogenic amines in dry sausages. *J. AOAC Int.* **1993**, *76*, 575–577. [PubMed]

69. Yoon, H.; Park, J.H.; Choi, A.; Hwang, H.J.; Mah, J.H. Validation of an HPLC analytical method for determination of biogenic amines in agricultural products and monitoring of biogenic amines in Korean fermented agricultural products. *Toxicol. Res.* **2015**, *31*, 299–305. [CrossRef] [PubMed]

70. Dadáková, E.; Křížek, M.; Pelikánová, T. Determination of biogenic amines in foods using ultra-performance liquid chromatography (UPLC). *Food Chem.* **2009**, *116*, 365–370. [CrossRef]

71. Saaid, M.; Saad, B.; Hashim, N.H.; Mohamed Ali, A.S.; Saleh, M.I. Determination of biogenic amines in selected Malaysian food. *Food Chem.* **2009**, *113*, 1356–1362. [CrossRef]

72. Anli, R.E.; Vural, N.; Yilmaz, S.; Vural, Ŷ.H. The determination of biogenic amines in Turkish red wines. *J. Food Compos. Anal.* **2004**, *17*, 53–62. [CrossRef]

73. Casal, S.; Oliveira, M.B.P.P.; Ferreira, M.A. Determination of biogenic amines in coffee by an optimized liquid chromatographic method. *J. Liq. Chromatogr. Relat. Technol.* **2002**, *25*, 2535–2549. [CrossRef]

74. Tamim, N.M.; Bennett, L.W.; Shellem, T.A.; Doerr, J.A. High-performance liquid chromatographic determination of biogenic amines in poultry carcasses. *J. Agric. Food Chem.* **2002**, *50*, 5012–5015. [CrossRef]

75. Triki, M.; Jiménez-Colmenero, F.; Herrero, A.M.; Ruiz-Capillas, C. Optimisation of a chromatographic procedure for determining biogenic amine concentrations in meat and meat products employing a cation-exchange column with a post-column system. *Food Chem.* **2012**, *130*, 1066–1073. [CrossRef]

76. Sánchez, J.A.; Ruiz-Capillas, C. Application of the simplex method for optimization of chromatographic analysis of biogenic amines in fish. *Eur. Food Res. Technol.* **2012**, *234*, 285–294. [CrossRef]

77. Zhao, Q.X.; Xu, J.; Xue, C.H.; Sheng, W.J.; Gao, R.C.; Xue, Y.; Li, Z.J. Determination of biogenic amines in squid and white prawn by high-performance liquid chromatography with postcolumn derivatization. *J. Agric. Food Chem.* **2007**, *55*, 3083–3088. [CrossRef]

78. Zotou, A.; Notou, M. Enhancing Fluorescence LC Analysis of Biogenic Amines in Fish Tissues by Precolumn Derivatization with Naphthalene-2,3-dicarboxaldehyde. *Food Anal. Method.* **2013**, *6*, 89–99. [CrossRef]

79. Vidal-Carou, M.C.; Lahoz-Portolés, F.; Bover-Cid, S.; Mariné-Font, A. Ion-pair high-performance liquid chromatographic determination of biogenic amines and polyamines in wine and other alcoholic beverages. *J. Chromatogr.* **2003**, *998*, 235–241. [CrossRef]

80. Lavizzari, T.; Teresa Veciana-Nogués, M.; Bover-Cid, S.; Mariné-Font, A.; Carmen Vidal-Carou, M. Improved method for the determination of biogenic amines and polyamines in vegetable products by ion-pair high-performance liquid chromatography. *J. Chromatogr.* **2006**, *1129*, 67–72. [CrossRef] [PubMed]

81. Latorre-Moratalla, M.L.; Bosch-Fusté, J.; Lavizzari, T.; Bover-Cid, S.; Veciana-Nogués, M.T.; Vidal-Carou, M.C. Validation of an ultra high pressure liquid chromatographic method for the determination of biologically active amines in food. *J. Chromatogr.* **2009**, *1216*, 7715–7720. [CrossRef] [PubMed]

82. Li, G.; Dong, L.; Wang, A.; Wang, W.; Hu, N.; You, J. Simultaneous determination of biogenic amines and estrogens in foodstuff by an improved HPLC method combining with fluorescence labeling. *LWT Food Sci. Technol.* **2014**, *55*, 355–361. [CrossRef]

83. Hlabangana, L.; Hernández-Cassou, S.; Saurina, J. Determination of biogenic amines in wines by ion-pair liquid chromatography and post-column derivatization with 1,2-naphthoquinone-4-sulphonate. *J. Chromatogr.* **2006**, *1130*, 130–136. [CrossRef] [PubMed]

84. Loukou, Z.; Zotou, A. Determination of biogenic amines as dansyl derivatives in alcoholic beverages by high-performance liquid chromatography with fluorimetric detection and characterization of the dansylated amines by liquid chromatography-atmospheric pressure chemical ionization mass spectrometry. *J. Chromatogr.* **2003**, *996*, 103–113. [CrossRef]

85. Lange, J.; Thomas, K.; Wittmann, C. Comparison of a capillary electrophoresis method with high-performance liquid chromatography for the determination of biogenic amines in various food samples. *J. Chromatogr. B Analyt. Technol. Biomed. Life Sci.* **2002**, *779*, 229–239. [CrossRef]

86. Nalazek-Rudnicka, K.; Wasik, A. Development and validation of an LC–MS/MS method for the determination of biogenic amines in wines and beers. *Monatshefte fur Chemie* **2017**, *148*, 1685–1696. [CrossRef]

87. Calbiani, F.; Careri, M.; Elviri, L.; Mangia, A.; Pistarà, L.; Zagnoni, I. Rapid assay for analyzing biogenic amines in cheese: Matrix solid-phase dispersion followed by liquid chromatography-electrospray-tandem mass spectrometry. *J. Agric. Food Chem.* **2005**, *53*, 3779–3783. [CrossRef]

88. Staruszkiewicz, W.F., Jr.; Bond, J.F. Gas chromatographic determination of cadaverine, putrescine, and histamine in foods. *J. Assoc. Off. Anal. Chem.* **1981**, *64*, 584–591. [PubMed]

89. Płotka-Wasylka, J.; Simeonov, V.; Namieśnik, J. An in situ derivatization - dispersive liquid-liquid microextraction combined with gas-chromatography - mass spectrometry for determining biogenic amines in home-made fermented alcoholic drinks. *J. Chromatogr.* **2016**, *1453*, 10–18. [CrossRef] [PubMed]

90. Cunha, S.C.; Faria, M.A.; Fernandes, J.O. Gas chromatography-mass spectrometry assessment of amines in port wine and grape juice after fast chloroformate extraction/derivatization. *J. Agric. Food Chem.* **2011**, *59*, 8742–8753. [CrossRef] [PubMed]

91. Fernandes, J.O.; Judas, I.C.; Oliveira, M.B.; Ferreira, I.M.P.L.V.; Ferreira, M.A. A GC-MS method for quantitation of histamine and other biogenic amines in beer. *Chromatographia* **2001**, *53*, S327–S331. [CrossRef]

92. Hungerford, J.M.; Walker, K.D.; Wekell, M.M.; LaRose, J.E.; Throm, H.R. Selective Determination of Histamine by Flow Injection Analysis. *Anal. Chem.* **1990**, *62*, 1971–1976. [CrossRef] [PubMed]

93. Hungerford, J.M.; Hollingworth, T.A.; Wekell, M.M. Automated kinetics-enhanced flow-injection method for histamine in regulatory laboratories: Rapid screening and suitability requirements. *Anal. Chim. Acta* **2001**, *438*, 123–129. [CrossRef]

94. Del Campo, G.; Gallego, B.; Berregi, I. Fluorimetric determination of histamine in wine and cider by using an anion-exchange column-FIA system and factorial design study. *Talanta* **2006**, *68*, 1126–1134. [CrossRef] [PubMed]

95. Niculescu, M.; Frébort, I.; Peč, P.; Galuszka, P.; Mattiasson, B.; Csöregi, E. Amine oxidase based amperometric biosensors for histamine detection. *Electroanalysis* **2000**, *12*, 369–375. [CrossRef]

96. Takagi, K.; Shikata, S. Flow injection determination of histamine with a histamine dehydrogenase-based electrode. *Anal. Chim. Acta* **2004**, *505*, 189–193. [CrossRef]

97. Santos, B.; Simonet, B.M.; Ríos, A.; Valcárcel, M. Direct automatic determination of biogenic amines in wine by flow injection-capillary electrophoresis-mass spectrometry. *Electrophoresis* **2004**, *25*, 3427–3433. [CrossRef] [PubMed]

98. Křížek, M.; Pelikánová, T. Determination of seven biogenic amines in foods by micellar electrokinetic capillary chromatography. *J. Chromatogr.* **1998**, *815*, 243–250. [CrossRef]

99. Zhang, N.; Wang, H.; Zhang, Z.X.; Deng, Y.H.; Zhang, H.S. Sensitive determination of biogenic amines by capillary electrophoresis with a new fluorogenic reagent 3-(4-fluorobenzoyl)-2-quinolinecarboxaldehyde. *Talanta* **2008**, *76*, 791–797. [CrossRef] [PubMed]

100. Sun, X.; Yang, X.; Wang, E. Determination of biogenic amines by capillary electrophoresis with pulsed amperometric detection. *J. Chromatogr.* **2003**, *1005*, 189–195. [CrossRef]

101. An, D.; Chen, Z.; Zheng, J.; Chen, S.; Wang, L.; Huang, Z.; Weng, L. Determination of biogenic amines in oysters by capillary electrophoresis coupled with electrochemiluminescence. *Food Chem.* **2015**, *168*, 1–6. [CrossRef] [PubMed]

102. Daniel, D.; dos Santos, V.B.; Vidal, D.T.R.; do Lago, C.L. Determination of biogenic amines in beer and wine by capillary electrophoresis-tandem mass spectrometry. *J. Chromatogr.* **2015**, *1416*, 121–128. [CrossRef] [PubMed]

foods

MDPI

Review

What We Know and What We Need to Know about Aromatic and Cationic Biogenic Amines in the Gastrointestinal Tract

Alberto Fernández-Reina [1], José Luis Urdiales [1,2,*] and Francisca Sánchez-Jiménez [1,2]

[1]   Departamento de Biología Molecular y Bioquímica, Facultad de Ciencias, Universidad de Málaga, 29071 Málaga, Spain; afernandezreina4@gmail.com (A.F.-R.); kika@uma.es (F.S.-J.)
[2]   CIBER de Enfermedades Raras & IBIMA, Instituto de Salud Carlos III, 29010 Málaga, Spain
*   Correspondence: jlurdial@uma.es; Tel.: +34-952-137-285

Received: 1 August 2018; Accepted: 29 August 2018; Published: 4 September 2018

Abstract: Biogenic amines derived from basic and aromatic amino acids (B/A-BAs), polyamines, histamine, serotonin, and catecholamines are a group of molecules playing essential roles in many relevant physiological processes, including cell proliferation, immune response, nutrition and reproduction. All these physiological effects involve a variety of tissue-specific cellular receptors and signalling pathways, which conforms to a very complex network that is not yet well-characterized. Strong evidence has proved the importance of this group of molecules in the gastrointestinal context, also playing roles in several pathologies. This work is based on the hypothesis that integration of biomedical information helps to reach new translational actions. Thus, the major aim of this work is to combine scientific knowledge on biomolecules, metabolism and physiology of the main B/A-BAs involved in the pathophysiology of the gastrointestinal tract, in order to point out important gaps in information and other facts deserving further research efforts in order to connect molecular information with pathophysiological observations.

Keywords: histamine; serotonin; catecholamines; polyamines; gastrointestinal tract; nutrition; inflammation; gastric cancer; bowel diseases; colon cancer

1. Introduction

Biogenic amines (BAs) are low molecular weight organic compounds synthetized in vivo by decarboxylation of L-amino acids or their derivatives, thus containing one or more amino groups [1]. BAs can be derived from L-basic amino acids, as for instance, histamine (HIS) derived from L-histidine, as well as putrescine (Put), agmatine (Agm), spermidine (Spd) and spermine (Spm) derived from L-arginine or L-ornithine, depending on the organism. Other BAs can also be synthetized from L-aromatic amino acids or derivatives in mammalian tissues, as is the case for serotonin (5-HT) and catecholamines (CAs) that have L-aromatic amino acids such as L-tryptophan and L-tyrosine as their precursors, respectively. Figure 1 shows chemical structures of BAs. Throughout this work, this set of biogenic amines derived from basic or aromatic L-amino acids are abbreviated as BA. We will focus our attention on the role of B/A-BAs, as they are the most important ones in the gastrointestinal context. Another important BA for the central nervous system (CNS), the gamma-aminobutyric acid (GABA), is derived from the amino acid L-glutamate. Many other BAs, outside the scope of this review, can also be synthetized in nature playing different roles along the phylogenetic scale (for instance, tyramine from L-tyrosine and cadaverine from L-lysine, among others) [2,3].

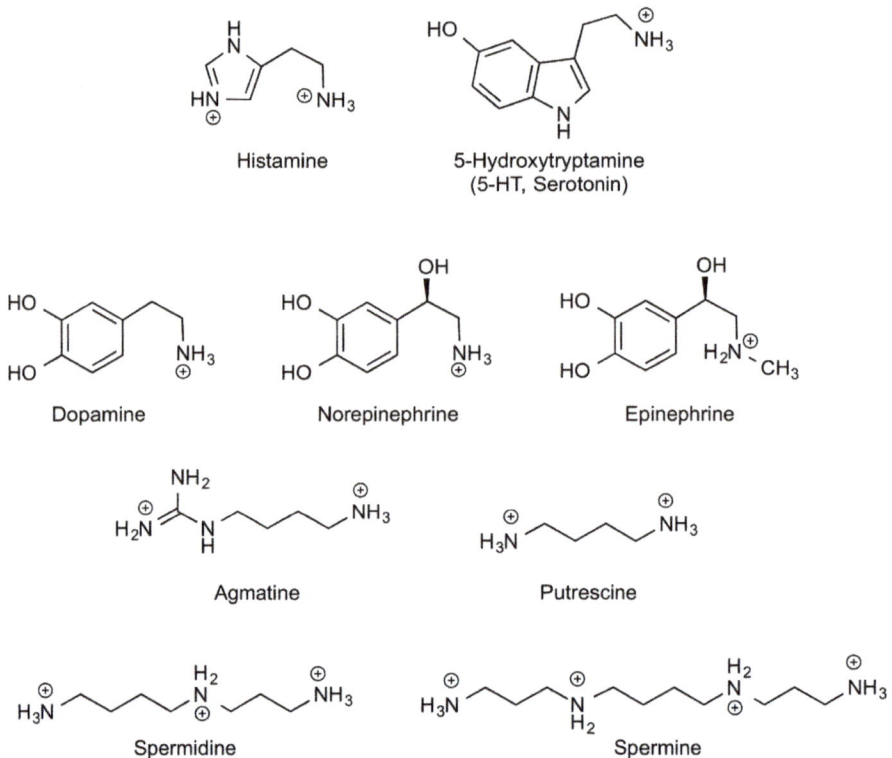

Figure 1. Chemical structures of B/A-BAs in their major forms at physiological pH. Histamine imidazole group is only partially protonated at pH 7 (pI $\approx$ 6).

All B/A-BA synthetic pathways include the alpha-decarboxylation of L-amino acids with cationic or aromatic side chains, or methylated or hydroxylated amino acid derivatives, as in the cases of 5-HT and CAs, respectively (Figure 2). In mammalian cells, B/A-BA synthesis involves the action of three pyridoxal 5′-phosphate (PLP)-dependent enzymes: ornithine decarboxylase (ODC, EC 4.1.1.17), histidine decarboxylase (HDC, EC 4.1.1.22) and aromatic L-amino acid decarboxylase (or DOPA decarboxylase, DDC, EC 4.1.1.28) [4–6]. In some cases, their common names used to derive from the precursor amino acid, as for HIS, that is synthetized from L-histidine, or 5-HT synthetized from L-tryptophan, but it is not a general rule. The metabolic origins of these BAs are shown in Figure 2 and further explained in the following sections.

Expressions of the involved PLP-decarboxylases—ODC, HDC and DDC—are cell-specific events, therefore linked to cell-specific developmental programs, for which we still ignore many involved factors. Both mammalian HDC and DDC share a high degree of homology; however mammalian ODC has a different evolutionary origin [7,8]. Nevertheless, all of them could compete for the cofactor PLP, in cases of vitamin B6 deficit or altered hepatic PLP metabolism (for instance, during aging [9]).

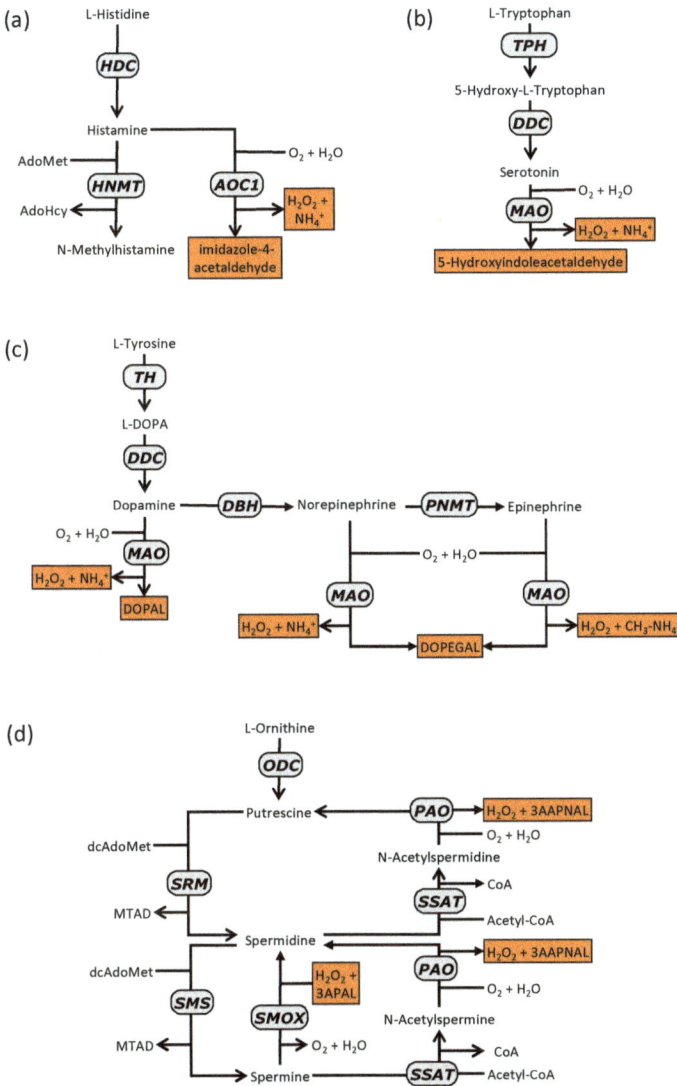

Figure 2. Aromatic and cationic BA synthesis and degradation pathways in mammalian cells. (**a**) histamine; (**b**) serotonin; (**c**) cathecolamines; (**d**) polyamines. Degradation products are depicted in orange boxes. Abbreviations (by alphabetical order): AdoMet, adenosylmethionine; AdoHcy, adenosylhomocysteine; AOC1; CoA, coenzyme A; diamine oxidase; DBH, dopamine β-hydroxylase; dcAdoMet, decarboxylated adenosylmethionine; ADDC, DOPA decarboxylase; DOPA, dihydroxyphenylalanine; DOPAL, 3,4-dihydroxyphenylacetaldehyde; DOPEGAL, 3,4-dihydroxyphenylglycoaldehyde; HDC, histidine decarboxylase; HNMT, histamine *N*-methyltransferase; MAO, monoamine oxidase; MTAD, methylthioadenosine; PAO, polyamine oxidase; PNMT, phenylethanolamine *N*-methyltransferase; SMOX, spermine oxidase; SMS, spermine synthase; SRM, spermidine synthase; SSAT, spermidine/spermine N^1-acetyltransferase; TH, tyrosine hydroxylase; TPH, tryptophan hydroxylase.

Another common fact is that BA degradation in vivo involves the action of amino oxidases. These reactions produce aldehydes (sometimes very toxic ones) and H_2O_2. A high oxidase activity could therefore cause local ROS and/or toxic aldehyde elevations. Amine oxidase specificities for each B/A-BA will be mentioned below. There are two families of amine oxidases, copper- or flavine-dependent oxidases [10,11]. These enzymes can be extra- or intracellular located and they also differ in the amine substrate specificities. For instance, the copper-dependent diamine oxidase (AOC1 or DAO, EC 1.4.3.22) can accept both HIS and Put as a substrate; these BAs come from different synthesis pathways. DDC products also share MAO activities (EC 1.4.3.4) [12]. Thus, degradation is also a process in which different BA metabolic pathways can eventually be confluent in the same physiological context.

In the following subsections, we will focus on descriptive overviews of the different B/A-BA specificities of their respective metabolic pathways and physiological functions in the gastrointestinal tract (GIT) system. It is a very complex physiological scenario still unveiled or confusing in many aspects. However, it is well known that BA metabolism in GIT can be highly decisive for health and quality of life, as occurring in the other physiological contexts mentioned above, therefore also deserving further biochemical and cellular research efforts to reach more efficient translational actions.

From a biochemical point of view, BAs were considered to be a part of secondary metabolism for many years, and consequently neglected in many general biochemistry textbooks. However, evidence has accumulated revealing important roles of these metabolites in mammalian pathophysiology. For instance, it is well known that HIS is an important mediator of the immune system, as well as a key biomolecule for correct gastric function [13–15] (Table 1). Nowadays, we can say that they are very important for human homeostasis, as they play important roles in the most important human physiological functions (neurotransmission, defence, digestion and nutrition, growth, apoptosis, and reproduction). Consequently, impairments in their metabolic (including signalling) pathways are related to many different pathological phenotypes and diseases.

Table 1. Nomenclature, precursors and main functions of the basic and aromatic amines involved in the gastrointestinal pathophysiology.

Common Names (Abbreviations) *	IUPAC Names	Precursor L-Amino Acids	Physiological Roles
Histamine (HIS)	2-(1H-Imidazol-4-yl)ethanamine	L-Histidine	Neurotransmitter. Immune mediator. Gastric acid secretion inducer.
Serotonin (5-HT)	3-(2-Aminoethyl)-1H-indol-5-ol	L-Triptophan	Neurotransmitter related to reward motivated behaviour. Modulator of vessel constriction and intestinal motility.
Catecholamines (CAs):			
Dopamine (DA)	4-(2-Aminoethyl)benzene-1,2-diol	L-Tyrosine	Blood pressure regulators. Modulators of nutrient absorption and intestinal motility.
Epinephrine	(R)-4-(1-Hydroxy-2-(methyl amino)ethyl)benzene-1,2-diol		
Norepinephrine	(R)-4-(2-amino-1-hydroxy ethyl)benzene-1,2-diol		
Polyamines (PAs):			
Putrescine (Put)	Butane-1,4-diamine	L-Ornithine	Essential for cell viability, proliferation and correct differentiation.
Spermidine (Spd)	N'-(3-aminopropyl)butane-1,4-diamine	L-Ornithine + L-Methionine	
Spermine (Spm)	N,N'-bis(3-aminopropyl)butane-1,4-diamine		
Agmatine (Agm)	2-(4-aminobutyl)guanidine	L-Arginine	Anti-apoptotic effects. Positive effects on brain, hepatic and renal functions.

Data from references [16–24]. *, Abbreviations used in the text.

BAs can be synthesized de novo by specific mammalian cell types, but can also have an exogenous origin [19]. Microbiota, as well as microorganisms taking part in food processing or contamination, can produce biogenic amines from dietary amino acids at different rates and with different structures to those synthetized by human cells, which can have physiological effects; for instance, the decarboxylation product of L-arginine, Agm [20]. Its endogenous synthesis is, at least, controversial [21]. BAs are also present in a huge variety of drinks and foods, especially those in which microbial activity takes place during storage or preparation, sometimes with negative consequences for human health; for instance, toxicity due to HIS overproduction in contaminated seafood (i.e., by *Morganella morganii* sp.) [22] or high levels of amines in cold cuts and fermented foods (lactic products, fermented vegetables, wine, beer, etc.) [23,24]. In addition, BAs could form carcinogenic nitrosamines in the presence of nitrites during food processing [25].

A full characterization of the physiological effects of exogenous BAs has always faced two big handicaps: the multiple difficulties to evaluate the degree in which dietary amines are absorbed by gut epithelium, and the complexity of the characterization of the BA metabolism capacities of our particular microbiota, as this factor can induce important changes in the BA concentrations available to gut epithelium.

The importance of B/A-BAs in our digestive systems has been observed throughout the 20th century. In spite of all these valuable pathophysiological data (thousands of indexed publications) available, many gaps in molecular information still exist with regard to the mechanisms involved in each case, delaying the progress towards more personalized and accurate solutions for digestive-related pathologies [26]. As a research group working on several Systems Biology initiatives [27,28] and BAs [1,29,30], our hypothesis is based on the concept that integration of information can reveal emergent information, offering light to new hypothesis and translational actions. As far as we know, there is no recent similar review devoted to gathering biochemical and pathophysiological information on B/A-BAs in the GIT. Thus, the major aim of this work is to present an overview of the known facts of biochemical and pathophysiological information on B/A-BAs in the GIT context. The objective is to point out interesting facts deserving further research in order to eliminate gaps of molecular information currently blocking or delaying translational possibilities for prevention, diagnosis, and/or intervention of gastrointestinal diseases.

2. Histamine Biochemistry and Physiology

2.1. Histamine Synthesis

HIS was one of the first discovered low molecular weight (111 Da) immune mediators at the beginning of the last century [17,31,32]. Its precursor, the semi essential amino acid L-histidine, can be at least partially endogenous or derived from dietary proteins [33]. In mammalian cells, HIS synthesis occurs by decarboxylation of L-histidine, which is catalysed by the enzyme named L-histidine decarboxylase (HDC) (Figure 2a). This activity is carried out by PLP-dependent enzymes in both Gram-negative bacteria and Metazoa [34]. However, in Gram positive bacteria potentially present in intestines (for instance, *Lactobacillus* sp.), the reaction is catalysed by a non-homologous pyruvoil-dependent enzyme [5,34,35]. In human tissue, HDC is only expressed in a short list of cell types. Among them, several immune differentiated cells (mast cells and basophils) [36–38], histaminergic neurons [18,39], and gastric enterochromaffin-like cells (ECL cells) [40] are able to both synthetize and store HIS. Transformed HIS producing cells can preserve or even increase their HIS-producing capacity, as in the case of malignant mastocytosis and several types of gastric cancer cells [41,42]. Other cells (for instance, macrophages, eosinophils, and platelets) can also synthetize HIS to any extent, being unable to store it in specialized vesicles [31,35].

A big gap of information still exists on mechanisms controlling cell type-specificities with respect to HDC expression, but it seems to be clear that epigenetic events play important roles (i.e., methylation/demethylation of CpG islands present in mammalian HDC gene promoter) [43,44].

It has been observed that HDC gene expression can increase in response to several stimuli such as gastrin, estrogens, or several interleukins (ILs) as IL-1, IL-3, IL-12 or IL-18. It depends on the specific receptors expressed by the target HIS-producing cells [35,45].

Alternative splicing events have been observed during mammalian (and human) HDC expression in HIS-producing cells [46,47]. The meaning of these aberrant messengers is still unknown. As the active protein is a dimer and taking into account that dimerization involves interaction of both N-terminus [34], some of the truncated sequences could act as natural HDC inhibitors.

In addition, the protein needs to be processed to reach the active conformation and is a very unstable enzyme [48–53]. Regulation of the enzyme processing and turnover can be important as a determinant of active HDC levels. However, HDC processing and maturation is a process not fully characterized. It seems to be clear that the monomer mature form must correspond to a 53–63 KDa fragment of the N-terminus of the primary translation product [49]. Nevertheless, the precise sequence of this fragment in vivo is not yet known.

The action mechanism of mammalian PLP-dependent L-amino acid decarboxylases has been previously described [54,55]. Briefly, it involves two transaldamination reactions from the PLP-enzyme complex to the L-amino acid-enzyme complex, which is decarboxylated in the substrate α-carboxylic group to form a covalent amine product-enzyme complex. This last complex suffers a second transaldimination reaction with PLP to recover the initial PLP-enzyme complex, thus releasing the amine product. Important changes in the global decarboxylase conformation have been observed for both mammalian HDC and DDC during catalysis [56,57]. The quaternary structure of HDC and DDC only differs in tautomeric forms of intermediates along the reaction, most probably due to slight differences in the active dimer conformation [54,55]. In fact, both enzymes can share substrates (i.e., L-histidine, but with different affinities) and inhibitors (for instance, epigallocathechine-3-gallate). This fact needs to be taken into account for design of specific inhibitors of any of these activities with pharmacological purposes.

2.2. Exogenous Histamine Synthesis

Important quantities of HIS can be present in some natural products, such as oranges or tomatoes. Fermented products (cheese, alcoholic drinks, fermented vegetables and fish) can also contain high quantities of HIS (and other BAs), as a result of the metabolic properties of the living organism involved in each case. Contamination during food processing or storage can also allow undesirable growth of HIS-producing microorganisms. Many efforts of EU COST actions (i.e., COST Action 917, 922 and BN0806) have been devoted to the study and control of BA levels in foods (see for instance, [58,59]) However, it is a very complex subject with many variables and a lot of uncertainty with regard to amine absorption mechanisms and traceability. This issue requires applying more holistic approaches, and the use of high-throughput technologies, in order to efficiently translate the knowledge to both new general nutritional recommendations and personalized diets.

In addition, GIT microbiota species can synthetize HIS (and many other BAs) by using PLP- or pyruvoyl-dependent enzymes. Thus, microbiota characterization should also be considered during personalized medicine initiatives of patients affected by BA-related diseases [26].

2.3. Histamine Degradation

In human tissues, HIS can be degraded by two different pathways (Figure 2a). The first one involves the intracellular *N*-methylation of HIS in its imidazole group catalysed by histamine *N*-methyl transferase (HNMT, EC 2.1.1.8) [60]. It is an ubiquitous enzyme expressed in liver and also in intestinal mucosa in a minor extent [61,62]. Its product, N-tele methyl-histamine, is a substrate of monoamine oxidase (MAO), which produces *N*-methylimidazole acetaldehyde. Finally, the enzyme aldehyde dehydrogenase (AD, EC 1.2.1.5) reduces this metabolite to *N*-methylimidazole acetic acid. This seems to be the major pathway in the brain [63]. However in GIT, the main pathway for HIS degradation involves the direct HIS oxidation by the action of human DAO producing imidazole acetaldehyde [35].

It is a copper-containing glycoprotein associated with cytosolic membrane and expressed in the stomach, duodenum, small intestine, and colon [61,62]. It can be released from membranes of their producing cells and is active in human serum [64]. It is also known as amyloride-binding protein-1 (AOC1) and histaminase.

2.4. Histamine Transport and Storage Mechanisms

In mammalian cells, HIS can be transported into epithelial cells throughout organic cations transporters (OCT 2 and 3), and the plasmatic membrane monoamine transporter (PMAT). OCT 2 and 3, and PMAT are located in the basolateral plasmatic membranes. OCT 3 has also been located in the luminal membranes of bronchial and small intestine epithelial cells [35,65].

Inside the cell, HIS can be transported through endosomal membranes by using the vesicular amine/proton antiporter systems named vesicular monoamine transporter 2 (VMAT2 or SLC18A2), which is also able to transport other monoamines such as DA, norepinephrine and 5-HT, and can be modulated by drugs such as amphetamines and cocaine [66].

With respect to storage, as mentioned previously, only a few cell types are able to store HIS. The major HIS-storing cells in a human body are mature mast cells, which can accumulate HIS, as well as 5-HT and even PAs into secretory granules, most probably derived from *trans*-Golgi vesicles [67]. Mast cells are infiltrated into mammalian epithelia, including GIT epithelia. Other important HIS storing cells in GIT are gastric ECL cells [68].

Three different mechanisms have been described for HIS secretion [35]:

1. Mast cell degranulation by immune stimuli. The presence of specific antigens induces IgE synthesis, inducing a high affinity binding between the specific IgE and IgE receptor known as FcεRI. This high affinity complex induces degranulation after further expositions to the antigen.
2. Cytokines can also induce degranulation. It is mediated by vesicular trafficking events involving fusion and/or content interchange between secretory granules and vesicles driven to exocytosis.
3. Constitutive HIS leakage due to non-active transport through cytosolic membranes or *trans*-Golgi vesicles driven to exocytosis.

2.5. Histamine Signalling and Physiological Functions

HIS could be considered the most pleiotropic amine, as it is involved in a wide spectrum of physiological processes concerning the most important function for a human being. It is a well-known immune mediator, as well as a neurotransmitter, thus involved in the most complex and still not fully characterized physiological capabilities. It also plays a key role in gastric acid secretion, and has also been described as a cell proliferation modulator, nutrition and cell proliferation being two essential functions for life [13].

HIS effects in different physiological scenarios are elicited by different HIS receptors. Four specific HIS receptors have been detected so far, namely H_1R, H_2R, H_3R and H_4R. All of them are members of the G-protein coupled receptor family. Their expressions are cell-type dependent and the elicited signals sometimes contradictory. Nevertheless, all of them somehow participate in GIT functions and homeostasis. Table 2 summarizes their specific characteristics.

Table 2. Molecular and functional properties of human histamine receptor types.

Properties	HIS Receptor 1 (H_1R)	HIS Receptor 2 (H_2R)	HIS Receptor 3 (H_3R)	HIS Receptor 4 (H_4R)
Chromosome	3	5	20	18
Molecular weight (KDa)	56	40	49	44
G protein signalling	$G\alpha_q$	$G\alpha_s$	$G_{i/o}$	$G_{i/o}$
Elicited signalling	PLC activation Increase of Ca^{2+} Production of NOS and cGMP	PKA activation Increase of cAMP PLC activation Increase of Ca^{2+}	Decrease of cAMP Inhibition of Ca^{2+} channels	Inhibition of cAMP Stimulation of MAP kinase phosphorylation
Expression	Brain, smooth muscle, skin, gastrointestinal and genitourinary tract, adrenal medulla, immune system and heart	Brain, smooth muscle, skin, gastrointestinal and genitourinary tract, adrenal medulla, immune system and heart	Widely found in brain and gastric mucosa	Inflammatory cells, dendritic cells and peripheral nerves
Physiological effects	Smooth muscle contraction Vasodilation and increase of vascular permeability	Inhibition of chemotaxis in basophils, gastric secretion of HCl and duodenal bicarbonate secretion	Release regulation of HIS (and other neurotransmitters) release from neurons Inhibition the secretion of gastric acid	Inflammatory processes such as allergies and asthma

Data from references [16–18]. Abbreviations; cAMP, 3′-5′-cyclic adenosine monophosphate; cGMP, 3′-5′-cyclic guanosyl monophosphate; HIS, histamine; MAP, mitogen-activated protein; NOS, nitric oxide synthase; PLC, phospholipase C.

H_1R and H_2R are the most ubiquitous receptors along the GIT. H_1R, H_2R and H_3R are located in gastric mucosa and their affinities for HIS are dependent on the expressed isoforms. H_4R can be expressed by inflammatory cells, as well as peripheral neurons associated with GIT. Specific HIS actions throughout the GIT are summarized below. H_4R was the most recent HIS receptor discovered, and is still not fully characterized, in spite of the multiple efforts made by international research groups [69]. Nevertheless, insights point out to its importance in GIT physiology, thus encouraging new actions to decipher this still veiled but important information for GIT pathophysiology. In fact, the receptor is proposed to be involved in gastric acid secretion, gastric mucosa defence, intestinal motility and secretion, visceral sensitivity, inflammation, immunity and gastric and colorectal carcinogenesis [70].

2.5.1. Histamine and Acid Gastric Secretion

Gastric acid secretion is regulated by different positive stimuli, such as acetylcholine, HIS and gastrin, and inhibitors such as somastostatin. Acetylcholine is a neurotransmitter coming from enteric neurons. HIS, gastrin and somatostatin are secreted by different endocrine cells infiltrated in GIT mucosa; they include ECL cells, G cells and D cells, respectively [71].

Figure 3 is a scheme of the balance between stimuli and inhibitors of gastric secretion. Briefly, on the one hand, binding of acetylcholine (from enteric neurons) to specific receptors stimulates parietal cells to secrete HCl; as well as gastrin (from gastric epithelium G cells), which binds to the cholecystokinin receptor 2 (CCK2 receptor) of ECL cells, thus inducing HIS secretion. As mentioned before, HIS is a stimulus for HCl secretion by parietal cells through the signalling pathway elicited by H_2R. On the other hand, circulating cholecystokinin (CCK) binds to CCK1 receptors of gastric D cells, thus stimulating somatostatin secretion [71]. Somatostatin directly inhibits acid secretion by parietal cells, as well as both HIS and gastrin secretion [72].

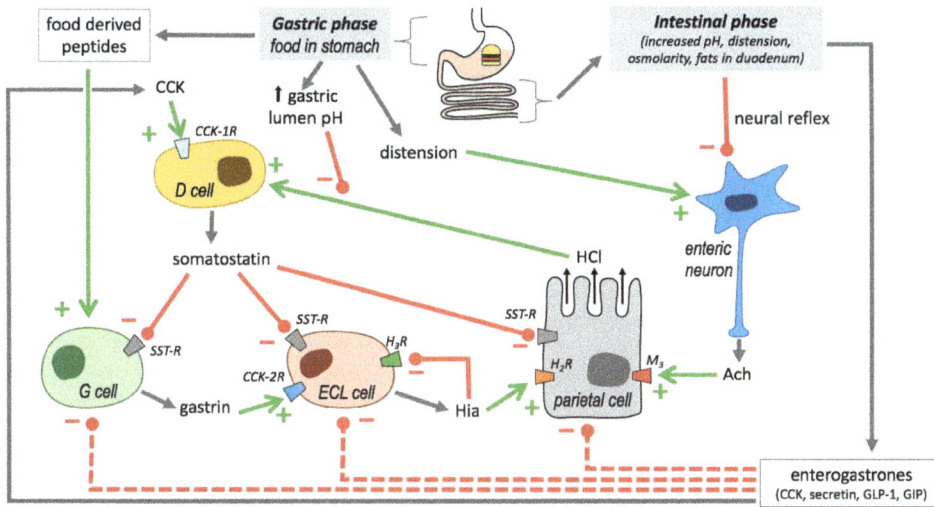

Figure 3. Balance between stimuli and inhibitors of gastric secretion. CCK, cholecystokinin; CCK_nR, different types of cholecystokinin receptors (1 or 2); D cell, somatostatin-releasing cell; ECL cell, enterochromaffin-like cell; G cell, gastrin-producing cell. GLP-1, glucagon-like peptide; GIP, gastric inhibitory polypeptide; H_nR, different types of histamine receptors; M_3, muscarinic acetylcholine receptor type 3; SST-R, somatostatin receptors. Products are represented by grey arrows, activations by green plus symbols and arrows, and inhibitions by minus symbols and red bars.

The balance between stimuli and inhibitors change throughout different phases involved in the process, including an intracranial phase, a gastric phase, and an intestinal phase. In the next paragraphs, we will focus on phases directly related to GIT.

In the gastric phase, the presence of food in the stomach induces acid gastric secretion by three different ways (Figure 3): the stomach distention caused by the food is detected by mechanoreceptors, which in turn induces neuronal reflexes for acetylcholine production; food derived-peptides and amino acids stimulate gastrin secretion by G cells; food increases gastric lumen pH, which is an inhibitory signal for somatostatin secretion [72].

When chyme reaches the duodenum, negative feedback mechanisms operate to reduce acid secretion (Figure 3). On the one hand, neuronal reflexes are activated, therefore blocking acetylcholine induced HCl secretion. On the other hand, in enteroendocrine cells, the synthesis of somatostatin synthesis activators (i.e., CCK, secretin, glucagon-like peptide and gastric inhibitory polypeptide) [73] are also promoted in different enteroendocrine cell types, which finally lead to gastrin, HIS and HCl secretion inhibition (Figure 3).

It has been proposed recently that HIS could also inhibit its own secretion through binding to H_3R present in ECL cells membranes [17]. Acting through other receptors, HIS has also been proposed as involved in the gastric vasodilatation and reactive hyperaemia produced in response to acid challenge (through H_1R), and the modulation of the gastric mucosal defence, the enteric neurotransmission and the feedback regulation of HIS release (through H_3R, and maybe also through H_4R) [73,74]. Nevertheless, the precise roles of H_4R in gastric physiology are still controversial [75].

2.5.2. Histamine and Immune Response in Gastrointestinal Tract

As mentioned before, mast cells (and basophyls) are the major producers of HIS. In these cells, HIS release can be induced by IgE, but also by cytokines, neuropeptides, growth factors, free radicals and anaphylotoxins [17], many of them potentially present in the GIT. Other eventually HIS-producing, but not HIS-storing cells, like lymphocytes, fibroblasts and macrophages are also present and interact with GIT epithelia [76,77].

HIS modulates immune response mainly through H_1R, H_2R and H_4R, depending on the receptor type expressed in each immune cell type. HIS elicited immune actions include the capability to modulate expression and/or activity of many cytokines and the complement system [78–80]. Again in turn, a cross regulation between cytokines and HIS seems to control GIT functions, as it has been proven that several cytokines such as TNF-α, IL-1 and IL-6 modulate HIS synthesis and secretion. Several components of complement systems like C3a, C4a and C5a (anaphylotoxins) have the capability to induce HIS release from mast cells and basophils [17]. During the last decade, interest of the role of H_4R in the GIT context has increased [81].

3. Serotonin Biochemistry and Physiology

3.1. Serotonin Synthesisn

5-HT is an L-tryptophan-derivative (Table 1). Meat, milk and fruit are the major sources of the essential amino acid precursor [82]. About 95% of the total 5-HT content in a human body is synthetized by GIT-associated cells (approximately 9/10 by intestinal enterochromaffin-cells (EC), and 1/10 by serotoninergic neurons located in the myenteric plexus. Only 5% of the 5-HT content in a human body is estimated to be synthetized in CNS [83].

The biosynthetic 5-HT pathway begins with the hydroxylation of the indole moiety C5 (Figure 2b) catalysed by the enzyme tryptophan hydroxylase (TPH, EC 1.14.16.4). It is a tetrameric non-heme iron-dependent monooxygenase that uses L-Trp and oxygen as substrates and tetrahydrobiopterin (BH_4) as the cofactor. The reaction occurs as two sequential half reactions: a reaction between the active site iron, oxygen, and the tetrahydropterin to produce a reactive $Fe^{IV}O$ intermediate and the hydroxylation of the amino acid by $Fe^{IV}O$ [84]. Two isoforms have been detected for this enzyme. TPH-2 expression is almost exclusive for neurons, and TPH-1 is expressed in other 5-HT-producing cell types [85].

The TPH product, 5-hydroxy-L-tryptophan (5-HTP), is the substrate of DDC that produces the amine 5-HT by decarboxylation of the 5-HTP α-carbon. This enzyme also decarboxylates other aromatic L-amino acids or derivatives; for instance, L-dihydroxyphenylalanine (L-DOPA) to produce DA [86]. In addition, it is also able to catalyse other reactions under different environmental circumstances or specific mutations (for instance, half-transaminations and oxidative deaminations) [87,88]. The mammalian enzyme is also a PLP-dependent enzyme, highly homologous to mammalian HDC, as mentioned above [26,89]. In fact, it is able to accept L-His as a substrate but with a much lower affinity than for 5-HTP or DOPA; however, the human DDC gene lacks the sequence encoding the carboxy-terminal fraction present in mammalian HDC, which is involved in mammalian HDC sorting to endoplasmic reticulum and activation [49]. This could suggest a different intracellular location for both enzymes. Mammalian HDC and DDC share the catalytic mechanism explained above for HDC [54,90]. However, at least in the case of the purified recombinant wild proteins, mammalian DDC seems to be a more efficient enzyme according to their respective catalytic constant values obtained in silico and in vitro [54,55]. In the case of DDC, slight modifications of the catalytic site environment seem to induce important changes in catalytic constant (k_{cat}) values [90]. Both enzymes (DDC and HDC) also share other structural properties related to enzyme stability and catalysis. For instance, the presence of PEST regions in the N-terminus of the monomers and a highly labile flexible loop, which is essential for conformation changes of the enzymes during PLP binding and for catalysis itself [6,53,88]. In the case of 5-HT biosynthesis, the limiting step is not decarboxylation but TPH activity.

It is noteworthy that HIS and 5-HT synthesis exhibit antagonist time-course patterns during differentiation of mouse bone marrow derived cells to mast cells in vitro, as well as opposite responses to PA inhibitors [67]. These results suggest a sort of regulatory coordination among all of these amine biosynthetic processes, which are not fully characterized yet, but should be taken into account in all of the pathophysiological scenarios where synthesis of these amines can be confluent, as GIT is.

3.2. Serotonin Degradation

5-HT degradation is catalysed mainly by any MAO activity (Figure 2b). MAO catalysis requires FAD as the cofactor to carry out an oxidative deamination of 5-HT, thus producing hydrogen peroxide and 5-hydroxy-3-indolacetaldehyde, which is rapidly processed to 5-hydroxy-3-indolacetic acid by the action of AD [91]. Human genome contains two different genes encoding MAO activities, namely MAO-A and MAO-B [12]. Both proteins are located in the external mitochondrial membrane [92]. MAO-A has a higher affinity by 5-HT as well as a wider expression spectrum. However, MAO-B is the only one detected in serotoninergic neurons [91]. Both MAO isozymes are expressed in GIT [61,62,93].

3.3. Serotonin Transport and Storage Mechanism

In GIT, 5-HT is mainly produced and secreted by the neuroendocrine enterochromaffin (EC) cells, located alongside the intestinal epithelium lining the lumen of the digestive tract. Recently, it was described that the sodium channel $Na_V1.3$ plays an important role for EC excitability and 5-HT release [94]. Once 5-HT is secreted by EC cells and binds to the specific receptors of the surrounding cells, it is removed from the interstitial space by the sodium-dependent 5-HT transporter (SERT), also named as the solute carrier family 6 member 4 (SLC6A4), which is expressed by GIT epithelial cells. SERT is a protein with 12-transmembrane fragments, which is able to transport 5-HT by a Na^+/K^+- and Cl-dependent mechanism [83,95]. Inside the GIT epithelial cells, 5-HP is rapidly degraded by MAO activity, as explained above [83,95,96].

In addition, postprandial 5-HT can also enter systemic circulation and is absorbed by platelets. Actually, most of the circulating 5-HT is stored in platelets, as these cells also express 5-HT transporters. SERT is negatively regulated by activation of tool-like receptors and several pro-inflammatory cytokines. On the contrary, other anti-inflammatory cytokines, such as IL-10, increase transporter activity. Treatment with specific SERT inhibitors leads to an increase of free 5-HT content, thus empowering 5-HT effects, not only in GIT but also in CNS [83].

3.4. Serotonin Signalling and Physiological Functions

It is well known that 5-HT has also been involved in very complex physiological processes such as being an essential neurotransmitter and paracrine molecule for brain-intestine crosstalk, commonly known as gut-brain axis [97]. The amine is involved in modulation of body temperature and circadian rhythm [98,99], as well as in cardiovascular activity, morphogenesis and cell proliferation [100,101]. In the GIT context, it has been described as a gastric motility and secretion, a nutrient absorption regulator and an immunoregulatory compound. Consequently, dysfunctions in 5-HT metabolism usually have very important negative consequences on human physiology including gut-brain communication [82,102]. 5-HT also elicits both motor and sensitive responses in the intestine by binding to different receptors expressed by mesenteric and mucosal neurons (Table 3).

Table 3. Molecular and functional properties of the best-known human 5-HT receptor subtypes important for GIT functions.

Properties	5-HT$_{1A}$ Receptors (5-HT$_{1A}$R)	5-HT$_{1D}$ Receptors (5-HT$_{1D}$R)	5-HT$_2$ Receptors (5-HT$_2$R)	5-HT$_3$ Receptors (5-HT$_3$R)	5-HT$_4$ Receptors (5-HT$_4$R)	5-HT$_7$ Receptors (5-HT$_7$R)
Chromosome	5	6	13/2/X	11 (A, B and C) and 3 (D and E)	5	10
Molecular weight (KDa)	421	390	471/481/458	Pentameric 478 (A); 441 (B); 447 (C); 279 (D); 471 (E)	387	445
G protein signalling	G$_{i/o}$	G$_{i/o}$	G$_{q/11}$	Activated by ligand binding and opening channels	G$_s$	G$_s$
Expression	Enteric neurons, substantia nigra, hippocampus	Enteric neurons, substantia nigra, basal ganglia	Stomach, fundus, caudate nucleus, cerebellum	Enteric, sympathetic and vagus nerves, area postrema	Enteric neurons (myenteric plexus), hippocampus	Smooth muscle, thalamus, hypothalamus and hippocampus
Physiological effects	Neuronal inhibition	Neuronal inhibition	Muscle contraction	Neuronal depolarization Increased neurotransmitter release	Muscle contraction Positive effects on cholinergic transmission.	Muscle relaxation

Data from references [95,103].

3.4.1. Regulation of GIT Smooth Muscle Contraction and Relaxation

5-HT is a regulator of both intestinal smooth muscle contraction and relaxation through the activation of enteric excitatory motor neurons and intrinsic inhibitory neurons, respectively [83,104]. The amine can bind 5-HT$_3$R and 5-HT$_4$R of excitatory cholinergic motor neurons, thus inducing acetylcholine release and smooth muscle contraction. However, 5-HT binding to 5-HT$_4$R, 5-HT$_{1A}$R, and/or the badly characterized 5-HT$_{1D}$R, present in inhibitory nitrergic motor neurons induces nitric oxide (NO) synthesis and consequently smooth muscle relaxation (Figure 4). In addition, 5-HT also participates in gastric muscle motility regulation [83,105].

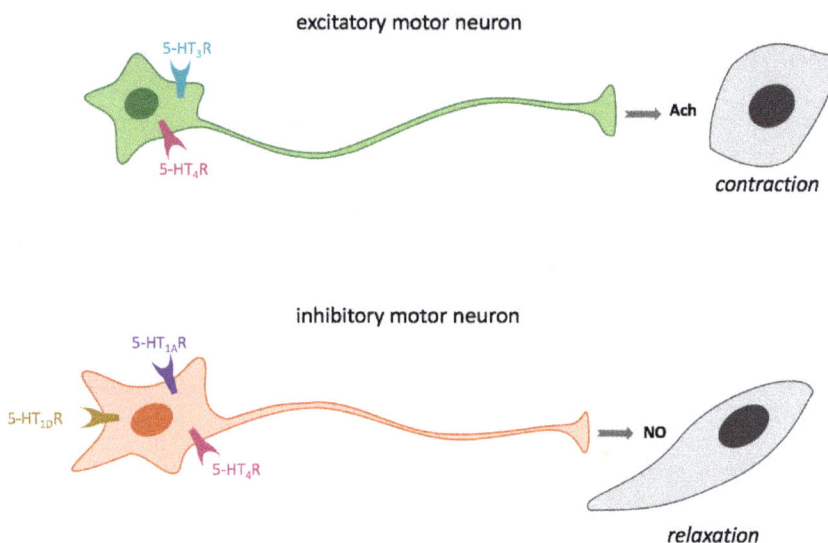

Figure 4. Regulation of GIT smooth muscle contraction and relaxation through 5-HT receptors. AcH, acetylcholine; 5-HT$_{1A}$R serotonin receptor type 1A; 5-HT$_{1D}$R; serotonin receptor type 1D, 5-HT$_3$R; serotonin receptor type 3, 5-HT$_4$R; serotonin receptor type 4.

3.4.2. Mucosal Sensory Transduction

EC cells secrete 5-HT in response to intraluminal pressure. The released amine can stimulate both the intrinsic primary afferent neurons (IPANs) located in submucosal and myenteric plexus and the extrinsic afferent neurons (vagal and spinal), through their binding to different receptors: 5-HT$_3$R, 5-HT$_4$R, 5-HT$_7$R; and 5-HT$_{1D}$R [83].

On the one hand, submucosal neurons release both acetylcholine and calcitonin gene-related peptide; however, myenteric neurons only release acetylcholine. Both neuron types are involved in the modulation of intestinal motility, secretion, and vasodilatation. Thus, submucosal neurons initiate peristaltic and secretory reflexes, and myenteric neurons start migratory contractions. On the other hand, spinal afferent neurons transmit signals related to digestive reflexes, satiety, and pain from the intestine to the CNS.

In addition, some authors claim an important role of neuronal 5-HT in promotion of development/survival of some classes of late-born enteric neurons, including dopaminergic neurons, which appear to innervate and activate the adult enteric nervous system [104].

3.4.3. Serotonin and Immune Response in GIT

It has been reported that 5-HT can also elicit pro-inflammatory responses in GIT that involve different transduction pathways most probably started by the amine binding to the 5-HT receptors expressed by dendritic cells located in *lamina propria*. Recently, this immuneregulatory role of 5-HT in GIT has been the subject of very relevant reviews on the topic [95,102].

4. Biochemistry and Physiology of Catecholamines

4.1. Synthesis of Catecholamines

DA, and their derivatives noradrenaline/norepinephrine and adrenaline/epinephrine, are the most important cathecolamines for human physiology (Table 1). All of them are synthetized from L-phenylalanine or L-tyrosine mainly from diet (Figure 2c). L-phenylalanine can be the substrate of phenylalanine hydroxylase (PAH) to produce L-tyrosine. The limiting step for DA synthesis is the enzyme tyrosine hydroxylase (TH, tyrosine 3-monooxygenase, EC 1.14.16.2), which introduces a hydroxyl group in the meta position of the cathecol ring of L-tyrosine to obtain L-3,4-dihydroxyphenylalanine or L-DOPA). This reaction requires Fe^{2+}, the cofactor BH$_4$ and O$_2$. TH is mainly expressed in neuroendocrine cells in both soluble and membrane bound forms. It is a highly stereo specific enzyme, although it can also act on L-phenylalanine [82,106]. Up to four different alternative TH mRNA spliced forms have been detected. The meaning of this variability is still uncertain [106].

L-DOPA is a substrate of the DDC mentioned in the previous section, producing DA. In fact, it has also been named as dopa decarboxylase in the literature [6,90]. La AADC or DDC is a ubiquitous enzyme expressed by cell types located in different organs like the adrenal medulla, kidney, liver, GIT and brain [61,106].

In addition to the above-mentioned pathway, there are two other alternative ways to produce DA (not shown in Figure 2). One way, L-tyrosine, can also be decarboxylated by AADC to produce tyramine, which is hydroxylated by a member of the cytochrome P450 family (family 2, subfamily D, or CYP2D). Nevertheless, L-phenylalanine can also be decarboxylated by AADC to produce phenyltyramine, which can be converted to DA by CYP2D [92].

In several peripheral tissues (mainly in adrenal medulla), two other further reactions can take place to produce norepinephrine and epinephrine. Firstly, the action of the enzyme dopamine β-hydroxylase (DBH, dopamine β-monooxygenase, EC 1.14.17.1) produces norepinephrine. This oxidase requires ascorbic acid as the electron donor. It is a highly antigenic homotetramer (Mr around 290 kD) with low substrate specificity [106].

Finally, the enzyme phenylethanolamine *N*-methyltransferase (PNMT, EC 2.1.1.28) catalyses the *N*-methylation of norepinephrine to produce epinephrine (Figure 1c). PNMT is a cytosolic enzyme that uses *S*-adenosylmethionine (SAM) as the methyl donor. It has a low substrate specificity that allows it to carry out the beta carbon methylation of a variety of amines. Its expression is mainly but not exclusively restricted to the suprarenal glands [106,107].

4.2. Degradation of Catecholamines

As occurring with other BAs, CAs can be the subject of oxidative deamination catalysed by MAO, thus producing H_2O_2 and aldehydes, 3,4-dihydroxyphenylacetaldehyde (DOPAL) from DA, and 3,4-dihydroxyphenylglycoaldehyde (DOPEGAL) from epinephrine and norepinephrine (Figure 2c). Both are instable compounds rapidly oxidized to dihydroxyphenylacetic acid and 3,4-dihydroxymandelic acid, respectively, by the action of AD [108,109]. Alternatively, the enzyme aldehyde reductase (AR, EC 1.1.1.21) can reduce DOPAL to 3,4-dihydroxyphenylethanol, and DOPEGAL to dihydroxyphenylglycol. The lack of a beta-hydroxyl group in DOPAL favors its oxidation by AD. Conversely, the presence of the β-hydroxyl group in DOPEGAL makes it a better substrate of AR [108].

Nevertheless, the major product of norepinephrine degradation in humans seems to be vanillylmandelic acid (VMA), produced mainly by a pathway that requires the consecutive actions of MAO, AR, catechol O-methyltransferase (COMT, EC 2.1.1.6) alcohol dehydrogenase (ADH, EC 1.1.1.1) and AD [108]. COMT also uses SAM as the methyl donor. Two different isoforms are encoded by a unique gene, the cytosolic isoform being the one present in glia and peripheral organs (for instance, liver and kidney) (not shown in Figure 2).

DA and their derivatives can be converted into other molecules in CNS and peripheral tissues before being excreted. The activity phenolsulfotransferase (EC 2.8.2.1) is able to produce dopamine-3-O-sulfate and dopamine-4-O-sulfate by transferring a sulfate group of 3′-phosphoadenosine-5′-fosfosulfate to any hydroxyl groups of the catechol (preferably to dopamine-3-O-sulfate). Moreover, the enzyme uridine diphosphoglucuronosyltransferase (EC 2.4.1.17) transfers glucuronic acid from UDP-glucuronic acid to both hydroxyl groups of the dopamine catechol ring. This enzyme is linked to the reticulum endoplasmic membrane [109].

4.3. Signalling and Physiological Functions of Catecholamines

CAs act as both neurotransmitters and hormones, depending on their targets in different tissues/organs [110]. In CNS, they are mainly involved in motor and emotional control, cognition, and memory [111].

In peripheral tissues, they also play important roles as modulators of blood pressure and renal excretion, as well as in the immune system and GIT functions [112]. DA receptors are G protein-coupled receptors (GPCRs) classified into 5 types (D1–D5) [113]. Heterodimerization among the subtypes and with other receptor type monomers have been reported, which results in a very complex cell-dependent signalling network. In addition, DA can also elicit physiological signalling through G-protein independent mechanisms (i.e., ion channels, tyrosine kinases and even arrestins [113].

Epinephrine and norepinephrine preferably bind to adrenergic receptors or adrenoceptors (all GCPRs) that are classified as α (1a,1b,1d and 2a,2b,2c) and β (1–3) subtypes [82,114]. In the intestine, the main epinephrine receptors are α1 and β2. However, in the case of norepinephrine, they are α1 and α2 (as well as β2, but in a minor degree). These differences determine the specific effects of the different CAs on absorption, blood flux, and motility in the intestine [82].

Regulation of Intestinal Blood Flux, Immunity and Motility

Norepinephrine binding to α-adrenergic receptors induces vasoconstriction and increases vascular resistance, thus reducing the blood flux in the intestine. Low epinephrine levels stimulate β receptors, inducing vasodilation and consequently an increase in blood flux. However, at high levels, it induces similar effects to norepinephrine [82]. DA preferentially binds to D1 at low concentrations, and to adrenoceptors β1, and even to α1, as the DA concentrations increase. Thus, low DA levels induce vasodilation and increase blood flux, but high DA levels can induce vasoconstriction and consequently abdominal flux blood decrease.

Recently, in 2017, Mittal et al. [82], summarized the major effects of CAs in GIT:

- *Nutrient absorption.* Both epinephrine and norepinephrine play important roles in nutrient absorption regulation. Epinephrine is able to induce a hyperglycemic response acting through β-adrenergic receptors, and it increases absorption of oligopeptides when bound to α-adrenoceptors.
- *Intestinal motility.* CAs binding to β-adrenoreceptors induces smooth muscle relaxation that lead to a global food transit delay. On the contrary, their bindings to α-adrenoreceptors stimulate intestinal smooth muscle contraction, and consequently gut motility and food transit.
- *CAs, immune system and GIT.* Recently, CAs, as well as 5-HT, have been described as regulators of the innate immune system, which can be related to food intolerance. In addition, it is also reported that these amines can influence the intestinal microbiota [115].

These effects are very interesting in the context of nutrition, therefore deserving further research [116]. Fortunately, the advances of high-throughput technologies can help us to reveal the bases of these complex but important subjects for healthy and personalized nutrition.

5. Biochemistry and Physiology of Polyamines

5.1. Synthesis of Polyamines

PAs synthetized by mammalian cells are aliphatic low molecular weight polycations with 2–4 protonated amino groups at physiological pH: Put, Spd and Spm. They are essential for all living organisms from Archaea to humans, as they modulate the most basic mechanism for life, macromolecular synthesis and structure and membrane dynamics. The diamine Put is the precursor of the triamine Spd and the tetramine Spm. In mammalian cells, Put is synthetized directly from α-decarboxylation of the amino acid L-ornithine. The aminopropyl groups of Spd (1) and Spm (2) are from decarboxylated SAM (dcAdoMet) (Figure 2d) [117]. Plants and microorganisms can synthetize other BAs with different lengths, positive charges and configurations; for instance, cadaverine (pentane-1,5-diamine), Agm, and other PAs synthetized by thermophilic microorganisms [118,119].

The first step for PA biosynthesis in mammalian cells is the hydrolysis of the guanidinium group of L-arginine catalysed by arginase (EC 3.5.3.1) activity (not shown in Figure 2d). Its product, L-ornihine, is the substrate of the PLP-dependent decarboxylase, ODC, a minor and instable protein, which is a limiting step of PA synthesis (Figure 2d). ODC product (1,4-butanodiamine) is commonly known as Put (Figure 2d). Eukaryotic ODC structure follows a model that belongs to the group IV of the PLP-dependent L-amino acid decarboxylases [7]. It needs to be a dimer to be active (≈a 102 kDa homodimer in mammals). The enzyme has one of the shortest half-lives known so far for mammalian proteins (10–50 min) and is located mainly in cytosol but it has also been detected in nucleus [120,121]. Its activity is highly regulated in response to different growth factors, oncogenes, trophic hormones, among other proliferative stimuli [122].

As mentioned before, dcAdoMet is required to synthetize higher PAs (Spd and Spm). Its synthesis involves the condensation of L-methionine and ATP to produce SAM, catalysed by any of the isoforms of *S*-adenosylmethionine synthetase or methionine adenosyltransferase (MAT, EC 2.5.1.6). SAM can then be decarboxylated by the action of *S*-adenosyl-L-methionine decarboxylase (SAMDC, EC 4.1.1.50), producing the nucleoside dcAdoMet, which acts as the aminopropyl donor for Spd and Spm synthesis (Figure 2d). SAMDC activity can also be a limiting step of the PA biosynthesis pathway. The mature enzyme suffers a post-translational maturation process, which renders the essential pyruvoyl prosthetic group [123]. Spd is synthetized by the transfer of the dcAdoMet aminopropyl moiety to the N^4 of Put through the action of spermidine synthase (SpdS, EC 2.5.1.16). Finally, the addition of a new dcAdoMet aminopropyl moiety to the N^8 of Spd gives rise to the tetramine Spm through the action of spermine synthase (SpmS, EC 2.5.1.22) [124].

The aminopropyl transferases SpdS and SpmS are homologus homodimers but with high substrate specificity. Steric restrictions avoid binding of Put to SpdS, as well as binding of Spd to SpmS. Nevertheless, both human enzymes contain two key Asp residues (Asp^{104} and Asp^{173} in SpdS, and Asp^{201} and Asp^{276} in SpmS), which are essential for the catalytic mechanism [119].

5.2. Degradation and Recycling of Polyamines

PA metabolism can be considered to be a very robust bicycle involving both metabolic branches (synthesis and degradation) (Figure 2d) [117]. Degradation involves a series of different amine oxidases. Spermine oxidase (SMO, EC 1.5.3.16) is a FAD-dependent oxidase able to directly transform Spm to Spd, 3-aminopropanal and H_2O_2 in the presence of H_2O and O_2. Several alternative splicing variants have been observed. It is highly inducible by PAs and their analogues, among other stimuli [124].

Al alternative pathway to convert Spm into Spd (as well as Spd into Put) requires the action of spermidine/spermine N^1-acetyltranferase (SSAT, EC 2.3.1.57). SSAT reaction using Spm and acetyl-CoA as the substrates transforms Spm into N^1-acetylspermine. This product may, in turn, follow two alternative pathways. It can be a substrate of the peroxisomal polyamine oxidase (PAO, EC 1.5.3.13) that produces the lower PA (Spd), 3-acetamidopropanal and H_2O_2 (Figure 2d). N^1-acetylspermine can even be acetylated again in its other terminal producing N^1,N^{12}-diacetylspermine. This metabolite can be a substrate for peroxisomal PAO to produce 3-acetamidopropanal, H_2O_2 and N^1-acetylspermidine. In a second reaction on N^1-acetylspermidine, PAO produces 3-acetamidopropanal, H_2O_2 and Put [125,126].

SSAT also can act on Spd to form N^1-acetylspermidine (and CoA). N^1-acetylspermidine, which is subsequently oxidized by PAO producing 3- the diamine Put, 3-acetamidopropanal and H_2O_2 [125]. The diamine Put can be further degraded by DAO [118], or alternatively recycled for higher PA synthesis. Acetylated PA are more easily excreted than their deacetylated counterparts, and their levels in urine have been used as biomarkers of elevated PA metabolism [127].

In summary, the net result of the action of SMO or the tandem SSAT plus PAO is the conversion of higher PAs or their acetylated versions into their respective lower poly- or diamine, which can be again recycled in the biosynthetic pathway (Figure 2d). It conforms two energy-consuming cycles being apparently futile. However, following the metabolic control theory, "futile cycles" confer high sensitivity for modulation of metabolic pathways that need to respond to regulatory stimuli in a coordinated way as a response (sometimes a compensation) to external alterations [128]. That indeed is the case for mammalian PA metabolism, as predicted by the mathematical model of mammalian PA metabolism and proven by its further validation [117,129]. This fact explains the difficulties experienced by many experimental groups when trying to deplete intracellular PA levels as an anticancer strategy [130,131].

5.3. Polyamine Transport Systems

PA import systems in mammalian cells are not yet fully characterized, in spite of the multiple efforts made by different group members of the PA research community; this fact is still one of the most important handicaps to control intracellular PA levels under pathological circumstances (for instance, cancer) [132,133].

Currently, three models have been proposed and reviewed by Poulin et al. [133]. A first model proposes the action of two permeases with different locations, one located in the cytosolic inner membrane (PMPP) and an H^+–coupled PA transporter located in vesicular membranes (VPA). A second mechanism (one step model) involves glypican-1, acting as a high affinity PA receptor. This binding could induce endocytosis leading to PA internalization, as described by Belting et al., for Spm transport [134]. The presence of NO and Ca^{++} in the endosomes would revert glypican-Spm binding. A third model proposes PA interaction with caveoline-1, which would also be reverted by NO. These mechanisms could explain the presence of higher PA in vesicles of several mammalian cell types. However, many doubts still remain concerning PA transport mechanisms through the different cellular compartments, as well as cell- and PA-specificities and regulation of each transport mechanism. Abdulhussein and Wallace recently reviewed this topic [135]. Specifically, Uemura et al., identified the amino acid transporter SLC3A2 as a Put export protein in colon cancer-derived cells [136] and also studied the specific characteristics of PA absorption by the intestinal tract, providing methods for PA transport analysis in the colon and the small intestine using membrane vesicles, culture cells, and mouse models [137].

5.4. Physiological Functions of Polyamines

PAs can be considered protonated amino groups kept together by aliphatic skeletons that impose specific distances among them. Thus, their positive charges and aliphatic chains can interact specifically with negative charges and hydrophobic residues/surfaces of other biomolecules (nucleic acid sequences, proteins and lipids) located at the correct distances, therefore modifying their conformations and consequently the functional properties of macromolecular structures. PAs are present and absolutely essential to keep cell viability in almost all living organisms. PA-DNA, PA-RNA and PA-membrane interactions and their conformational consequences can be reproduced and studied in vitro (in an abiotic environment) working with their purified components (PA, polynucleotides and/or lipids). This knowledge on specific molecular PA interactions gives rise to new hypotheses [138–140]. On the one hand, it is clear that some specific binding modes have been detected (among an immense quantity of possibilities for PA-biomolecules interactions) with physiological consequences or applications (for instance, nanotechnology applied to drug delivery) [141]. On the other hand, it is tempting to hypothesize that their interaction with both nucleic acids and membranes was of absolutely essential value from the beginning of life on Earth [142].

The interactive properties of PAs allow them to modulate a long list of processes involved in cell cycle progression and gene expression as DNA condensation, replication fidelity, RNA secondary and tertiary structure stabilization, translation initiation, elongation and fidelity, and posttranslational modification of proteins, among others [143]. The importance of the mentioned physiological functions modulated by PAs explains that their metabolism is strictly regulated, as well as very robust. Nevertheless, imbalance in PA levels is associated with a long list of human diseases that includes aberrant cell growth and differentiation and/or abnormal protein expression and folding; for instance, cancer, GIT and neurodegenerative diseases [144], as will be mentioned further on. Nevertheless, many molecular questions still remain unsolved on the molecular bases of the cellular functions of PAs [145], thus delaying the biotechnological applications of this yet unveiled knowledge.

As the other above mentioned BAs, PAs also play important roles in GIT physiology [146,147]. Synthetized de novo or uptaken from the intestinal lumen intestinal, they promote two different processes described for intestinal mucous reparation: DNA-independent cell migration and replacement of damaged cells by cell proliferation [148].

It is also proven that PAs are important for the correct biochemical and morphological maturation of intestine during the postnatal period. The benefits are dose and PA-specific (being Spm > Spd > Put). In addition, PAs have also been proposed as being involved in the correct immunological system development in neonatal intestine [146,149].

5.5. The Particular Case of Agmatine

Agm is a member of the PA family, as it is the product of L-arginine decarboxylation. It can be converted into Put through the action of a liver agmatinase activity, a hydrolase that removes the agmatine guanidinium moiety. It has finally been clarified that human cells express active agmatinase, but not an active arginine decarboxylase (ADC, EC 4.1.1.19) [21]. However, some bacteria present in human microbiota have, in fact, active ADC, so producing Agm that can be absorbed by intestinal epithelium. Agm protects mitochondrial functions and confers resistance to cellular apoptosis [150], being able to modulate several processes such as hepatic regeneration and renal function [151], among other proposed physiological functions. For instance, it is able to block *N*-methylaspartate receptor receptor in brain areas related to learning and memory [152], as well as modulate mental stress [153,154]. Thus, it seems to be a good candidate to participate in gut-brain axis. As HIS and Put, Agm is also a DAO substrate to produce γ-butiramide that is finally converted into γ-guanidinobutirate in CNS [39]. It can also compete with the other diamines for binding to OCT transporters [155].

6. Biogenic Amines and Microbiota-Intestine Crosstalk

In addition to BAs from diet and the endogenous BA metabolism of GIT cells, there is another important contributor to BA levels in the intestine: the intestinal microbiota metabolism.

On the one hand, intestinal microbiota species, like any other living organism, are able to synthesize PA [156]. These PAs can be uptaken by intestinal epithelia, thus contributing to cellular growth and tissue renewal, especially in the colon [137,156].

On the other hand, microbiota species can also synthetize any of the other BAs mentioned in this text. For instance, a PLP-dependent HDC homologous to the mammalian enzyme can be expressed by Gram negative Enterobacteria. On the contrary, Gram positive bacteria (for instance, *Lactobacillus sp.*) can express a non-homologous pyruvoil-dependent HDC [157]. The physiological effects of HIS produced by microbiota are controversial and need more research to be fully understood [15]. For instance, in spite of the general idea of a deleterious role of diet HIS on human health (see next section), it has been reported that HIS synthetized by the probiotic *Lactobacillus reuteri* acts as a positive immune regulator acting through H_2R [158].

In addition, some bacteria potentially taking part in human microbiota can synthetize other BAs, different for the mentioned endogenous ones, and can also be bioactive for human physiology (for instance, tyramine) [159]. In general, as all BAs are described as neurotransmitters, neuroendocrine factors or neuromodulators, amines synthetized by microbiota may interact with host signals establishing a microbiota-endocrinology system crosstalk that is a part of the larger microbiota-gut-brain axis, with consequences in health and diseases [160].

Moreover, other products of the microbiota metabolism different from amines can regulate endogenous BA metabolism. For instance, bile acids and short-chain fatty acids, affect 5-HT synthesis that, in turn, directly or indirectly regulates gut motility [161] and enteric neuroimmune mechanisms [102]. A protective role of probiotics against histamine-mediated colon carcinogenesis have been recently reported [162], as well as against gastric cancer by modulating PA metabolism [163].

Summarizing, the so called microbiota-gut-brain axis is a very interesting but extremely complex open subject of current biomedicine that absolutely requires the help of new approaches from system biology and high-throughput technologies.

7. What Is Known about Biogenic Amines Roles in Human Gastro Intestinal Pathologies?

It is clear that alterations of elements of the BA metabolism, including transport and signalling pathways, are involved in a wide diversity of GIT pathologies. However, further efforts are needed to fully characterize the specific aberrant element(s) and/or mechanism(s) responsible for each pathological consequence. In the next subsection, we will summarize the current state of the art for the better known relationships between BA-related elements and human GIT diseases.

7.1. Gastric Diseases

7.1.1. Peptic Ulcers

Peptic Ulcers are mucosal erosions induced by gastric secretions. In addition to those located in gastric mucosa (gastric ulcers), similar damages can appear at the entrance of duodenum (duodenal ulcers), or even in a minor percentage in the oesophagus or other intestinal segments [164]. Gastric ulcers are usually located along the minor curvature, particularly in the corpus-antrum transitional mucosa, their prevalence being higher in over 40 year-old humans. Duodenal ulcers are located between the lower part of the stomach and the start of the intestine (that is, the intestinal area exposed to gastric acid) with the highest frequency being between 20–50 year-old humans [70].

Abdominal pain is the most common symptom of peptic ulcer. Swelling, loss of appetite, nausea and/or indigestion are also usual symptoms. Associated complications include bleeding, perforation and stenosis. Bleeding is the most common complication among peptic ulcer patients (15–20%). Reciprocally, 40% of humans suffering from upper GIT bleeding are peptic ulcer patients [70].

Peptic ulcers are classified depending on their anatomical location [70,165,166]. When they are located in the stomach, they evolve as atrophic gastritis. Initially, inflammation induces parietal cell apoptosis, which in turn induces gastric acid hyposecretion and mucosal atrophy. Chronic inflammation and mucosal atrophy lead to gastric ulcers and eventually to gastric cancer.

Other individuals present gastritis located in the pylorus area. In this case, excessive quantities of gastric acid are usually produced, which lead to duodenal ulcers. The inflammatory response induces cytokine secretion that finally induces dysregulation of endocrine cells located in this area. Thus, G cells are stimulated to overproduce gastrin, while somatostatin secretion is inhibited. As a consequence, HIS synthesis and secretion increase, which in turn promotes proliferation and stimulation of parietal cells, leading to gastric acid hypersecretion (Figure 3). These facts explain the treatment of the ulcers with proton pump inhibitors and H_2R antagonists [167,168].

Infection by *Helicobacter pylori* is a very common origin of peptic ulcers. Nevertheless, other risk factors as alcohol and smoking abuse, as well as a continual use of several drugs (for instance, nonsteroidal anti-inflammatory drugs or NSAIDs) have been described. *H. pylori* is a microaerophilic flagellated bacteria able to colonize human gastric mucosa. It is a highly common infection that can take place for tens of years. This Gram-negative bacteria is considered an important pathogenic agent associated with several human pathologies like chronic gastritis, gastrointestinal ulcers, and neoplasms such as gastric adenocarcinomas and gastric mucosa-associated lymphomas (Table 4) [169].

H. pylori induces HDC expression, and consequently an increase in endogenous HIS synthesis, which leads to an inflammatory response including an increased presence of neutrophils and lymphocytes, which in turn produces different cytokines and chemokines (for instance, IL-1, IL-6, TNF-α and IFN-γ). Gene polymorphisms detected in some of these human cytokine genes are proposed as being involved in resistance or susceptibility to *H. pylori* [75,170].

Table 4. Features associated with *H. pylori* infections and subsequent inflammation located in different parts of the gastrointestinal tract.

Location	Acid Secretion	Gastric Features and Histology	Intestinal Features and Histology	Pathology
Stomach (pan-gastritis)	Hyposecretion	Chronic inflammation and parietal cell apoptosis Atrophy Intestinal metaplasia	Normal	Gastric ulcer Gastric cancer
Pylorus area	Hypersecretion	Chronic inflammation and increased gastrin released Inhibition of somatostatin Increase parietal cell stimulation	Gastric metaplasia Active chronic inflammation	Duodenal ulcer

Data from reference [16–18].

Different signalling mechanisms have been described to explain the pathological consequences of *H. pylori* infection. On the one hand, there is a CAG (cytotoxin-associated gene)-dependent pathway (involving signal transduction elements like Rho GTPases, PKA, MKK4 and JNK, and the transcription factors AP-1 and NF-κB), which result in synthesis and secretion of cytokines acting as an innate immune response. On the other hand, MAP kinase pathway (involving Raf-1, ERK, MEK) is activated by a CAG-independent mechanism, which finally results in activation of BP1 and BP2.

These transcription factors act as inducers on the HDC promoter, which lead to an increase in gastric acid synthesis [171].

7.1.2. Gastric Cancer

Gastric cancer presents a high morbidity and mortality, being described as one of the most common worldwide malignant neoplasms. Nevertheless, its prevalence has decreased in the last decades in most European countries probably due to changes in lifestyle such as smoking reduction and *H. pylori* control [172].

Chronic inflammation produced by the pathogen can result in changes in the normal architecture of gastric mucosa, destruction of gastric glands, parietal cell loss, decreased gastric acid secretion, intestinal metaplasia, and finally gastric cancer. Thus, World Health Organization estimates that chronic *H. pylori* infection increases the risk of gastric cancer by 10 times, considering it as a class I carcinogen [70]. In fact, a reduced gastric acid secretion is not only a predisposition to gastric ulcers, but also to gastric cancer. Low levels of gastric acid reduce vitamin C absorption and allow an excessive growth of salivary and intestinal bacteria in the stomach, which can promote carcinogenesis [165].

It has been observed that treatments with gastrin/CCK-2 receptor antagonists reduce parietal cell apoptosis and inhibit gastric atrophy. Similar results were obtained with an irreversible H_2R antagonist (i.e., cimetidine); suggesting that both receptors could be involved in gastric cell apoptosis and carcinogenesis [173]. However, a prospective, double blind trial carried out among hundreds of gastric cancer patients by the British stomach cancer group did not result in any increase of survival [174].

Recent research on H_4R physiological roles reveals interesting information on the involvement of the receptor in the relationship between the immune system and GIT carcinogenesis [70].

As PAs (mainly Spd and Spm) are also essential for gastric cancer progression, some authors claim that new probiotic-based anticancer strategies are able to reduce endogenous PA levels and gastric cancer growth as a result [163].

7.2. Intestinal Diseases

7.2.1. Irritable Bowel Syndrome

Formerly known as spastic colon, irritable bowel syndrome (IBS) is a prevalent disorder that is characterized by alterations in intestinal secretions and motility, mainly affecting the colon. The symptoms (cramps, abdominal pain, intestinal habit alterations, and food intolerances) can appear during childhood or in adults [175,176]. Different subtypes have been described. The IBS-D subtype is characterized by diarrhoea; IBS-C is characterized by constipation; and IBS-M presents both intestinal alterations [175].

Several research groups have proposed the involvement of 5-HT-related elements (enzymes, transporters and receptors) in the pathology. For instance, results of several studies point out genetic variants of the gene encoding the 5-HT transport SERT (chromosome 17) that could predispose to IBS [177]. However, other results are not conclusive and sometimes contradictory [178]. Thus, it is one of the complex BA-GIT relationships needing further investigation.

Different treatments are prescribed for IBS patients depending on the severity of the symptoms. Diet adjustments and behavioaral education are usually enough for mild and moderate symptoms. However, for most severe forms, multidisciplinary approaches including pharmacotherapy are required. For IBS-D, treatment with antidiarrhoeal drugs, such as loperamide, can be necessary. In the case of women with severe symptoms, the $5\text{-}HT_3R$ antagonist alosetron can be used while the $5\text{-}HT_4R$ agonist prucalopride is used to relieve constipation in IBS-patients [179].

Other drug discovery initiatives are trying to develop effective inhibitors of tryptophan hydroxylase 1 (TPH1, the isoform expressed in GIT), which are unable to pass the blood-brain barrier, as an alternative to reduce 5-HT synthesis by EC cells, and consequently reduce/avoid their deleterious effects induced by dysregulation of the gastrointestinal serotonergic system (for instance IBS and carcinoid syndrome) [180,181].

7.2.2. Inflammatory Bowel Diseases

Crohn disease (CD) and ulcerative colitis (UC) are inflammatory bowel diseases (IBD) with different characteristics. Nevertheless, both diseases occur with alternating periods of remission and relapse. UC is characterized by inflammation with the presence of superficial colon mucosa ulcers that usually originate in the colon and then progress towards upper colon sections. UC symptoms include diarrhoea, cramping, and rectal bleeding. CD is characterized by a discontinuous pattern along the intestinal tract and can present larger ulcerations and sometimes granulomas; abdominal pain, diarrhoea, weight loss and bleeding being its most common symptoms [45].

Different research groups have observed a HIS secretion increase in jejunum of CD patients and high HIS levels in the intestinal mucosa of UC patients. In addition, levels of the HIS degradation product *N*-methylhistamine are elevated in the urine of both disease patients, which suggest a more active HIS metabolism (synthesis and degradation) in the intestine with respect to control individuals. The results indicate that degranulation of mast cells infiltrated along the intestinal tract must be involved in these diseases. However, there is a lack of information about molecular details of the signalling mechanisms responsible for the HIS effects on IBD evolution, thus blocking the development of efficient intervention strategies [75,182]. It has been proven that the chronic use of H_2R receptor antagonists increases the risk of more severe CD, suggesting a protective role of HIS on the intestinal mucosa when acting through an H_2R receptor [45]. IBD intervention acting through H_4R receptor has also been recently proposed [183,184]. In fact, several studies highlight the potential of H_4 receptor targeted therapy in the treatment of various gastrointestinal disorders such as IBD, IBS and cancer [184].

DAO has been suggested as an IBD marker, but it is a controversial subject [64,185]. Currently, IBD treatment consists of diet adjustment, psychological support, and eventually surgery. Recently, it has been proven that microbiota is usually altered in IBD patients. Working with a mouse model of intestinal inflammation, it has been demonstrated that the probiotic *Lactobacillus reuteri* is able to reduce intestinal inflammation by a HIS signalling-dependent mechanism, thus acting as a preventive factor for cancer risk associated with chronic inflammation [186]. Microbiota can also have a positive role in IBD patients, as microbial-derived metabolites (for instance, bile acids and short-chain fatty acids) can regulate intestinal 5-HT synthesis, and consequently intestinal motility, which opens new perspectives for probiotic-based strategies [161].

7.2.3. Intestinal Neoplasias

Colon cancer is considered the most common GIT cancer. It is a multifactorial disease and its etiology can combine genetic inherited factors, exposition to environmental risk factors (including diet), as well as other endogenous circumstances such as chronic intestinal inflammation. At present, it is one of the most frequent human cancers, besides being one of the principal causes of death among human cancer patients [187]. As in other cancer types, colon cancer biopsies present increased activities of PA-synthesis key enzymes and elevated PA content up to 10-15-fold with respect to the levels observed in normal colon epithelium. Thus, together with inflammation, PAs are considered to be markers of colon carcinogenesis [188].

It is known that several oncogenes and suppressor genes that regulate specific phases of colon carcinogenesis are also involved in PA metabolism regulation. Under normal conditions, the suppressor gene adenomatous polyposis coli (APC) repress MYC transcription, a family of transcription factors required for cell proliferation. MYC overexpression is related to uncontrolled proliferation and progression of carcinogenic process in different cancer types, including colon cancer. Members of MYC family are inducers of ODC transcription [189]. In addition, APC up-regulate the expression of ODC antizymes, a protein family acting as ODC inhibitors, as they bind to ODC monomer blocking the active quaternary conformation of the enzyme, thus targeting the ODC monomer for an antizyme-dependent and ubiquitin-independent proteasomal degradation. When APC is deleted or inactivated by mutation during the early stages of carcinogenesis (as occurring in adenomatous polyposis patients), MYC APC-induced repression is lost, and consequently ODC and PA synthesis are upregulated. In addition, the upregulation of ODC antizyme expression is lost, which leads to an increase in ODC turnover [190].

Active KRAS oncogene downregulates the peroxisome proliferator-activated receptor gamma (PPARγ), which has been proposed as a marker for colorectal cancer survival [191]. Transcription of the key enzyme for PA degradation (SSAT) is upregulated by PPARγ response elements (PPREs) present in SSAT promoter [192]. In advanced stages of colon cancer, oncogenic mutations in KRAS lead to permanent KRAS activation, which in turn downregulates PPARγ and, consequently, SSAT expression and PA degradation [193]. These regulatory mechanisms explain the molecular bases of the relationship between oncogenic events and PA elevation in colon cancer patients. As mentioned in previous sections, elevation of PA levels helps replication and macromolecular synthesis of the transformed cells and confers other advantages for cancer progression.

In addition to the endogenous regulation of PA metabolism, it is also worth mentioning that diet and microbiota are also potential PA sources [194]. The lack of effective PA absorption inhibitors, as well as the robustness of PA metabolism, is blocking the success of antitumoural strategies based on PA depletion in different cancer types [190]. In colon cancer, treatment depends on the progression stage. Surgery can be enough during the first stages; then, chemotherapy or immunotherapy must be required. What about chemoprevention? The irreversible ODC inhibitor DFMO is able to act on both Enterobacteria and mammalian ODC activities. As PAs are essential for colon cancer progression, DFMO could be effective as a chemopreventive agent for putative familiar colon cancer patients. This was the hypothesis claimed by Gerner and Meyskens several years ago, and validated with positive results when administered as a combined therapy with NSAIDs, the latest acting as PPARγ inducers [188,190].

Working with HDC knocked out mice under treatment with probiotics (*Lactobacillus reuteri*), results obtained by Gao et al. [162] indicate that luminal HIS produced by gut microbiota could suppress inflammation-associated colon carcinogenesis.

As mentioned above (Table 1), HIS is an immune system regulator playing important roles in immune cell development, span lives, and functions. These effects can involve any of the four receptors (mainly H_1R and H_4R) [37,195]. HIS has also been described as a cell proliferation regulator of several cancer types (for instance, breast cancer, GIT cancer, leukaemia, lung cancer, lymphoma) [196]. These effects can be different depending on the HIS receptors expressed by the different cell types. In addition, communication between the immune system and cancer is a dynamic process involving different immune cells; for instance, macrophages, monocytes, mast cells and regulatory B cells and T cells [197,198]. Consequently, HIS effects on carcinogenesis and cancer progression is a very complex but interesting topic, which also deserves further research efforts. Recent results support the therapeutic potential of H_4R ligand in several cancer types, including colon cancer [198].

Evolution of Zollinger-Ellison syndrome, a rare GIT pathology caused by the presence of gastrin-secreting tumours in pancreas and/or intestine, involves both synthesis and secretion of HIS [29].

8. Conclusions and Future Prospects

Previous sections demonstrate that BAs play important physiological roles in the entire GIT. Consequently, aberrant functions of the metabolic pathways are involved in the most important gastrointestinal pathologies. HIS metabolism seems to be mainly important in gastric physiopathology, as well as in inflammatory intestinal diseases. 5-HT plays major pathophysiological roles in the intestine, as an immune modulator and regulator of the intestinal smooth muscle contraction/relaxation. Its involvement in inflammatory diseases needs further clarification. CAs are modulators of intestinal absorption, blood flux and motility and they have also been proposed as immune modulators in the GIT system. Some of these functions may be regulated by 5-HT [104]. This is a very interesting hypothesis that would also require more research efforts to be fully validated. PAs, being essential biomolecules for cell growth, are important for both epithelial reparation and proliferation. Thus, they are beneficial for a healthy intestinal epithelium and have been described as both immune and epithelial permeability modulators in GIT [199], but also proposed as a promising target for colon cancer chemoprevention. From a phenomenological point of view, all B/A-BAs have been described as modulators of both immunity and epithelial cell growth in the GIT, but many of the underlying molecular mechanisms are not fully characterized, yet. In any case, these results point to the GIT as an interesting scenario to study BA metabolic and functional interplay.

In spite of the specialization of the amine effects along the different segments of the GIT, in some cases, their metabolic routes are coincident in a given GIT segment, so they can share/compete for common elements, thus establishing a crosstalk among their metabolism and consequently their physiological missions. For instance, enzymes such as decarboxylases (i.e., DDC), amine oxidases (i.e., MAO and DAO), cofactors (i.e., PLP, BH_4), and metabolites (i.e., SAM), among others. Moreover, at least in neurons, heteroreceptor complexes have also been detected among HIS and DA receptors [200], and both Spd and HIS are ligands and modulate *N*-methylaspartate receptor activity [201]. In mouse mast cells, synthesis of PA, HIS and 5-HT seem to be antagonistic processes in the mast cell differentiation process. Thus, the pathophysiological consequences of this cross-talk among BA metabolic elements still present many gaps and open questions (mentioned throughout this review) in the GIT context, which deserve deeper molecular characterization, as this information could provide valuable insights useful for new diagnosis and intervention initiatives in the gastroenterology field as well as more personalized nutritional advices and preventive actions.

It is clear that the pathophysiological effects of B/A-BAs in GIT is a very complex issue that interacts with and is modified by a very extensive list of endogenous and exogenous factors, from metabolic interactions with other immune and/or neuroendocrine compounds/systems to the influx of diet and microbiota composition. The full characterization of the entire involved interactions still requires filling many gaps on specific biochemical and molecular details. This research objective should be helped by systemic approaches to provide integrative views (and even predictive models) of the multiple pathophysiological processes associated with BAs in the GIT. In fact, several independent groups have proposed to approach the problem, or subsets of the problem, by using integrative high-throughput and Systems Biology strategies currently successfully used with many other complex biological systems [1,202].

The translational benefits of the final objective are clear taking into account that the topic involves three of the most important but complex physiological systems for a human being, neurotransmission/neuroendocrine, immune and digestive systems. Consequently, life quality and/or span life of many human beings can depend on research advances in the topic. Thus, it should be considered among the research priorities not only for nutrition but for biomedicine, in general. In addition, different companies have developed a wide spectrum of drugs capable of modulating different elements of BA metabolism and signalling. The usefulness and/or efficiency of these compounds (or their analogues/derivatives) will probably increase when a deeper degree of integrative knowledge about the molecular basis and the roles of all B/A-BAs in the GIT system is achieved.

Author Contributions: Conceptualization, A.F.-R., J.L.U. and F.S.-J.; Investigation, A.F.-R., J.L.U. and F.S.-J.; Writing-Original Draft Preparation, A.F.-R. and F.S.-J.; Writing-Review & Editing, J.L.U. and F.S.-J.; Visualization, A.F.-R., J.L.U. and F.S.-J.; Supervision, J.L.U. and F.S.-J.; Funding Acquisition, F.S.-J.

Funding: The present work was funded by Consejería de Economía, Innovación, Ciencia y Empleo, Junta de Andalucía (PAIDI group BIO267).

Acknowledgments: Thanks are due to Consejería de Economía, Innovación, Ciencia y Empleo, Junta de Andalucía. This work also takes part of the activities of the group in "CIBER de enfermedades raras" (CIBERER), and IBIMA, both institutions being part of Instituto de Salud Carlos III (MINECO, Spain).

Conflicts of Interest: The authors declare no conflict of interest. The funders had no role in the design of the study; in the collection, analyses, or interpretation of data; in the writing of the manuscript, and in the decision to publish the results.

References

1. Sánchez-Jiménez, F.; Ruiz-Perez, M.V.; Urdiales, J.L.; Medina, M.A. Pharmacological potential of biogenic amine-polyamine interactions beyond neurotransmission. *Br. J. Pharmacol.* **2013**, *170*, 4–16. [CrossRef] [PubMed]
2. Okada, K.; Hidese, R.; Fukuda, W.; Niitsu, M.; Takao, K.; Horai, Y.; Umezawa, N.; Higuchi, T.; Oshima, T.; Yoshikawa, Y.; et al. Identification of a novel aminopropyltransferase involved in the synthesis of branched-chain polyamines in hyperthermophiles. *J. Bacteriol.* **2014**, *196*, 1866–1876. [CrossRef] [PubMed]
3. Suzzi, G.; Torriani, S. Editorial: Biogenic amines in foods. *Front. Microbiol.* **2015**, *6*, 472. [CrossRef] [PubMed]
4. Bodmer, S.; Imark, C.; Kneubühl, M. Biogenic amines in foods: Histamine and food processing. *Inflamm. Res.* **1999**, *48*, 296–300. [CrossRef] [PubMed]
5. Moya-Garcia, A.A.; Pino-Angeles, A.; Gil-Redondo, R.; Morreale, A.; Sanchez-Jimenez, F. Structural features of mammalian histidine decarboxylase reveal the basis for specific inhibition. *Br. J. Pharmacol.* **2009**, *157*, 4–13. [CrossRef] [PubMed]
6. Giardina, G.; Montioli, R.; Gianni, S.; Cellini, B.; Paiardini, A.; Voltattorni, C.B.; Cutruzzolà, F. Open conformation of human DOPA decarboxylase reveals the mechanism of PLP addition to Group II decarboxylases. *Proc. Natl. Acad. Sci. USA* **2011**, *108*, 20514–20519. [CrossRef] [PubMed]
7. Sandmeier, E.; Hale, T.I.; Christen, P. Multiple evolutionary origin of pyridoxal-5'-phosphate-dependent amino acid decarboxylases. *Eur. J. Biochem.* **1994**, *221*, 997–1002. [CrossRef] [PubMed]
8. Sánchez-Jiménez, F.; Moya-García, A.A.; Pino-Ángeles, A. New structural insights to help in the search for selective inhibitors of mammalian pyridoxal 5'-phosphate-dependent histidine decarboxylase. *Inflamm. Res.* **2006**, *55*, S55–S56. [CrossRef] [PubMed]
9. Fonda, M.L.; Eggers, D.K.; Mehta, R. Vitamin B-6 metabolism in the livers of young adult and senescent mice. *Exp. Gerontol.* **1980**, *15*, 457–463. [CrossRef]
10. Jalkanen, S.; Salmi, M. Cell surface monoamine oxidases: Enzymes in search of a function. *EMBO J.* **2001**, *20*, 3893–3901. [CrossRef] [PubMed]
11. Finney, J.; Moon, H.-J.; Ronnebaum, T.; Lantz, M.; Mure, M. Human copper-dependent amine oxidases. *Arch. Biochem. Biophys.* **2014**, *546*, 19–32. [CrossRef] [PubMed]
12. Edmondson, D.E.; Binda, C.; Mattevi, A. Structural insights into the mechanism of amine oxidation by monoamine oxidases A and B. *Arch. Biochem. Biophys.* **2007**, *464*, 269–276. [CrossRef] [PubMed]
13. Rodríguez-López, R.; Morales, M.; Sánchez-Jiménez, F. Histamine and its receptors as a module of the biogenic amine diseasome. In *Histamine Receptors*; Springer International Publishing: Cham, Switzerland, 2016; pp. 173–214.
14. Schubert, M.L. Gastric acid secretion. *Curr. Opin. Gastroenterol.* **2016**, *32*, 452–460. [CrossRef] [PubMed]
15. Barcik, W.; Wawrzyniak, M.; Akdis, C.A.; O'Mahony, L. Immune regulation by histamine and histamine-secreting bacteria. *Curr. Opin. Immunol.* **2017**, *48*, 108–113. [CrossRef] [PubMed]
16. Schneider, E.; Rolli-Derkinderen, M.; Arock, M.; Dy, M. Trends in histamine research: New functions during immune responses and hematopoiesis. *Trends Immunol.* **2002**, *23*, 255–263. [CrossRef]
17. Peters, L.J.; Kovacic, J.P. Histamine: Metabolism, physiology, and pathophysiology with applications in veterinary medicine. *J. Vet. Emerg. Crit. Care* **2009**, *19*, 311–328. [CrossRef] [PubMed]
18. Panula, P.; Chazot, P.L.; Cowart, M.; Gutzmer, R.; Leurs, R.; Liu, W.L.S.; Stark, H.; Thurmond, R.L.; Haas, H.L. International union of basic and clinical pharmacology. XCVIII. Histamine receptors. *Pharmacol. Rev.* **2015**, *67*, 601–655. [CrossRef] [PubMed]
19. Gardini, F.; Özogul, Y.; Suzzi, G.; Tabanelli, G.; Özogul, F. Technological factors affecting biogenic amine content in foods: A review. *Front. Microbiol.* **2016**, *7*, 1218. [CrossRef] [PubMed]
20. Piletz, J.E.; Aricioglu, F.; Cheng, J.-T.; Fairbanks, C.A.; Gilad, V.H.; Haenisch, B.; Halaris, A.; Hong, S.; Lee, J.E.; Li, J.; et al. Agmatine: Clinical applications after 100 years in translation. *Drug Discov. Today* **2013**, *18*, 880–893. [CrossRef] [PubMed]

21. López-Contreras, A.J.; López-Garcia, C.; Jiménez-Cervantes, C.; Cremades, A.; Peñafiel, R. Mouse ornithine decarboxylase-like gene encodes an antizyme inhibitor devoid of ornithine and arginine decarboxylating activity. *J. Biol. Chem.* **2006**, *281*, 30896–30906. [CrossRef] [PubMed]

22. Biji, K.B.; Ravishankar, C.N.; Venkateswarlu, R.; Mohan, C.O.; Gopal, T.K.S. Biogenic amines in seafood: A review. *J. Food Sci. Technol.* **2016**, *53*, 2210–2218. [CrossRef] [PubMed]

23. Morgan, D.M.L.; White, A.; Sánchez-Jiménez, F.; Bardócz, S. *COST 917—Biogenically Active Amines in Food. Volume IV, First General Workshop*; Office for Official Publicationsn of European Communities: Luxemburg, 2000.

24. Wallace, H.M.; Hughes, A. *COST Action 922. Health Implications of Dietary Amines*; Office for Official Publicationsn of European Communities: Luxemburg, 2004.

25. Naila, A.; Flint, S.; Fletcher, G.; Bremer, P.; Meerdink, G. Control of biogenic amines in food-existing and emerging approaches. *J. Food Sci.* **2010**, *75*, R139–R150. [CrossRef] [PubMed]

26. Sanchez-Jiménez, F.; Pino-Ángeles, A.; Rodríguez-López, R.; Morales, M.; Urdiales, J.L. Structural and functional analogies and differences between histidine decarboxylase and aromatic L-amino acid decarboxylase molecular networks: Biomedical implications. *Pharmacol. Res.* **2016**, *114*, 90–102. [CrossRef] [PubMed]

27. Rodríguez-López, R.; Reyes-Palomares, A.; Sánchez-Jiménez, F.; Medina, M.Á. PhenUMA: A tool for integrating the biomedical relationships among genes and diseases. *BMC Bioinform.* **2014**, *15*, 375. [CrossRef] [PubMed]

28. Reyes-Palomares, A.; Bueno, A.; Rodríguez-López, R.; Medina, M.Á.; Sánchez-Jiménez, F.; Corpas, M.; Ranea, J.A.G. Systematic identification of phenotypically enriched loci using a patient network of genomic disorders. *BMC Genom.* **2016**, *17*, 232. [CrossRef] [PubMed]

29. Pino-Ángeles, A.; Reyes-Palomares, A.; Melgarejo, E.; Sánchez-Jiménez, F. Histamine: An undercover agent in multiple rare diseases? *J. Cell. Mol. Med.* **2012**, *16*, 1947–1960. [CrossRef] [PubMed]

30. Ruiz-Pérez, M.V.; Medina, M.Á.; Urdiales, J.L.; Keinänen, T.A.; Tuomo, A.; Sánchez-Jiménez, F. Polyamine metabolism is sensitive to glycolysis inhibition in human neuroblastoma cells. *J. Biol. Chem.* **2015**, *290*, 6106–6119. [CrossRef] [PubMed]

31. Watanabe, T.; Ohtsu, H. L-histidine decarboxylase as a probe in studies on histamine. *Chem. Rec.* **2002**, *2*, 369–376. [CrossRef] [PubMed]

32. Stark, H. *Histamine H4 Receptor: A Novel Drug Target for Immunoregulation and Inflammation*; Versita: Berlin, Germany, 2013.

33. Nakanishi, T.; Kekuda, R.; Fei, Y.J.; Hatanaka, T.; Sugawara, M.; Martindale, R.G.; Leibach, F.H.; Prasad, P.D.; Ganapathy, V. Cloning and functional characterization of a new subtype of the amino acid transport system N. *Am. J. Physiol. Cell Physiol.* **2001**, *281*, C1757–C1768. [CrossRef] [PubMed]

34. Moya-Garcia, A.A.; Medina, M.A.; Sánchez-Jiménez, F. Mammalian histidine decarboxylase: From structure to function. *Bioessays* **2005**, *27*, 57–63. [CrossRef] [PubMed]

35. Schwelberger, H.G.; Ahrens, F.; Fogel, W.A.; Sánchez-Jiménez, F. Histamine metabolism. In *Histamine H4 Receptor: A Novel Drug Target in Immunoregulation and Inflammation*; Stark, H., Ed.; Versita: Berlin, Germany, 2013; pp. 63–102.

36. Metcalfe, D.D. Mast cells and mastocytosis. *Blood* **2008**, *112*, 946–956. [CrossRef] [PubMed]

37. Ennis, M.; Ciz, M.; Dib, K.; Friedman, S.; Gangwar, R.S.; Gibbs, B.F.; Levi-Schaffer, F.; Lojek, A.; Migalovich-Sheikhet, H.; O'Mahony, L.; et al. Histamine receptors and inflammatory cells. In *Histamine H4 Receptor: A Novel Drug Target in Immunoregulation and Inflammation*; Stark, H., Ed.; Versita: London, UK, 2013; pp. 103–144.

38. Dwyer, D.F.; Barrett, N.A.; Austen, K.F.; Immunological Genome Project Consortium. Expression profiling of constitutive mast cells reveals a unique identity within the immune system. *Nat. Immunol.* **2016**, *17*, 878–887. [CrossRef] [PubMed]

39. Fabbri, R.; Furini, C.R.G.; Passani, M.B.; Provensi, G.; Baldi, E.; Bucherelli, C.; Izquierdo, I.; de Carvalho Myskiw, J.; Blandina, P. Memory retrieval of inhibitory avoidance requires histamine H1 receptor activation in the hippocampus. *Proc. Natl. Acad. Sci. USA* **2016**, *113*, E2714–E2720. [CrossRef] [PubMed]

40. Bernsand, M.; Ericsson, P.; Bjorkqvist, M.; Zhao, C.-M.; Hakanson, R.; Norlen, P. Submucosal microinfusion of endothelin and adrenaline mobilizes ECL-cell histamine in rat stomach, and causes mucosal damage: A microdialysis study. *Br. J. Pharmacol.* **2003**, *140*, 707–717. [CrossRef] [PubMed]

41. Krauth, M.-T.T.; Agis, H.; Aichberger, K.J.; Simonitsch-Klupp, I.; Müllauer, L.; Mayerhofer, M.; Böhm, A.; Horny, H.-P.P.; Valent, P. Immunohistochemical detection of histidine decarboxylase in neoplastic mast cells in patients with systemic mastocytosis. *Hum. Pathol.* **2006**, *37*, 439–447. [CrossRef] [PubMed]

42. Osefo, N.; Ito, T.; Jensen, R.T. Gastric acid hypersecretory states: Recent insights and advances. *Curr. Gastroenterol. Rep.* **2009**, *11*, 433–441. [CrossRef] [PubMed]

43. Kuramasu, A.; Saito, H.; Suzuki, S.; Watanabe, T.; Ohtsu, H. Mast cell-/basophil-specific transcriptional regulation of human L-histidine decarboxylase gene by CpG methylation in the promoter region. *J. Biol. Chem.* **1998**, *273*, 31607–31614. [CrossRef] [PubMed]

44. Correa-Fiz, F.; Reyes-Palomares, A.; Fajardo, I.; Melgarejo, E.; Gutiérrez, A.; García-Ranea, J.A.; Medina, M.A.; Sánchez-Jiménez, F. Regulatory cross-talk of mouse liver polyamine and methionine metabolic pathways: A systemic approach to its physiopathological consequences. *Amino Acids* **2011**, *42*, 577–595. [CrossRef] [PubMed]

45. Smolinska, S.; Jutel, M.; Crameri, R.; O'Mahony, L. Histamine and gut mucosal immune regulation. *Allergy* **2014**, *69*, 273–281. [CrossRef] [PubMed]

46. Mamune-Sato, R.; Yamauchi, K.; Tanno, Y.; Ohkawara, Y.; Ohtsu, H.; Katayose, D.; Maeyama, K.; Watanabe, T.; Shibahara, S.; Takishima, T. Functional analysis of alternatively spliced transcripts of the human histidine decarboxylase gene and its expression in human tissues and basophilic leukemia cells. *Eur. J. Biochem.* **1992**, *209*, 533–539. [CrossRef] [PubMed]

47. Abrighach, H.; Fajardo, I.; Sánchez-Jiménez, F.; Urdiales, J.L. Exploring polyamine regulation by nascent histamine in a human-transfected cell model. *Amino Acids* **2010**, *38*, 561–573. [CrossRef] [PubMed]

48. Olmo, M.T.; Urdiales, J.L.; Pegg, A.E.; Medina, M.A.; Sánchez-Jiménez, F. In vitro study of proteolytic degradation of rat histidine decarboxylase. *Eur. J. Biochem.* **2000**, *267*, 1527–1531. [CrossRef] [PubMed]

49. Fleming, J.V.; Fajardo, I.; Langlois, M.R.; Sanchez-Jimenez, F.; Wang, T.C. The C-terminus of rat L-histidine decarboxylase specifically inhibits enzymic activity and disrupts pyridoxal phosphate-dependent interactions with L-histidine substrate analogues. *Biochem. J.* **2004**, *381*, 769–778. [CrossRef] [PubMed]

50. Furuta, K.; Nakayama, K.; Sugimoto, Y.; Ichikawa, A.; Tanaka, S. Activation of histidine decarboxylase through post-translational cleavage by caspase-9 in a mouse mastocytoma P-815. *J. Biol. Chem.* **2007**, *282*, 13438–13446. [CrossRef] [PubMed]

51. Olmo, M.T.; Rodríguez-Agudo, D.; Medina, M.A.; Sánchez-Jiménez, F. The pest regions containing C-termini of mammalian ornithine decarboxylase and histidine decarboxylase play different roles in protein degradation. *Biochem. Biophys. Res. Commun.* **1999**, *257*, 269–272. [CrossRef] [PubMed]

52. Rodriguez-Agudo, D.; Olmo, M.T.; Sanchez-Jimenez, F.; Medina, M.A. Rat histidine decarboxylase is a substrate for m-calpain in vitro. *Biochem. Biophys. Res. Commun.* **2000**, *271*, 777–781. [CrossRef] [PubMed]

53. Pino-Angeles, A.; Morreale, A.; Negri, A.; Sánchez-Jiménez, F.; Moya-García, A.A. Substrate uptake and protein stability relationship in mammalian histidine decarboxylase. *Proteins* **2010**, *78*, 154–161. [CrossRef] [PubMed]

54. Olmo, M.T.; Sanchez-Jimenez, F.; Medina, M.A.; Hayashi, H. Spectroscopic analysis of recombinant rat histidine decarboxylase. *J. Biochem.* **2002**, *132*, 433–439. [CrossRef] [PubMed]

55. Moya-García, A.A.; Ruiz-Pernía, J.; Martí, S.; Sánchez-Jiménez, F.; Tuñón, I. Analysis of the decarboxylation step in mammalian histidine decarboxylase. A computational study. *J. Biol. Chem.* **2008**, *283*, 12393–12401. [CrossRef] [PubMed]

56. Rodríguez-Caso, C.; Rodríguez-Agudo, D.; Moya-García, A.A.; Fajardo, I.; Medina, M.A.; Subramaniam, V.; Sánchez-Jiménez, F. Local changes in the catalytic site of mammalian histidine decarboxylase can affect its global conformation and stability. *Eur. J. Biochem.* **2003**, *270*, 4376–4387. [CrossRef] [PubMed]

57. Fleming, J.V.; Sánchez-Jiménez, F.; Moya-García, A.A.; Langlois, M.R.; Wang, T.C. Mapping of catalytically important residues in the rat L-histidine decarboxylase enzyme using bioinformatic and site-directed mutagenesis approaches. *Biochem. J.* **2004**, *379*, 253–261. [CrossRef] [PubMed]

58. Morgan, D.M.L.; Milovic, V.; Krizek, M.; White, A. *COST Action 917. Biogenically Active Amines in Food. Volume V. Polyamines and Tumor Growth, Biologically Active Amines in Food Processing and Amines Produced by Bacteria*; European Commission: Luxemburg, 2001.

59. Wallace, H.M. Health implications of dietary amines: An overview of COST Action 922 (2001–2006). *Biochem. Soc. Trans.* **2007**, *35*, 293–294. [CrossRef] [PubMed]

60. Schwelberger, H.G. Histamine *N*-methyltransferase (HNMT) enzyme and gene. In *Histamine: Biology and Medical Aspects*; Falus, A., Grosman, N., Darvas, Z., Eds.; SpringMed Publishing: Budapest, Hungary, 2004; pp. 53–59.

61. Fagerberg, L.; Hallström, B.M.; Oksvold, P.; Kampf, C.; Djureinovic, D.; Odeberg, J.; Habuka, M.; Tahmasebpoor, S.; Danielsson, A.; Edlund, K.; et al. Analysis of the human tissue-specific expression by genome-wide integration of transcriptomics and antibody-based proteomics. *Mol. Cell. Proteom.* **2014**, *13*, 397–406. [CrossRef] [PubMed]

62. Duff, M.O.; Olson, S.; Wei, X.; Garrett, S.C.; Osman, A.; Bolisetty, M.; Plocik, A.; Celniker, S.E.; Graveley, B.R. Genome-wide identification of zero nucleotide recursive splicing in Drosophila. *Nature* **2015**, *521*, 376–379. [CrossRef] [PubMed]

63. Prell, G.D.; Green, J.P. Measurement of histamine metabolites in brain and cerebrospinal fluid provides insights into histaminergic activity. *Agents Actions* **1994**, *41*, C5–C8. [CrossRef] [PubMed]

64. Song, W.-B.; Lv, Y.-H.; Zhang, Z.-S.; Li, Y.-N.; Xiao, L.-P.; Yu, X.-P.; Wang, Y.-Y.; Ji, H.-L.; Ma, L. Soluble intercellular adhesion molecule-1, D-lactate and diamine oxidase in patients with inflammatory bowel disease. *World J. Gastroenterol.* **2009**, *15*, 3916–3919. [CrossRef] [PubMed]

65. Koepsell, H.; Lips, K.; Volk, C. Polyspecific organic cation transporters: Structure, function, physiological roles, and biopharmaceutical implications. *Pharm. Res.* **2007**, *24*, 1227–1251. [CrossRef] [PubMed]

66. Eiden, L.E.; Weihe, E. VMAT2: A dynamic regulator of brain monoaminergic neuronal function interacting with drugs of abuse. *Ann. N. Y. Acad. Sci.* **2011**, *1216*, 86–98. [CrossRef] [PubMed]

67. Garcia-Faroldi, G.; Rodriguez, C.E.; Urdiales, J.L.; Perez-Pomares, J.M.; Davila, J.C.; Pejler, G.; Sanchez-Jimenez, F.; Fajardo, I. Polyamines are present in mast cell secretory granules and are important for granule homeostasis. *PLoS ONE* **2010**, *5*, e15071. [CrossRef] [PubMed]

68. Andersson, K.; Chen, D.; Håkanson, R.; Mattsson, H.; Sundler, F. Enterochromaffin-like cells in the rat stomach: Effect of alpha-fluoromethylhistidine-evoked histamine depletion. A chemical, histochemical and electron-microscopic study. *Cell Tissue Res.* **1992**, *270*, 7–13. [CrossRef] [PubMed]

69. Thurmond, R.L. The histamine H4 receptor: From orphan to the clinic. *Front. Pharmacol.* **2015**, *6*, 65. [CrossRef] [PubMed]

70. Kusters, J.G.; van Vliet, A.H.M.; Kuipers, E.J. Pathogenesis of Helicobacter pylori infection. *Clin. Microbiol. Rev.* **2006**, *19*, 449–490. [CrossRef] [PubMed]

71. Chen, D.; Aihara, T.; Zhao, C.-M.; Håkanson, R.; Okabe, S. Differentiation of the Gastric Mucosa I. Role of histamine in control of function and integrity of oxyntic mucosa: Understanding gastric physiology through disruption of targeted genes. *Am. J. Physiol. Gastrointest. Liver Physiol.* **2006**, *291*, G539–G544. [CrossRef] [PubMed]

72. Waldum, H.L.; Hauso, Ø.; Fossmark, R. The regulation of gastric acid secretion—Clinical perspectives. *Acta Physiol.* **2014**, *210*, 239–256. [CrossRef] [PubMed]

73. Coruzzi, G.; Adami, M.; Pozzoli, C.; de Esch, I.J.P.; Smits, R.; Leurs, R. Selective histamine H3 and H4 receptor agonists exert opposite effects against the gastric lesions induced by HCl in the rat stomach. *Eur. J. Pharmacol.* **2011**, *669*, 121–127. [CrossRef] [PubMed]

74. Rydning, A.; Lyng, O.; Falkmer, S.; Grønbech, J.E. Histamine is involved in gastric vasodilation during acid back diffusion via activation of sensory neurons. *Am. J. Physiol. Gastrointest. Liver Physiol.* **2002**, *283*, G603–G611. [CrossRef] [PubMed]

75. Ahmad, J.; Misra, M.; Rizvi, W.; Kumar, A. Histamine aspects in acid peptic diseases and cell proliferation. In *Biomedical Aspects of Histamine. Current Perspectives*; Shahid, M., Khardori, N., Khan, R.A., Tripathi, T., Eds.; Springer: Dordrecht, The Netherlands, 2010; pp. 175–198.

76. Sander, L.E.; Lorentz, A.; Sellge, G.; Coëffier, M.; Neipp, M.; Veres, T.; Frieling, T.; Meier, P.N.; Manns, M.P.; Bischoff, S.C. Selective expression of histamine receptors H1R, H2R, and H4R, but not H3R, in the human intestinal tract. *Gut* **2006**, *55*, 498–504. [CrossRef] [PubMed]

77. Schneider, E.; Leite-de-moraes, M.; Dy, M. Histamine, immune cells and autoimmunity. *Adv. Exp. Med. Biol.* **2010**, *709*, 81–94. [PubMed]

78. Xie, H.; He, S.-H. Roles of histamine and its receptors in allergic and inflammatory bowel diseases. *World J. Gastroenterol.* **2005**, *11*, 2851–2857. [CrossRef] [PubMed]

79. Gutzmer, R.; Diestel, C.; Mommert, S.; Köther, B.; Stark, H.; Wittmann, M.; Werfel, T. Histamine H4 receptor stimulation suppresses IL-12p70 production and mediates chemotaxis in human monocyte-derived dendritic cells. *J. Immunol.* **2005**, *174*, 5224–5232. [CrossRef] [PubMed]

80. Gutzmer, R.; Mommert, S.; Gschwandtner, M.; Zwingmann, K.; Stark, H.; Werfel, T. The histamine H4 receptor is functionally expressed on T(H)2 cells. *J. Allergy Clin. Immunol.* **2009**, *123*, 619–625. [CrossRef] [PubMed]

81. Coruzzi, G.; Adami, M.; Pozzoli, C. Role of histamine H4 receptors in the gastrointestinal tract. *Front. Biosci.* **2012**, *4*, 226–239. [CrossRef]

82. Mittal, R.; Debs, L.H.; Patel, A.P.; Nguyen, D.; Patel, K.; O'Connor, G.; Grati, M.; Mittal, J.; Yan, D.; Eshraghi, A.A.; et al. Neurotransmitters: The critical modulators regulating gut-brain axis. *J. Cell. Physiol.* **2017**, *232*, 2359–2372. [CrossRef] [PubMed]

83. Sikander, A.; Rana, S.V.; Prasad, K.K. Role of serotonin in gastrointestinal motility and irritable bowel syndrome. *Clin. Chim. Acta* **2009**, *403*, 47–55. [CrossRef] [PubMed]

84. Roberts, K.M.; Fitzpatrick, P.F. Mechanisms of tryptophan and tyrosine hydroxylase. *IUBMB Life* **2013**, *65*, 350–357. [CrossRef] [PubMed]

85. O'Mahony, S.M.; Clarke, G.; Borre, Y.E.; Dinan, T.G.; Cryan, J.F. Serotonin, tryptophan metabolism and the brain-gut-microbiome axis. *Behav. Brain Res.* **2015**, *277*, 32–48. [CrossRef] [PubMed]

86. Hayashi, H.; Tsukiyama, F.; Ishii, S.; Mizuguchi, H.; Kagamiyama, H. Acid-base chemistry of the reaction of aromatic L-amino acid decarboxylase and dopa analyzed by transient and steady-state kinetics: Preferential binding of the substrate with its amino group unprotonated. *Biochemistry* **1999**, *38*, 15615–15622. [CrossRef] [PubMed]

87. Bertoldi, M.; Voltattorni, C.B. Dopa decarboxylase exhibits low pH half-transaminase and high pH oxidative deaminase activities toward serotonin (5-hydroxytryptamine). *Protein Sci.* **2001**, *10*, 1178–1186. [CrossRef] [PubMed]

88. Bertoldi, M.; Gonsalvi, M.; Contestabile, R.; Voltattorni, C.B. Mutation of tyrosine 332 to phenylalanine converts dopa decarboxylase into a decarboxylation-dependent oxidative deaminase. *J. Biol. Chem.* **2002**, *277*, 36357–36362. [CrossRef] [PubMed]

89. Ruiz-Pérez, M.V.; Pino-Ángeles, A.; Medina, M.A.; Sánchez-Jiménez, F.; Moya-García, A.A. Structural perspective on the direct inhibition mechanism of EGCG on mammalian histidine decarboxylase and DOPA decarboxylase. *J. Chem. Inf. Model.* **2012**, *52*, 113–119. [CrossRef] [PubMed]

90. Montioli, R.; Cellini, B.; Dindo, M.; Oppici, E.; Voltattorni, C.B. Interaction of human Dopa decarboxylase with L-Dopa: Spectroscopic and kinetic studies as a function of pH. *BioMed Res. Int.* **2013**, *2013*, 161456. [CrossRef] [PubMed]

91. Bortolato, M.; Chen, K.; Shih, J.C. The degradation of serotonin: Role of MAO. *Handb. Behav. Neurosci.* **2010**, *21*, 203–218. [CrossRef]

92. Meiser, J.; Weindl, D.; Hiller, K. Complexity of dopamine metabolism. *Cell Commun. Signal.* **2013**, *11*, 34. [CrossRef] [PubMed]

93. Nagatsu, T. Progress in monoamine oxidase (MAO) research in relation to genetic engineering. *Neurotoxicology* **2004**, *25*, 11–20. [CrossRef]

94. Strege, P.R.; Knutson, K.; Eggers, S.J.; Li, J.H.; Wang, F.; Linden, D.; Szurszewski, J.H.; Milescu, L.; Leiter, A.B.; Farrugia, G.; et al. Sodium channel NaV1.3 is important for enterochromaffin cell excitability and serotonin release. *Sci. Rep.* **2017**, *7*, 15650. [CrossRef] [PubMed]

95. Mawe, G.M.; Hoffman, J.M. Serotonin signalling in the gut—Functions, dysfunctions and therapeutic targets. *Nat. Rev. Gastroenterol. Hepatol.* **2013**, *10*, 473–486. [CrossRef] [PubMed]

96. Gershon, M.D. 5-HT (serotonin) physiology and related drugs. *Curr. Opin. Gastroenterol.* **2000**, *16*, 113–120. [CrossRef] [PubMed]

97. Greenwood-van Meerveld, B. Importance of 5-hydroxytryptamine receptors on intestinal afferents in the regulation of visceral sensitivity. *Neurogastroenterol. Motil.* **2007**, *19*, 13–18. [CrossRef] [PubMed]

98. Versteeg, R.I.; Serlie, M.J.; Kalsbeek, A.; la Fleur, S.E. Serotonin, a possible intermediate between disturbed circadian rhythms and metabolic disease. *Neuroscience* **2015**, *301*, 155–167. [CrossRef] [PubMed]

99. Shortall, S.E.; Spicer, C.H.; Ebling, F.J.P.; Green, A.R.; Fone, K.C.F.; King, M.V. Contribution of serotonin and dopamine to changes in core body temperature and locomotor activity in rats following repeated administration of mephedrone. *Addict. Biol.* **2016**, *21*, 1127–1139. [CrossRef] [PubMed]

100. Arreola, R.; Becerril-Villanueva, E.; Cruz-Fuentes, C.; Velasco-Velázquez, M.A.; Garcés-Alvarez, M.E.; Hurtado-Alvarado, G.; Quintero-Fabian, S.; Pavón, L. Immunomodulatory Effects Mediated by Serotonin. *J. Immunol. Res.* **2015**, *2015*, 354957. [CrossRef] [PubMed]

101. Maggiorani, D.; Manzella, N.; Edmondson, D.E.; Mattevi, A.; Parini, A.; Binda, C.; Mialet-Perez, J. Monoamine oxidases, oxidative stress, and altered mitochondrial dynamics in cardiac ageing. *Oxid. Med. Cell. Longev.* **2017**, *2017*, 3017947. [CrossRef] [PubMed]

102. Margolis, K.G.; Gershon, M.D. Enteric neuronal regulation of intestinal inflammation. *Trends Neurosci.* **2016**, *39*, 614–624. [CrossRef] [PubMed]

103. Filip, M.; Bader, M. Overview on 5-HT receptors and their role in physiology and pathology of the central nervous system. *Pharmacol. Rep.* **2009**, *61*, 761–777. [CrossRef]

104. Li, Z.; Chalazonitis, A.; Huang, Y.-Y.; Mann, J.J.; Margolis, K.G.; Yang, Q.M.; Kim, D.O.; Côté, F.; Mallet, J.; Gershon, M.D. Essential roles of enteric neuronal serotonin in gastrointestinal motility and the development/survival of enteric dopaminergic neurons. *J. Neurosci.* **2011**, *31*, 8998–9009. [CrossRef] [PubMed]

105. Soyer, T.; Aktuna, Z.; Reşat Aydos, T.; Osmanoğlu, G.; Korkut, O.; Akman, H.; Cakmak, M. Esophageal and gastric smooth muscle activity after carbon dioxide pneumoperitoneum. *J. Surg. Res.* **2010**, *161*, 278–281. [CrossRef] [PubMed]

106. Flatmark, T. Catecholamine biosynthesis and physiological regulation in neuroendocrine cells. *Acta Physiol. Scand.* **2000**, *168*, 1–17. [CrossRef] [PubMed]

107. Wu, Q.; McLeish, M.J. Kinetic and pH studies on human phenylethanolamine *N*-methyltransferase. *Arch. Biochem. Biophys.* **2013**, *539*, 1–8. [CrossRef] [PubMed]

108. Eisenhofer, G.; Kopin, I.J.; Goldstein, D.S. Catecholamine metabolism: A contemporary view with implications for physiology and medicine. *Pharmacol. Rev.* **2004**, *56*, 331–349. [CrossRef] [PubMed]

109. Daubner, S.C.; Le, T.; Wang, S. Tyrosine hydroxylase and regulation of dopamine synthesis. *Arch. Biochem. Biophys.* **2011**, *508*, 1–12. [CrossRef] [PubMed]

110. Tank, A.W.; Lee Wong, D. Peripheral and Central Effects of Circulating Catecholamines. *Compr. Physiol.* **2014**, *5*, 1–15. [CrossRef]

111. Kobayashi, K. Role of catecholamine signaling in brain and nervous system functions: New insights from mouse molecular genetic study. *J. Investig. Dermatol. Symp. Proc.* **2001**, *6*, 115–121. [CrossRef] [PubMed]

112. Arreola, R.; Alvarez-Herrera, S.; Pérez-Sánchez, G.; Becerril-Villanueva, E.; Cruz-Fuentes, C.; Flores-Gutierrez, E.O.; Garcés-Alvarez, M.E.; de la Cruz-Aguilera, D.L.; Medina-Rivero, E.; Hurtado-Alvarado, G.; et al. Immunomodulatory effects mediated by dopamine. *J. Immunol. Res.* **2016**, *2016*, 3160486. [CrossRef] [PubMed]

113. Beaulieu, J.-M.; Espinoza, S.; Gainetdinov, R.R. Dopamine receptors—IUPHAR review 13. *Br. J. Pharmacol.* **2015**, *172*, 1–23. [CrossRef] [PubMed]

114. Elenkov, I.J.; Wilder, R.L.; Chrousos, G.P.; Vizi, E.S. The sympathetic nerve—An integrative interface between two supersystems: The brain and the immune system. *Pharmacol. Rev.* **2000**, *52*, 595–638. [PubMed]

115. Rizzetto, L.; Fava, F.; Tuohy, K.M.; Selmi, C. Connecting the immune system, systemic chronic inflammation and the gut microbiome: The role of sex. *J. Autoimmun.* **2018**, in press. [CrossRef] [PubMed]

116. Natale, G.; Ryskalin, L.; Busceti, C.L.; Biagioni, F.; Fornai, F. The nature of catecholamine-containing neurons in the enteric nervous system in relationship with organogenesis, normal human anatomy and neurodegeneration. *Arch. Ital. Biol.* **2017**, *155*, 118–130. [PubMed]

117. Rodriguez-Caso, C.; Montañez, R.; Cascante, M.; Sanchez-Jimenez, F.; Medina, M.A. Mathematical modeling of polyamine metabolism in mammals. *J. Biol. Chem.* **2006**, *281*, 21799–21812. [CrossRef] [PubMed]

118. Pegg, A.E. Mammalian polyamine metabolism and function. *IUBMB Life* **2009**, *61*, 880–894. [CrossRef] [PubMed]

119. Pegg, A.E. Functions of Polyamines in Mammals. *J. Biol. Chem.* **2016**, *291*, 14904–14912. [CrossRef] [PubMed]

120. Perez-Leal, O.; Merali, S. Regulation of polyamine metabolism by translational control. *Amino Acids* **2012**, *42*, 611–617. [CrossRef] [PubMed]

121. Kahana, C. Protein degradation, the main hub in the regulation of cellular polyamines. *Biochem. J.* **2016**, *473*, 4551–4558. [CrossRef] [PubMed]

122. Pegg, A.E. Regulation of ornithine decarboxylase. *J. Biol. Chem.* **2006**, *281*, 14529–14532. [CrossRef] [PubMed]

123. Pegg, A.E. *S*-Adenosylmethionine decarboxylase. *Essays Biochem.* **2009**, *46*, 25–45. [CrossRef] [PubMed]

124. Pegg, A.E.; Casero, R.A. Current status of the polyamine research field. In *Polyamines. Methods and Protocols*; Pegg, A.E., Casero, R.A., Eds.; Humana Press: Newyork, NK, USA, 2011; pp. 3–35.

125. Casero, R.A.; Marton, L.J. Targeting polyamine metabolism and function in cancer and other hyperproliferative diseases. *Nat. Rev. Drug Discov.* **2007**, *6*, 373–390. [CrossRef] [PubMed]

126. Casero, R.A.; Pegg, A.E. Polyamine catabolism and disease. *Biochem. J.* **2009**, *421*, 323–338. [CrossRef] [PubMed]

127. Cecco, L.; Antoniello, S.; Auletta, M.; Cerra, M.; Bonelli, P. Pattern and concentration of free and acetylated polyamines in urine of cirrhotic patients. *Int. J. Biol. Mark.* **1992**, *7*, 52–58.

128. Qian, H.; Beard, D.A. Metabolic futile cycles and their functions: A systems analysis of energy and control. *Syst. Biol.* **2006**, *153*, 192–200. [CrossRef]

129. Reyes-Palomares, A.; Montañez, R.; Sánchez-Jiménez, F.; Medina, M.A.; Sanchez-Jimenez, F.; Medina, M.A. A combined model of hepatic polyamine and sulfur amino acid metabolism to analyze S-adenosyl methionine availability. *Amino Acids* **2012**, *42*, 597–610. [CrossRef] [PubMed]

130. Muth, A.; Madan, M.; Archer, J.J.; Ocampo, N.; Rodriguez, L.; Phanstiel, O. Polyamine transport inhibitors: Design, synthesis, and combination therapies with difluoromethylornithine. *J. Med. Chem.* **2014**, *57*, 348–363. [CrossRef] [PubMed]

131. Uimari, A.; Keinänen, T.A.; Karppinen, A.; Woster, P.; Uimari, P.; Jänne, J.; Alhonen, L. Spermine analogue-regulated expression of spermidine/spermine N1-acetyltransferase and its effects on depletion of intracellular polyamine pools in mouse fetal fibroblasts. *Biochem. J.* **2009**, *422*, 101–109. [CrossRef] [PubMed]

132. Soulet, D.; Gagnon, B.; Rivest, S.; Audette, M.; Poulin, R. A fluorescent probe of polyamine transport accumulates into intracellular acidic vesicles via a two-step mechanism. *J. Biol. Chem.* **2004**, *279*, 49355–49366. [CrossRef] [PubMed]

133. Poulin, R.; Casero, R.A.; Soulet, D. Recent advances in the molecular biology of metazoan polyamine transport. *Amino Acids* **2011**, *42*, 711–723. [CrossRef] [PubMed]

134. Belting, M.; Mani, K.; Jönsson, M.; Cheng, F.; Sandgren, S.; Jonsson, S.; Ding, K.; Delcros, J.-G.; Fransson, L.-A. Glypican-1 is a vehicle for polyamine uptake in mammalian cells: A pivital role for nitrosothiol-derived nitric oxide. *J. Biol. Chem.* **2003**, *278*, 47181–47189. [CrossRef] [PubMed]

135. Abdulhussein, A.A.; Wallace, H.M. Polyamines and membrane transporters. *Amino Acids* **2014**, *46*, 655–660. [CrossRef] [PubMed]

136. Uemura, T.; Yerushalmi, H.F.; Tsaprailis, G.; Stringer, D.E.; Pastorian, K.E.; Hawel, L.; Byus, C.V.; Gerner, E.W. Identification and characterization of a diamine exporter in colon epithelial cells. *J. Biol. Chem.* **2008**, *283*, 26428–26435. [CrossRef] [PubMed]

137. Uemura, T.; Gerner, E.W. Polyamine transport systems in mammalian cells and tissues. *Methods Mol. Biol.* **2011**, *720*, 339–348. [CrossRef] [PubMed]

138. Ruiz-Chica, J.; Medina, M.A.; Sánchez-Jiménez, F.; Ramírez, F.J. Fourier transform Raman study of the structural specificities on the interaction between DNA and biogenic polyamines. *Biophys. J.* **2001**, *80*, 443–454. [CrossRef]

139. Finger, S.; Schwieger, C.; Arouri, A.; Kerth, A.; Blume, A. Interaction of linear polyamines with negatively charged phospholipids: The effect of polyamine charge distance. *Biol. Chem.* **2014**, *395*, 769–778. [CrossRef] [PubMed]

140. Lightfoot, H.L.; Hall, J. Endogenous polyamine function—The RNA perspective. *Nucleic Acids Res.* **2014**, *42*, 11275–11290. [CrossRef] [PubMed]

141. Thomas, T.J.; Tajmir-Riahi, H.A.; Thomas, T. Polyamine-DNA interactions and development of gene delivery vehicles. *Amino Acids* **2016**, *48*, 2423–2431. [CrossRef] [PubMed]

142. Acosta-Andrade, C.; Artetxe, I.; Lete, M.G.; Monasterio, B.G.; Ruiz-Mirazo, K.; Goñi, F.M.; Sánchez-Jiménez, F. Polyamine-RNA-membrane interactions: From the past to the future in biology. *Colloids Surf. B Biointerfaces* **2017**, *155*, 173–181. [CrossRef] [PubMed]

143. Igarashi, K.; Kashiwagi, K. Modulation of protein synthesis by polyamines. *IUBMB Life* **2015**, *67*, 160–169. [CrossRef] [PubMed]

144. Ramani, D.; De Bandt, J.P.; Cynober, L. Aliphatic polyamines in physiology and diseases. *Clin. Nutr.* **2014**, *33*, 14–22. [CrossRef] [PubMed]

145. Miller-Fleming, L.; Olin-Sandoval, V.; Campbell, K.; Ralser, M. Remaining mysteries of molecular biology: The role of polyamines in the cell. *J. Mol. Biol.* **2015**, *427*, 3389–3406. [CrossRef] [PubMed]

146. Murphy, G.M. Polyamines in the human gut. *Eur. J. Gastroenterol. Hepatol.* **2001**, *13*, 1011–1014. [CrossRef] [PubMed]

147. Yuan, Q.; Ray, R.M.; Viar, M.J.; Johnson, L.R. Polyamine regulation of ornithine decarboxylase and its antizyme in intestinal epithelial cells. *Am. J. Physiol. Gastrointest. Liver Physiol.* **2001**, *280*, G130–G138. [CrossRef] [PubMed]

148. Seiler, N.; Raul, F. Polyamines and the intestinal tract. *Crit. Rev. Clin. Lab. Sci.* **2007**, *44*, 365–411. [CrossRef] [PubMed]

149. Deloyer, P.; Peulen, O.; Dandrifosse, G. Dietary polyamines and non-neoplastic growth and disease. *Eur. J. Gastroenterol. Hepatol.* **2001**, *13*, 1027–1032. [CrossRef] [PubMed]

150. Arndt, M.A.; Battaglia, V.; Parisi, E.; Lortie, M.J.; Isome, M.; Baskerville, C.; Pizzo, D.P.; Ientile, R.; Colombatto, S.; Toninello, A.; et al. The arginine metabolite agmatine protects mitochondrial function and confers resistance to cellular apoptosis. *Am. J. Physiol. Cell Physiol.* **2009**, *296*, C1411–C1419. [CrossRef] [PubMed]

151. Satriano, J.; Isome, M.; Casero, R.A.; Thomson, S.C.; Blantz, R.C. Polyamine transport system mediates agmatine transport in mammalian cells. *Am. J. Physiol. Cell Physiol.* **2001**, *281*, C329–C334. [CrossRef] [PubMed]

152. Yang, X.C.; Reis, D.J. Agmatine selectively blocks the *N*-methyl-D-aspartate subclass of glutamate receptor channels in rat hippocampal neurons. *J. Pharmacol. Exp. Ther.* **1999**, *288*, 544–549. [PubMed]

153. Molderings, G.J.; Heinen, A.; Menzel, S.; Lübbecke, F.; Homann, J.; Göthert, M. Gastrointestinal uptake of agmatine: Distribution in tissues and organs and pathophysiologic relevance. *Ann. N. Y. Acad. Sci.* **2003**, *1009*, 44–51. [CrossRef] [PubMed]

154. Halaris, A.; Plietz, J. Agmatine: Metabolic pathway and spectrum of activity in brain. *CNS Drugs* **2007**, *21*, 885–900. [CrossRef] [PubMed]

155. Winter, T.N.; Elmquist, W.F.; Fairbanks, C.A. OCT2 and MATE1 provide bidirectional agmatine transport. *Mol. Pharm.* **2011**, *8*, 133–142. [CrossRef] [PubMed]

156. Sugiyama, Y.; Nara, M.; Sakanaka, M.; Gotoh, A.; Kitakata, A.; Okuda, S.; Kurihara, S. Comprehensive analysis of polyamine transport and biosynthesis in the dominant human gut bacteria: Potential presence of novel polyamine metabolism and transport genes. *Int. J. Biochem. Cell Biol.* **2017**, *93*, 52–61. [CrossRef] [PubMed]

157. Landete, J.M.; De las Rivas, B.; Marcobal, A.; Muñoz, R. Updated molecular knowledge about histamine biosynthesis by bacteria. *Crit. Rev. Food Sci. Nutr.* **2008**, *48*, 697–714. [CrossRef] [PubMed]

158. Ganesh, B.P.; Hall, A.; Ayyaswamy, S.; Nelson, J.W.; Fultz, R.; Major, A.; Haag, A.; Esparza, M.; Lugo, M.; Venable, S.; et al. Diacylglycerol kinase synthesized by commensal *Lactobacillus reuteri* diminishes protein kinase C phosphorylation and histamine-mediated signaling in the mammalian intestinal epithelium. *Mucosal. Immunol.* **2018**, *11*, 380–393. [CrossRef] [PubMed]

159. Williams, B.B.; Van Benschoten, A.H.; Cimermancic, P.; Donia, M.S.; Zimmermann, M.; Taketani, M.; Ishihara, A.; Kashyap, P.C.; Fraser, J.S.; Fischbach, M.A. Discovery and characterization of gut microbiota decarboxylases that can produce the neurotransmitter tryptamine. *Cell Host Microbe* **2014**, *16*, 495–503. [CrossRef] [PubMed]

160. Westfall, S.; Lomis, N.; Kahouli, I.; Dia, S.Y.; Singh, S.P.; Prakash, S. Microbiome, probiotics and neurodegenerative diseases: Deciphering the gut brain axis. *Cell. Mol. Life Sci.* **2017**, *74*, 3769–3787. [CrossRef] [PubMed]

161. Ge, X.; Pan, J.; Liu, Y.; Wang, H.; Zhou, W.; Wang, X. Intestinal crosstalk between microbiota and serotonin and its impact on gut motility. *Curr. Pharm. Biotechnol.* **2018**, in press. [CrossRef] [PubMed]

162. Gao, C.; Ganesh, B.P.; Shi, Z.; Shah, R.R.; Fultz, R.; Major, A.; Venable, S.; Lugo, M.; Hoch, K.; Chen, X.; et al. Gut microbe-mediated suppression of inflammation-associated colon carcinogenesis by luminal histamine production. *Am. J. Pathol.* **2017**, *187*, 2323–2336. [CrossRef] [PubMed]

163. Russo, F.; Linsalata, M.; Orlando, A. Probiotics against neoplastic transformation of gastric mucosa: Effects on cell proliferation and polyamine metabolism. *World J. Gastroenterol.* **2014**, *20*, 13258–13272. [CrossRef] [PubMed]

164. Ramakrishnan, K.; Salinas, R.C. Peptic ulcer disease. *Am. Fam. Physician* **2007**, *76*, 1005–1012. [PubMed]

165. Calam, J.; Baron, J.H. ABC of the upper gastrointestinal tract: Pathophysiology of duodenal and gastric ulcer and gastric cancer. *BMJ* **2001**, *323*, 980–982. [CrossRef] [PubMed]

166. Lai, L.H.; Sung, J.J.Y. Helicobacter pylori and benign upper digestive disease. *Best Pract. Res. Clin. Gastroenterol.* **2007**, *21*, 261–279. [CrossRef] [PubMed]

167. Konturek, S.J.; Konturek, P.C.; Brzozowski, T.; Konturek, J.W.; Pawlik, W.W. From nerves and hormones to bacteria in the stomach; Nobel prize for achievements in gastrology during last century. *J. Physiol. Pharmacol.* **2005**, *56*, 507–530. [PubMed]

168. Singh, V.; Gohil, N.; Ramírez-García, R. New insight into the control of peptic ulcer by targeting the histamine H2 receptor. *J. Cell. Biochem.* **2018**, *119*, 2003–2011. [CrossRef] [PubMed]

169. Safavi, M.; Sabourian, R.; Foroumadi, A. Treatment of Helicobacter pylori infection: Current and future insights. *World J. Clin. Cases* **2016**, *4*, 5–19. [CrossRef] [PubMed]

170. Figueiredo, C.A.; Marques, C.R.; dos Santos Costa, R.; da Silva, H.B.F.; Alcantara-Neves, N.M. Cytokines, cytokine gene polymorphisms and *Helicobacter pylori* infection: Friend or foe? *World J. Gastroenterol.* **2014**, *20*, 5235–5243. [CrossRef] [PubMed]

171. Wessler, S.; Höcker, M.; Fischer, W.; Wang, T.C.; Rosewicz, S.; Haas, R.; Wiedenmann, B.; Meyer, T.F.; Naumann, M. Helicobacter pylori activates the histidine decarboxylase promoter through a mitogen-activated protein kinase pathway independent of pathogenicity island-encoded virulence factors. *J. Biol. Chem.* **2000**, *275*, 3629–3636. [CrossRef] [PubMed]

172. Wadhwa, R.; Song, S.; Lee, J.-S.; Yao, Y.; Wei, Q.; Ajani, J.A. Gastric cancer-molecular and clinical dimensions. *Nat. Rev. Clin. Oncol.* **2013**, *10*, 643–655. [CrossRef] [PubMed]

173. Takaishi, S.; Cui, G.; Frederick, D.M.; Carlson, J.E.; Houghton, J.; Varro, A.; Dockray, G.J.; Ge, Z.; Whary, M.T.; Rogers, A.B.; et al. Synergistic inhibitory effects of gastrin and histamine receptor antagonists on Helicobacter-induced gastric cancer. *Gastroenterology* **2005**, *128*, 1965–1983. [CrossRef] [PubMed]

174. Langman, M.J.; Dunn, J.A.; Whiting, J.L.; Burton, A.; Hallissey, M.T.; Fielding, J.W.; Kerr, D.J. Prospective, double-blind, placebo-controlled randomized trial of cimetidine in gastric cancer. *Br. J. Cancer* **1999**, *81*, 1356–1362. [CrossRef] [PubMed]

175. Garvin, B.; Wiley, J.W. The role of serotonin in irritable bowel syndrome: Implications for management. *Curr. Gastroenterol. Rep.* **2008**, *10*, 363–368. [CrossRef] [PubMed]

176. Crowell, M.D. Role of serotonin in the pathophysiology of the irritable bowel syndrome. *Br. J. Pharmacol.* **2004**, *141*, 1285–1293. [CrossRef] [PubMed]

177. Foley, S.; Garsed, K.; Singh, G.; Duroudier, N.P.; Swan, C.; Hall, I.P.; Zaitoun, A.; Bennett, A.; Marsden, C.; Holmes, G.; et al. Impaired uptake of serotonin by platelets from patients with irritable bowel syndrome correlates with duodenal immune activation. *Gastroenterology* **2011**, *140*, 1434–1443. [CrossRef] [PubMed]

178. Mawe, G.M.; Coates, M.D.; Moses, P.L. Intestinal serotonin signalling in irritable bowel syndrome. *Aliment. Pharmacol. Ther.* **2006**, *23*, 1067–1076. [CrossRef] [PubMed]

179. Konturek, P.C.; Brzozowski, T.; Konturek, S.J. Stress and the gut: Pathophysiology, clinical consequences, diagnostic approach and treatment options. *J. Physiol. Pharmacol.* **2011**, *62*, 591–599. [PubMed]

180. Liu, Q.; Yang, Q.; Sun, W.; Vogel, P.; Heydorn, W.; Yu, X.-Q.; Hu, Z.; Yu, W.; Jonas, B.; Pineda, R.; et al. Discovery and characterization of novel tryptophan hydroxylase inhibitors that selectively inhibit serotonin synthesis in the gastrointestinal tract. *J. Pharmacol. Exp. Ther.* **2008**, *325*, 47–55. [CrossRef] [PubMed]

181. Matthes, S.; Bader, M. Peripheral Serotonin Synthesis as a New Drug Target. *Trends Pharmacol. Sci.* **2018**, *39*, 560–572. [CrossRef] [PubMed]

182. Fogel, W.A.; Lewiński, A.; Jochem, J. Histamine in idiopathic inflammatory bowel diseases—Not a standby player. *Folia Med. Cracov.* **2005**, *46*, 107–118. [PubMed]

183. Neumann, D.; Seifert, R. The therapeutic potential of histamine receptor ligands in inflammatory bowel disease. *Biochem. Pharmacol.* **2014**, *91*, 12–17. [CrossRef] [PubMed]

184. Deiteren, A.; De Man, J.G.; Pelckmans, P.A.; De Winter, B.Y. Histamine H_4 receptors in the gastrointestinal tract. *Br. J. Pharmacol.* **2015**, *172*, 1165–1178. [CrossRef] [PubMed]

185. Xie, Q.; Gan, H.-T. Controversies about the use of serological markers in diagnosis of inflammatory bowel disease. *World J. Gastroenterol.* **2010**, *16*, 279–280. [CrossRef] [PubMed]

186. Thomas, H. IBD: Probiotics for IBD: A need for histamine? *Nat. Rev. Gastroenterol. Hepatol.* **2016**, *13*, 62–63. [CrossRef] [PubMed]

187. Milano, A.F.; Singer, R.B. The cancer mortality risk project—Cancer mortality risks by anatomic site: Part 1—Introductory overview; part II—Carcinoma of the Colon: 20-Year mortality follow-up derived from 1973-2013 (NCI) SEER*Stat Survival Database. *J. Insur. Med.* **2017**, *47*, 65–94. [CrossRef] [PubMed]

188. Babbar, N.; Gerner, E.W. Targeting polyamines and inflammation for cancer prevention. *Recent Results Cancer Res.* **2010**, *188*, 49–64. [CrossRef]

189. Shen, P.; Pichler, M.; Chen, M.; Calin, G.A.; Ling, H. To WNT or Lose: The Missing Non-Coding Linc in Colorectal Cancer. *Int. J. Mol. Sci.* **2017**, *18*, 2003. [CrossRef] [PubMed]

190. Gerner, E.W.; Meyskens, F.L. Combination chemoprevention for colon cancer targeting polyamine synthesis and inflammation. *Clin. Cancer Res.* **2009**, *15*, 758–761. [CrossRef] [PubMed]

191. Ogino, S.; Shima, K.; Baba, Y.; Nosho, K.; Irahara, N.; Kure, S.; Chen, L.; Toyoda, S.; Kirkner, G.J.; Wang, Y.L.; et al. Colorectal cancer expression of peroxisome proliferator-activated receptor gamma (PPARG, PPARgamma) is associated with good prognosis. *Gastroenterology* **2009**, *136*, 1242–1250. [CrossRef] [PubMed]

192. Babbar, N.; Ignatenko, N.A.; Casero, R.A.; Gerner, E.W. Cyclooxygenase-independent induction of apoptosis by sulindac sulfone is mediated by polyamines in colon cancer. *J. Biol. Chem.* **2003**, *278*, 47762–47775. [CrossRef] [PubMed]

193. Gerner, E.W.; Meyskens, F.L. Polyamines and cancer: Old molecules, new understanding. *Nat. Rev. Cancer* **2004**, *4*, 781–792. [CrossRef] [PubMed]

194. Vargas, A.J.; Ashbeck, E.L.; Thomson, C.A.; Gerner, E.W.; Thompson, P.A. Dietary polyamine intake and polyamines measured in urine. *Nutr. Cancer* **2014**, *66*, 1144–1153. [CrossRef] [PubMed]

195. Simon, T.; László, V.; Falus, A. Impact of histamine on dendritic cell functions. *Cell Biol. Int.* **2011**, *35*, 997–1000. [CrossRef] [PubMed]

196. Martinel Lamas, D.J.; Rivera, E.S.; Medina, V.A. Histamine H$_4$ receptor: Insights into a potential therapeutic target in breast cancer. *Front. Biosci.* **2015**, *7*, 1–9.

197. Palucka, A.K.; Coussens, L.M. The Basis of Oncoimmunology. *Cell* **2016**, *164*, 1233–1247. [CrossRef] [PubMed]

198. Medina, V.A.; Coruzzi, G.; Martinel Lamas, D.J.; Massari, N.; Adami, M.; Levi-Schaffer, F.; Ben-Zimra, M.; Schwelberger, H.G.; Rivera, E.S. Histamine in cancer. In *Histamine H4 Receptor: A Novel Drug Target in Immunoregulation and Inflammation*; Stark, H., Ed.; Versita: London, UK, 2013; pp. 259–308.

199. Barilli, A.; Rotoli, B.M.; Visigalli, R.; Ingoglia, F.; Cirlini, M.; Prandi, B.; Dall'Asta, V. Gliadin-mediated production of polyamines by RAW264.7 macrophages modulates intestinal epithelial permeability in vitro. *Biochim. Biophys. Acta* **2015**, *1852*, 1779–1786. [CrossRef] [PubMed]

200. Rodríguez-Ruiz, M.; Moreno, E.; Moreno-Delgado, D.; Navarro, G.; Mallol, J.; Cortés, A.; Lluís, C.; Canela, E.I.; Casadó, V.; McCormick, P.J.; et al. Heteroreceptor complexes formed by dopamine d1, histamine h3, and n-methyl-d-aspartate glutamate receptors as targets to prevent neuronal death in Alzheimer's disease. *Mol. Neurobiol.* **2017**, *54*, 4537–4550. [CrossRef] [PubMed]

201. Burban, A.; Faucard, R.; Armand, V.; Bayard, C.; Vorobjev, V.; Arrang, J.-M. Histamine potentiates N-methyl-D-aspartate receptors by interacting with an allosteric site distinct from the polyamine binding site. *J. Pharmacol. Exp. Ther.* **2010**, *332*, 912–921. [CrossRef] [PubMed]

202. Hornung, B.; Martins Dos Santos, V.A.P.; Smidt, H.; Schaap, P.J. Studying microbial functionality within the gut ecosystem by systems biology. *Genes Nutr.* **2018**, *13*, 5. [CrossRef] [PubMed]

foods

MDPI

Review

Biogenic Amine Production by Lactic Acid Bacteria: A Review

Federica Barbieri [1], Chiara Montanari [1], Fausto Gardini [1,2] and Giulia Tabanelli [1,2,*]

[1] Interdepartmental Center for Industrial Agri-Food Research, University of Bologna, 47521 Cesena, Italy; federica.barbieri16@unibo.it (F.B.); chiara.montanari8@unibo.it (C.M.); fausto.gardini@unibo.it (F.G.)
[2] Department of Agricultural and Food Sciences, University of Bologna, 40126 Bologna, Italy
* Correspondence: giulia.tabanelli2@unibo.it; Tel.: +39-347-032-8294

Received: 26 November 2018; Accepted: 2 January 2019; Published: 7 January 2019

Abstract: Lactic acid bacteria (LAB) are considered as the main biogenic amine (BA) producers in fermented foods. These compounds derive from amino acid decarboxylation through microbial activities and can cause toxic effects on humans, with symptoms (headache, heart palpitations, vomiting, diarrhea) depending also on individual sensitivity. Many studies have focused on the aminobiogenic potential of LAB associated with fermented foods, taking into consideration the conditions affecting BA accumulation and enzymes/genes involved in the biosynthetic mechanisms. This review describes in detail the different LAB (used as starter cultures to improve technological and sensorial properties, as well as those naturally occurring during ripening or in spontaneous fermentations) able to produce BAs in model or in real systems. The groups considered were enterococci, lactobacilli, streptococci, lactococci, pediococci, oenococci and, as minor producers, LAB belonging to *Leuconostoc* and *Weissella* genus. A deeper knowledge of this issue is important because decarboxylase activities are often related to strains rather than to species or genera. Moreover, this information can help to improve the selection of strains for further applications as starter or bioprotective cultures, in order to obtain high quality foods with reduced BA content.

Keywords: biogenic amines; decarboxylase enzymes; lactic acid bacteria; starter cultures

1. Biogenic Amine Toxicity and Physiological Role in Microorganisms

A large number of metabolites, exerting both beneficial and detrimental properties for human health, can be synthetized by microorganisms. Among these, amino acid derivatives produced during bacterial growth and fermentation can interact with human physiology in several ways, showing health-modulating potential [1]. This group includes bioactive compounds such as biogenic amines (BAs), which are responsible for adverse effects and are involved in several pathogenic syndromes [1]. In fact, ingestion of food containing high BA amounts is a risk for consumer health since these compounds can cause headache, heart palpitations, vomiting, diarrhea and hypertensive crises [2–4]. However, their toxic effect depends on the type of BA, on individual sensitivity or allergy and on the consumption of monoaminooxidase inhibitory drugs or ethanol, which interact with aminooxidase enzymatic systems responsible for the detoxification process of exogenous BAs [5,6].

Due to the severity of symptoms they may cause, histamine and tyramine are the most dangerous BAs and are responsible for symptomatology known as "scombroid fish poisoning" and "cheese reaction," respectively [3,7]. The "scombroid fish poisoning", often due to the consumption of fish such as tuna, sardines, anchovies, mackerel, etc., consists in flushing of face, neck and upper arms, oral numbness and/or burning, headache, heart palpitations, asthma attacks, hives, gastrointestinal symptoms, and difficulties in swallowing [8]. Tyramine intoxication is known as "cheese reaction" because this BA is the most frequently found in cheese and it can causes dietary-induced migraine, increased cardiac output, nausea, vomiting, respiratory disorders and elevated blood glucose [7,9].

As far as other BAs, the presence of high level of 2-phenylethylamine, putrescine, cadaverine, agmatine, spermine and spermidine can lead to toxicity. Moreover, they can potentiate the effects of histamine and tyramine toxicity by inhibiting their metabolizing enzymes [10].

Although the consumption of food containing large amounts of BAs can have toxicological consequences, there is no specific legislation regarding the presence of BAs in foods, with the exception of fishery products, for which the maximum acceptable level of histamine is defined [11]. However, recently, EFSA conducted a qualitative risk assessment concerning BA in fermented foods in the European Union, indicating concentrations that could induce adverse effects in consumers [12].

According to their chemical structures, BAs can be classified as aromatic (tyramine and 2-phenylethylamine), aliphatic (putrescine, cadaverine, spermine and spermidine) and heterocyclic (histamine and tryptamine) (see Table 1) and they are analogous to those naturally found in fresh food products, which exert a physiological role associated with cell growth and proliferation [13,14].

The exogenous BAs derive from bacterial decarboxylation of the corresponding amino acids through decarboxylase enzymes. Histamine and cadaverine can be formed by converting histidine and lysine via histidine decarboxylase (HDC) and via lysine decarboxylase (LDC), respectively. Tyrosine is converted in tyramine by tyrosine decarboxylase (TDC), which can act also on phenylalanine obtaining 2-phenylethylamine. This latter aromatic BA is produced by TDC with a lower efficiency with respect to tyramine and it is accumulated when tyrosine is almost completely depleted [15–17]. The formation of these BAs is based on one-step decarboxylation reactions of their respective amino acids and requires systems for amino acid active transport such as antiporter protein in exchange for the resulting BA. Putrescine can be accumulated with a single-step decarboxylation pathway by ornithine decarboxylase (ODC), common in Gram negative bacteria (such as enterobacteria and pseudomonads) or Lactic Acid Bacteria (LAB) deriving from wine environment [16,18–20]. However, this BA can also be formed through agmatinase pathway, which directly converts agmatine to urea and putrescine, or by agmatine deiminase (AgDI) pathway, common in LAB, which transforms arginine to agmatine by arginine decarboxylase. Subsequently, agmatine is converted to putrescine by the agmatine deiminase system, consisting of three enzymes: agmatine deiminase, putrescine carbamoyltransferase and carbamate kinase [21]. The biosynthesis of higher polyamines (spermine and spermidine) proceeds with complex pathways starting from putrescine released from ornithine or agmatine [22,23].

The decarboxylative pathways are activated for several physiological reasons. In fact, decarboxylation of amino acids is coupled with an electrogenic antiport system that can counteract intracellular acidification [24,25]. Therefore, BA accumulation can represent a cellular defense mechanism to withstand acid stress and it has been demonstrated that the transcription of many decarboxylase genes is induced by low pH and improves cell performances in acid conditions [9,17,19,20,26,27]. Moreover, the transfer of a net positive charge outside the cell can generate a proton motive force, leading to cell membrane energization and bringing supplementary energy. It has been demonstrated that the decarboxylase pathway can support the primary metabolism in environmental critical conditions [26–28]. This function can be particularly important for microorganisms lacking a respiratory chain, such as most LAB [29]. del Rio et al. [30] demonstrated that AgDI pathway promotes the growth of *Lactococcus lactis* after nutrient depletion.

Table 1. Biogenic amines, precursors, decarboxylase enzyme and their producers lactic acid bacteria found in fermented foods.

Biogenic Amine	Amino Acid Precursor	Classification	Decarboxylase Enzyme or Pathway in LAB	Lactic Acid Bacteria Producing Species	References
Histamine	Histidine	Heterocyclic	Histidine decarboxylase (HDC)	E. faecium, E. faecalis, L. sakei, L. curvatus, L. parabuchneri, L. buchneri, L. plantarum, L. brevis, L. casei, L. paracasei, L. vaginalis, L. reuteri, L. hilgardii, L. mali, L. rhamnosus, L. paracollinoides, L. rossiae, L. helveticus, S. thermophilus, O. oeni, P. parvulus, Leuc. mesenteroides, W. cibaria, W. confusa, W. paramesenteroides, T. muriaticus, T. halophilus	[24,31–46]
Tyramine	Tyrosine	Aromatic	Tyrosine decarboxylase (TDC)	E. faecium, E. faecalis, E. durans, E. hirae, E. casseliflavus, E. mundtii, L. sakei, L. curvatus, L. plantarum, L. brevis, L. buchneri, L. casei, L. paracasei, L. reuteri, L. hilgardii, L. homohiochii, L. delbrueckii subsp. bulgaricus, S. thermophilus, S. macedonicus, Lc. lactis, Leuc. mesenteroides, W. cibaria, W. confusa, W. paramesenteroides, W. viridescens, C. divergens, C. maltaromaticum, C. gallinarum, T. halophilus, Sporolactobacillus sp.	[9,17,32,47–58]
2-phenylethylamine	Phenylalanine	Aromatic	Tyrosine decarboxylase (TDC)	E. faecium, E. faecalis, E. durans, E. hirae, E. casseliflavus, E. mundtii, L. brevis, Lc. lactis, Leuc. mesenteroides, C. divergens	[9,15–17,32,42,47,59]
Cadaverine	Lysine	Aliphatic	Lysine decarboxylase (LDC)	E. faecium, E. faecalis, L. curvatus, L. brevis, L. casei, L. paracasei, S. thermophilus, Pediococcus spp., Leuc. mesenteroides, T. halophilus	[32,47,56,60–63]
Putrescine	Arginine	Aliphatic	Ornithine decarboxylase (ODC)	E. faecium, E. faecalis, E. durans, E. hirae, E. casseliflavus, L. sakei, L. curvatus, L. buchneri, L. plantarum, L. brevis, L. paracasei, L. mali, L. rhamnosus, L. rossiae, L. homohiochii, Lc. lactis, S. thermophilus, S. mutans, P. parvulus, O. oeni, T. halophilus	[20,30,32,54,56,60,64–72]
	Agmatine	Aliphatic	Agmatine deiminase (AgDI)	E. faecalis, E. faecium, E. durans, E. hirae, E. mundtii, L. curvatus, L. plantarum, L. brevis, S. thermophilus, S. mutans, Lc. lactis, O. oeni, P. parvulus, P. pentosaceus, Leuc. mesenteroides, W. halotolerans, C. divergens, C. maltaromaticum, C. gallinarum	

Several Gram negative and Gram positive bacteria are able to produce BAs. Spoilage bacteria belonging to enterobacteria and pseudomonads can accumulate histamine, putrescine and cadaverine [47,64,73–75]. For this reason, BA content has been related to poor hygienic quality of non-fermented foods, being associated with a massive growth of decarboxylase positive spoilage microorganisms, and several authors proposed BA content as a microbial quality index [75,76]. Decarboxylase activity has been described also in Gram positive microbial groups, such as staphylococci, *Bacillus* spp. and, especially, LAB, considered the most efficient tyramine producers [9,48]. Moreover, the ability to produce histamine, cadaverine and putrescine by bacteria belonging to LAB have been reported [21,64,77]. According to some authors, also yeasts and moulds are implicated in BA accumulation, even if with a controversial role [78–80]. It is important to point out that the capability to produce BAs is generally a strain-specific characteristic, with strong variability in aminobiogenetic potential between different strains belonging to the same species.

2. Role of LAB in Fermented Food BA Content and Their Decarboxylase Clusters Genetic Organization

BA content in fermented foods is of great interest not only for its potential health concerns but also from an economic point of view. On the other hand, the presence of small concentrations of these compounds in fermented foods is unavoidable. In fact, the BA content in these products can range from concentrations below 20 mg/kg for alcoholic and no-alcoholic beverages, fermented vegetables and soy products, up to several hundred mg/kg for some sausages and cheeses [12]. The presence of different BAs is dependent on the precursor availability due to proteolysis during ripening. Moreover, the presence of decarboxylase positive non-starter microbiota, deriving from raw material and productive environment, often leads to high BA concentrations in fermented foods, especially in those obtained without the use of starter cultures [47,75,81,82]. In addition to precursor availability and the presence of BA producing microorganisms, the accumulation of these compounds depends on various intrinsic, environmental and technological factors, recently revised by Gardini et al. [83].

Decarboxylase activity is often expressed independently of cell viability and these enzymes maintain their activity after cell lysis also in harsh environmental conditions [31,49,84,85]. Moreover, once produced, BAs are stable to heat treatment, freezing, and smoking [86].

Dairy products, especially ripened cheeses, have been associated with foodborne intoxications due to their high content of BAs, such as tyramine, histamine, putrescine, and 2-phenylethylamine [32,47]. In any case, BA content varies between different types of cheeses and even different sections of the same cheese [87]. In fermented meats the most prevalent BAs are tyramine, cadaverine, putrescine, and, with minor extent, histamine and their levels strongly vary among different types of products and [75,88–90]. The presence of these compounds in such products depends on low quality processing conditions favoring contamination and on the presence of autochthonous microbiota with decarboxylase potential [82]. Also in alcoholic beverages BAs (mainly histamine, tyramine, putrescine and cadaverine) can be formed through microbial activity during production and storage [60,61]. The presence of BAs has been reported also in fermented vegetables, such as sauerkraut or table olives, where the presence of aminobiogenic spoilage microorganisms can result in high putrescine, cadaverine and tyramine content [91,92]. Abundant amounts of histamine have been detected in fermented fish products [93].

Even if the production of diamines is usually attributed to spoiling Gram negative bacteria, such as enterobacteria and pseudomonads [82], LAB are considered mainly responsible for BA production in fermented foods [47,94]. Although starter cultures are accurately selected for the absence of decarboxylase activity, non-controlled autochthonous LAB involved in ripening process can contribute to BA accumulation. These non-starter LAB (NSLAB) consist mainly of mesophilic facultative or obligate heterofermentative bacteria, which exert a crucial role in maturation phenomena such as the development of flavor [95,96]. These bacteria show good adaptation to unfavorable growth conditions and can survive for long period after sugar depletion, thanks to their ability to obtain energy for growth and survival from other substrates, among which amino acids [97–100]. Moreover, the adaptation to some ecological niches has required the capability to resist acid stresses, activating bacterial mechanisms able to counteract low pH. In stress conditions occurring in fermented foods during ripening, NSLAB encode specific genetic mechanisms that lead to stress responses producing physiological changes among which decarboxylation reactions acquire important roles thanks to the maintenance of pH homeostasis [24,25,101,102]. In fact, expression (by transcriptional induction) and/or activation (by catalytic modulation) of amino acid decarboxylation systems in LAB are reported to be adaptive responses to energy depletion but also strategies to counteract acid stress [103]. The presence of the decarboxylase genes involved in the production of BAs are mostly strain dependent rather than species specific, highlighting the occurrence of horizontal gene transfer between strains as part of a mechanism of survival and adaptation to specific environments [16,33,50]. Recently, the genes belonging to BA biosynthetic pathways in LAB have been identified and the genetic organization of decarboxylase clusters has been reviewed [9,34,64,65]. Generally, enzymes responsible for specific amino acid decarboxylation are organized in clusters in which some genes are always present, i.e., the specific amino acid decarboxylase and the corresponding antiporter permease.

The first tyrosine decarboxylase locus (*tdc*) described in bacteria was found in *Enterococcus faecalis* JH2-2 [104]. This cluster has been annotated also in the genome sequence of other LAB [16,51,52,105–107]. Marcobal et al. [9] evidenced for all tyramine biosynthetic loci a high similarity in both gene sequence and organization, since this locus usually contains the genes encoding tyrosine decarboxylase (*tyrDC*), tyrosyl tRNA synthetase (*tyrS*, located upstream the *tyrDC* gene), putative tyrosine/tyramine permease (*tyrP*, located downstream the *tyrDC* gene) and a Na^+/H^+ antiporter (*nhaC*) [47]. The similar organization of different *tdc* clusters, their distribution, and their high similarity of sequence suggest a horizontal transfer of this cluster from a common source [106]. However, different strains can have different transcriptional organizations of the *tdc* gene cluster, as demonstrated by reverse transcription polymerase chain reaction (PCR) analyses. In fact, the four complete Open Reading Frame (ORF) can be co-transcribed [53] or *tyrS* can be transcribed independently and not included in the catabolic operon [27].

The LAB histidine decarboxylases belong to pyruvoil-dependent decarboxylases group and the encoding histidine decarboxylase gene (*hdcA*) has been identified in several LAB species [33–35,85,108–113]. The histidine decarboxylase gene clusters (*hdc*) of Gram positive bacteria usually comprise the decarboxylase gene *hdcA* and the histidine/histamine antiporter gene *hdcP*. Frequently, an *hdcB* gene, involved in the conversion of the histidine decarboxylase proenzyme to the active decarboxylase can be found [114]. Moreover, for lactobacilli, a histidyl-tRNA synthetase (*hisS*) gene has also been described [35]. The transcriptional studies demonstrated that these genes are located on an operon transcribed as a polycistronic mRNA. However, some authors demonstrated that the antiporter gene is transcribed as a monocistronic RNA and that transcriptional termination structures are present in the intergenic regions of histamine operon in *Lactobacillus buchneri* [111]. Rossi et al. [85] found that *hdcA* gene of *Streptococcus thermophilus* PRI60 was genetically different from the *hdcA* genes sequenced in other LAB, in agreement with the findings of Calles-Enríquez et al. [35], who reported that *hdc* cluster of *S. thermophilus* was more closely related to genera such as *Clostridium* and *Staphylococcus* than other LAB. Another interesting feature of *hdc* gene is its possibility to be located on a plasmid [34]. Lucas et al. [33,36] found that *Lactobacillus hilgardii* 0006, *Tetragenococcus muriaticus*, and *Oenococcus oeni* strains showed 99 to 100% identical *hdcA*- and *hdcB*-encoded proteins, highlighting the presence of a plasmid-encoded histidine decarboxylase system recently transferred horizontally between bacteria. Furthermore, they found that the *hdc* gene cluster, responsible for histamine production in *L. hilgardii* IOEB 0006, was located on an 80-kb plasmid that proved to be unstable. In fact, the capability to form histamine was lost in relation to the growth conditions.

Depending on the producer bacterium, genes/enzymes involved and the ecological niche from which it originates, two different metabolic routes have been described in LAB for the biosynthesis of putrescine [20,64,115]. The first is a decarboxylation system consisting of an ornithine decarboxylase (ODC) and an ornithine/putrescine exchanger. These enzymes are encoded by a gene cluster containing two adjacent genes: (i) *speC* encoding a biosynthetic/constitutive form of the ODC enzyme and (ii) *potE* encoding the transmembrane substrate/product exchanger protein [19,20,116]. Gram positive bacteria, however, have been infrequently reported to possess an ODC enzyme and putrescine-producing LAB strains via the ODC pathway are essentially, although not exclusively, derived from wine environment, belonging to the species *Lactobacillus saerimneri*, *Lactobacillus brevis* [19,20], *Lactobacillus mali* [18], and *O. oeni* [66]. In contrast, the agmatine deiminase (AgDI) pathway is relatively frequent in LAB and it is even considered a species trait in some enterococci [54]. This pathway consists of a more complex system, comprising AgDI, a putrescine transcarbamylase, a carbamate kinase, and an agmatine/putrescine exchanger [65,101]. Five genes are grouped in the agmatine deiminase cluster (*AgDI*): the regulator gene *aguR* and the metabolic genes *aguB*, *aguD*, *aguA* and *aguC* (*aguBDAC*). Linares et al. [117] reported that *aguR* is constitutively transcribed from its promoter (P*aguR*) while the catabolic genes are co-transcribed in a single mRNA from the *aguB* promoter (P*aguB*) in a divergent orientation. These pathway genes were occasionally detected in a putative acid resistance locus in LAB species [101]. In this locus, the *AgDI* genes are found adjacent to the genes associated with the tyrosine decarboxylase pathway on the chromosome [53], suggesting the presence of genes for high-alkalinizing routes (such as amino acid decarboxylases) in LAB genome.

3. Main LAB Involved in BA Production in Fermented Foods

All fermented foods are subjected to the risk of BA contamination. Although LAB are considered GRAS (Generally Regarded As Safe) organisms, they can have the capability to produce toxic compounds as BAs. In particular, in fermented foods, NSLAB can accumulate BAs and strains of lactobacilli, enterococci, lactococci, pediococci, streptococci, and leuconostocs have been associated with high levels of these compounds [118]. Genetic studies have revealed that many of these strains harbor genes or operons coding for decarboxylating enzymes or other pathways implicated in BA biosynthesis [9,64].

Hereafter, the main LAB genera associated with fermented products and involved in BA production in vitro or in situ are described.

3.1. Enterococcus

The *Enterococcus* genus has not been yet classified as safe for human consumption since it neither is recommended for the Qualified Presumption of Safety (QPS) list nor have GRAS status. Most of the species harbor a series of virulence factors and antibiotic resistance and they have been associated with several infections, having the ability to mediate gene transfer with different genetic elements, including plasmids, phages and conjugative transposons [119,120]. The role of enterococci in fermented foods remains controversial. They show remarkable ecological adaptability and ability to grow in adverse conditions. Due to their tolerance to salt and low pH, they are highly adapted to several food systems and they are also involved in the fermentation process of traditional cheeses and dry sausages [121]. Moreover, some *Enterococcus* strains show probiotic features [122] or can improve sensorial properties of dairy products when added as adjunct starters, taking part to flavour generation through proteolytic and lipolytic activities and the accumulation of C4 metabolites such as diacetyl, acetoin or 2, 3-butanediol [123,124]. In addition, their ability to biosynthesize bacteriocins with a wide-range effectiveness on pathogenic and spoilage bacteria is known [125].

Nevertheless, enterococci presence in fermented foods has been associated with the production of BAs (mainly tyramine) and this activity has been reported for strains belonging to different species isolated from meat, cheese, fish, wine and human faeces [54,126–132]. However, not all the strains able to decarboxylate tyrosine were characterized by the same phenotypic potential in relation to the kinetics of tyramine accumulation [15] (Table 1).

Enterococci have been recognized as important part of the natural microbiota in many artisanal cheeses and, in some cases, they can predominate over lactobacilli and lactococci [133]. Usually, enterococci are not present in starter cultures and thus all species of this genus isolated from cheese samples represent contaminating microbial communities, and can include aminobiogenic strains. The most common species found in milk are *Enterococcus faecium*, *Enterococcus durans* and *E. faecalis* but, even if with minor extent, *Enterococcus casseliflavus* may also be isolated [123] and mostly of the strains belonging to these species and isolated from cheese have been identified as tyramine producers [132]. Several authors found a relation between the enterococci counts and the concentrations of tyramine [134–137] and putrescine [67] in dairy products.

Burdychova and Komprda [138] detected tyraminogenic isolates from cheese belonging to *E. durans*, *E. faecium*, *E. faecalis* and *E. casseliflavus* species. Rea et al. [139] studied the effect of six strains of *E. faecalis*, *E. faecium*, *E. durans* and *E. casseliflavus* species on tyramine production in Cheddar cheese during manufacturing and ripening and found that all strains, except *E. casseliflavus*, produced this BA, with *E. durans* responsible for the highest concentration after 9 months of ripening at 8 °C. Enterococcal strains isolated from an Italian cheese and from raw goat milk showed high decarboxylase activity with tyrosine and phenylalanine as substrates [59,136]. Kalhotka et al. [140] investigated the decarboxylase activity of enterococci isolated from goat milk and found that all the tested strains, identified as *Enterococcus mundtii*, *E. faecium* and *E. durans*, showed significant tyrosine and arginine decarboxylase activity, in relation to temperature and time of incubation. Martino et al. [141] studied safety features of four enterococcal strains isolated from a regional Argentinean cheese founding that these strains possessed *tdc* gene cluster, even if only two of four strains gave a positive result in Bover-Cid and Holzapfel decarboxylase screening medium [142]. These authors hypothesized the possibility that this pathway was not active, although all the strains possessed the complete decarboxylase cluster.

The presence of enterococci able to produce BAs is a relevant food issue also in meat products, despite their recognized role in the development of sensory properties of fermented products particularly in sausage [143,144]. In fact, enterococci are constituents of the natural microbiota of raw meat and of many fermented meat products [145], with *E. faecium* and *E. faecalis* being the predominant species, followed by *Enterococcus hirae*, *E. durans* and *E. mundtii* [122]. For this reason, dry fermented sausages can easily accumulate high levels of BAs, especially tyramine, putrescine and cadaverine [82]. In contrast, histamine is usually scarcely found in fermented sausages [146].

Landeta et al. [147] found that 79% of *E. faecium* strains isolated from Spanish dry-cured sausages were able to produce tyramine and that some strains were PCR-positive for the presence of the tyrosine decarboxylase gene, but were not able to accumulate this BA, due to the absence of gene expression. These results were in agreement with those obtained by Komprda et al. [37] who reported that 88% of enterococcal strains isolated in ripened fermented sausages and belonging to *E. faecium* and *E. faecalis* species, possessed *tdc* sequences. These authors found also that 71% of enterococcal isolates had *hdc* gene sequence, assuming that the decarboxylation pathway (producing proton motive force) gives the strains a competitive advantage in nutrient-depleted conditions and acidic environments, such as fermented sausages at the end of ripening. The potential of different indigenous enterococci to contribute to BA formation in spontaneously fermented game meat sausages has been reported also by Maksimovic et al. [148], who found that 100% of *E. durans* and about 7% of *E. casseliflavus* possessed *tdc* genes. Iacumin et al. [62] indicated enterococci able to accumulate large amount of BAs as responsible for spoilage in goose sausages produced in the north of Italy. In fact, despite the addition of starter, enterococci grew during ripening and produced a large amount of BAs. This ability was confirmed in vitro, since all the isolates (n = 100), belonging to the species *E. faecium* and *E. faecalis*, were able to decarboxylate amino acids and produce BAs. In particular, all the strains produced histamine, and 60 out of 70 *E. faecium* and 25 out 30 *E. faecalis* strains produced cadaverine and 10 isolates belonging to both species produced tyramine.

Enterococci has been reported as mainly responsible for tyramine accumulation in wine during malolactic fermentation, together with some *Lactobacillus* species [18,127,149,150]. These latter authors isolated *E. faecium* strains during malolactic fermentation of red wine and demonstrated that, although all the isolates harbored decarboxylase genes, only five strains were able to survive under the harsh conditions found in wine (high ethanol content and low pH), leading to a higher concentration of BAs in samples, including tyramine, histamine and 2-phenylethylamine.

E. faecium and *E. faecalis* have been considered responsible also for BA production in fermented soybean food [151] and in tofu [152].

Although aminobiogenic capability is reported to be strain dependent, Ladero et al. [54] suggested that tyramine and putrescine biosynthesis is a species level trait in *E. faecalis*. In fact, independently of the origin, several strains have been identified as BA producers. Moreover, PCR results demonstrated that the same genetic organization was present in all the tested strains and their decarboxylase clusters were independently located in the chromosome, with flanking regions showing within-species homogeneity.

In *E. faecalis*, putrescine is formed from agmatine by the AgDI pathway, which is repressed by carbon source, suggesting a role in the energy production [153]. Perez et al. [154] studied the possible co-regulation among TDC and AgDI pathways in *E. faecalis*. They investigated firstly the tyrosine effect on the *tdc* cluster transcription of *E. faecalis* by microarray experiment, highlighting, in the presence of tyrosine, an over-expression of *tdcA*, *tdcP*, and *nhac-2* genes and a repression of *tyrS*. Bargossi et al. [15,155] have also demonstrated the same effect in other *E. faecalis* strains. Moreover, Perez et al. [154] showed that tyrosine induced putrescine biosynthesis genes, as confirmed by reverse transcription quantitative PCR (RT-qPCR) results. On the other hand, this effect was not observed in the mutant strain, which was unable to decarboxylate tyrosine and produce tyramine, showing that *tdc* cluster was involved in the tyrosine induction of putrescine biosynthesis.

Recently, some authors demonstrated that also *E. mundtii* possesses the capability to produce both tyramine and 2-phenilethylammine [107]. The genetic organization indicated that the tyramine-forming pathway in *E. mundtii* is similar to that found in phylogenetically closer enterococcal species, such as *E. faecium*, *E. hirae* and *E. durans*. The gene Na^+/H^+ antiporter (*nhaC*) that usually follows *tyrP* was missing. However, the analysis of the available data on *E. mundtii* genome revealed the presence of a further region that includes two genes encoding for an additional pyridoxal phosphate (PLP)-dependent decarboxylase and an amino acid permease, correlated with the tyrosine decarboxylating potential of this species.

In any case, tyramine is often accumulated by enterococci in high amounts already during the late exponential growth, before stationary phase, suggesting that this decarboxylation activity is not necessarily a response to starvation or nutrient depletion, and no competition between sugar catabolism and amino acid decarboxylation was observed [15,17]. In particular, these latter authors tested the ability to accumulate tyramine and 2-phenylethylamine by two strains of *E. faecalis* and two strains *E. faecium* in two culture media added or not with tyrosine. They demonstrated that, although all the tested enterococcal strains possessed a TDC pathway, they differed in BA accumulation level and in the expression rate of *tdc* gene, underlining the extremely variable decarboxylating potential of strains belonging to the same species, suggesting strain-dependent implications in food safety.

Environmental factors such as pH, temperature and NaCl concentrations can affect BA production in enterococci and several studies on decarboxylase activity of *Enterococcus* spp. in different conditions have been carried out. Gardini et al. [156] investigated the combined effects of temperature, pH and NaCl concentration on tyramine production by the strain *E. faecalis* EF37, finding that production of tyramine was mainly dependent on cell number. Moreover, these authors reported that this strain was able to accumulate also 2-phenylethylamine. A study regarding EF37 *tyrDC* expression revealed that stress could induce greater tyrosine decarboxylase activity, suggesting that suboptimal environmental conditions could lead to a higher tyrosine production, not necessarily associated with cell growth. This could be explained with the physiological role of this biochemical pathways associated with the survival of LAB in hostile environments [157]. Acidic conditions favored tyramine production in an *E. durans* BA-producing strain isolated from cheese [158] and in *E. faecium* [16,26], demonstrating the role of tyrosine decarboxylation in pH homeostasis. On the other hand, transcriptional studies of the *tdc* cluster in *E. durans* 655 showed a pH regulation of tyramine biosynthesis, being the gene expression quantification during the exponential phase induced by high concentrations of tyrosine, under acidic conditions [159].

Bargossi et al. [105] investigated the diversity of tyramine production capability of two *E. faecalis* and two *E. faecium* strains in buffered systems in relation to their genetic characteristics and to pH,

NaCl concentration and incubation temperature, comparing the results with those obtained with a purified tyrosine decarboxylase under the same conditions. They found that TDC activity was greatly heterogeneous within the enterococci, being *E. faecalis* EF37 the most efficient in tyramine accumulation. This heterogeneity depended on different genetic determinants, regulation mechanisms and environmental factors, above all incubation temperature.

A reduced transcription of genes involved in tyramine production was observed in the presence of 6.5% of NaCl in *E. faecalis* [160]. Also Bargossi et al. [105] showed that *E. faecalis* partially reduced its tyraminogenic potential passing from 0 to 5% of NaCl but the decarboxylation activity did not change significantly increasing NaCl concentration up to 15%. On the contrary, Liu et al. [161] demonstrated that NaCl stress can upregulate the expression of *tyrDC* and *tyrP* to improve the tyramine production of a single *E. faecalis* strain under certain conditions.

3.2. Lactobacillus

Lactobacilli are reported to be strong BA producers in different fermented foods [94].

In fermented sausages, beside to enterococci, the main tyramine producers among LAB are strains belonging to *Lactobacillus curvatus* species, which is, together with *Lactobacillus sakei*, the predominant *Lactobacillus* species in fermented meat products [162,163]. In fact, the majority of *L. curvatus* strains isolated from meat were reported to be tyramine producers [126]. However, Bover-Cid et al. [126] reported also some strains of *Lactobacillus paracasei*, *L. brevis*, and *L. sakei* isolated from pork-fermented meat as tyramine forming. Pereira et al. [164] demonstrated tyrosine and ornithine decarboxylase activities in *Lactobacillus homohiochii* and *L. curvatus* isolated from a Portuguese traditional dry fermented sausage.

Freiding et al. [165] screened *L. curvatus* strains from different origins, finding strain dependent tyrosine decarboxylase activity. Moreover, although *L. sakei* is usually described as non- aminogenic, histidine decarboxylase activity in one *L. sakei* strain has been evidenced [81,126,166].

Lactobacillus parabuchneri and *L. buchneri*, present as contaminants in fermented meat products, can produce histamine [167].

LAB populations were isolated from dry fermented sausages produced with different starters and using two spice mixtures in different process time by Kompdra et al. [37]. Tyrosine-decarboxylase and histidine-decarboxylase DNA sequence was identified in 44% and of 16% of lactobacilli isolates, respectively. In particular, several *Lactobacillus plantarum*, *L. brevis* and *Lactobacillus casei/paracasei* strains were identified as tyramine and histamine producers in the sausages analysed.

Although several microorganisms in cheese, including Gram negative bacteria, are able to produce BAs, *Lactobacillus* species, such as *Lactobacillus helveticus*, *L. buchneri* and *L. curvatus*, can be responsible for their accumulation in such products [38,138,168]. For example, specific strains of *L. buchneri* and *L. parabuchneri* harbor the histidine decarboxylase enzyme and can develop high levels of histamine, even at refrigerate temperature [39,169]. Wüthrich et al. [167] analysed several *L. parabuchneri* strains isolated from cheeses, finding some histamine positive among them. Moreover, these authors determined the complete genome of a histamine positive strain, showing that *hdc* gene cluster is located in a genomic island, transferred within the *L. parabuchneri* species. Diaz et al. [40] isolated, for the first time, 25 histamine-producing *Lactobacillus vaginalis* strains and sequenced *hdc* gene cluster and its flanking regions for a representative strain (*L. vaginalis* IPLA11050). These authors suggested that *hdc* locus was localized in the chromosome and, being the flanking regions the same in all histamine-producing *L. vaginalis* tested strains, histamine production has been suggested to be a species level trait. In addition, the organization of the examined genes was the same described for *Lactobacillus reuteri* [41], *L. buchneri* [111] and *L. hilgardii* [33] but differed to that of *S. thermophilus*.

L. brevis tyramine-producing strains have been isolated from cheeses by several authors [170,171] and this feature has been described as a strain-level trait (perhaps horizontally acquired) in *L. brevis*. Pachlová et al. [172] assessed the development of BA content in model cheese samples individually inoculated with two BA producing NSLAB strains of *L. curvatus* subsp. *curvatus* and *L. paracasei*, demonstrating the ability of these strains to accumulate tyramine up to 200 mg/kg in real dairy products during a 90 days ripening period. Yilmaz and Görkmen [173] demonstrated the capability to produce tyramine by a *L. plantarum* strain in yogurt and highlighted a possible indirect effect of *Lactobacillus delbrueckii* subsp. *bulgaricus* on accumulation of tyramine in the yoghurts, due to its synergistic interactions with tyraminogenic LAB strains.

It has been reported that non-starter *L. brevis* and *L. curvatus* are able to produce both tyramine and putrescine [174]. Although the ODC pathway has been described in several LAB, including strains of *L. brevis* [20], this pathway is not commonly used by dairy bacteria [64,67,175]. Ladero et al. [174] confirmed this aspect, showing that the detected putrescine-producing lactobacilli used AgDI pathway. Lucas et al. [101] found that *L. brevis* IOEB 9809 produced putrescine from agmatine but not from arginine, indicating the lack of a pathway converting arginine into agmatine. Moreover, it has been suggested that in *L. brevis* the AgDI genetic determinants are linked to those of the TDC pathway and are located in an acid resistance mechanism locus, probably acquired by horizontal gene transfer [20].

Several native lactobacilli, together with *O. oeni* and *Pediococcus parvulus* strains, are responsible for BA accumulation in wine. Their formation in wine depends on several conditions such as precursor amounts and the presence of specific decarboxylase-positive species and strains [176,177] and they are produced mainly during malolactic fermentation, particularly due to the presence of *L. brevis* and *L. hilgardii* [178–180]. Landete et al. [42] reported aminobiogenic potential of LAB isolated from wine samples, evidencing *L. mali* strains able to produce histamine, *L. brevis* strains able to accumulate tyramine and 2-phenylethylamine and a *L. hilgardii* strain showing histamine, tyramine, 2-phenylethylamine and putrescine production ability. On the other hand, the HDCs of *L. hilgardii* isolated from wine are well documented [33]. The enhancing effects of lower pH on histamine production (as responses to acidic stress) was observed in *L. brevis* [181]. Henríquez-Aedo et al. [182] reported that *Lactobacillus rhamnosus* was unexpectedly the predominant species in the vinification process of Chilean Cabernet Sauvignon wines and that it was mainly responsible for histamine accumulation in the products, presenting a significantly higher BA formation capability with respect to *O. oeni* isolated from the same samples. Arena and Manca de Nadra [21] studied a *L. plantarum* strain able to produce putrescine from arginine and ornithine while Moreno-Arribas et al. [183] found two wine strains of *L. buchneri* able to form putrescine via ornithine decarboxylase.

In wines, tyrosine decarboxylase has been associated with *Lactobacillus* spp., particularly *L. brevis* strains [84]. The same authors purified this pyridoxal 5P-phosphate dependent enzyme and showed it was highly substrate-specific for L-tyrosine and had an optimum pH of 5 [184]. The ability of a *L. plantarum* strain isolated from a red wine to produce tyramine from peptides containing tyrosine, especially during the late exponential growth phase, has been demonstrated [55] and *tdc* genes shared 98% identity with those in *L. brevis* consistent with horizontal gene transfer from *L. brevis* to *L. plantarum*. Arena et al. [185] assessed the expression of *L. brevis* IOEB 9809 *tdc* and *aguA1* genes during wine fermentation and evaluated the effect of substrate availability and pH on it, as well as on BA production, showing that the strain was able to produce both tyramine and putrescine. In addition, qRT-PCR analysis suggested a strong influence of substrate availability on the expression of BA pathway genes while less evident was pH influence. Afterwards, Lucas and Lonvaud-Funel [186] and Lucas et al. [53] reported for the same strain the complete *tdc* sequences, describing four complete genes (*tyrS*, *tyrDC*, *tyrP* and *nhaC*).

The BA production ability in different *Lactobacillus* strains, isolated from wine and cider, and their metabolic pathway were explored by Constantini et al. [68]. Their results demonstrated that most of the *L. brevis* analyzed harbor both *AgDI* and *tdc* genes and were tyramine and putrescine producers. Interestingly, these authors detected *hdc* genes in a *L. casei* strain isolated from cider.

Beer spoilage LAB showed several metabolic strategies to grow in nutrient poor environment, with acidic pH and hop presence among which BA production, contributing to energy supply and pH homeostasis, has been highlighted [187]. In particular, heterofermentative *L. brevis* strains accumulated tyramine and ornithine while *Lactobacillus lindneri* and *Lactobacillus paracollinoides* beer spoiling agent displayed ornithine and histamine production, respectively. Strains belonging to this latter species have been indicated as new potential histamine- and putrescine-producers in cider analysed by Ladero et al. [188]. Also Lorencová et al. [63] demonstrated that *L. brevis* strains isolated from beer can be a tyramine source in these products. The same authors showed the possibility to produce BAs by some probiotic strains belonging to *L. rhamnosus* species, opening a serious concern about the need to investigate the decarboxylation activity of probiotic or functional cultures before their use.

Recently, the capability to produce putrescine of a *Lactobacillus rossiae* strain, previously isolated from sourdough, has been reported [69]. This species is widely distributed in this fermented food [189] but the possibility of *L. rossiae* sourdough strains to produce BAs was not previously shown. In fact, only a strain of this species isolated from a wine starter has been described as histamine producer [190]. del Rio et al. [69] showed that *L. rossiae* strain accumulated this BA via the ODC pathway and the genetic organization and transcriptional analysis of the gene cluster identified the *odc* and *potE* genes forming an operon that is transcriptionally regulated by ornithine in a dose-dependent manner. Moreover, putrescine production via the ODC system improved the survival of *L. rossiae* by counteracting the cytoplasm acidification when the cells were subjected to acidic conditions, providing a biochemical defense mechanism against acidic environments. For this reason, this strain could easily produce putrescine during the fermentation process and the potential presence in sourdough of other BA-producing microorganisms cannot be ruled out.

The first description of ODC system in LAB was reported for *L. saerimneri* 30a [191]. Further investigation on this strain evidenced the presence of a unique genomic organization in which *odc* does not have an adjacent specific transporter gene but a three-component decarboxylase system with a lysine decarboxylase gene (*aadc*) and a promiscuous amino acid-amine transporter gene (*aat*), appearing atypical from those of other LAB [192].

3.3. Streptococcus

Although certain *Streptococcus* species are responsible for many disease (i.e., meningitis, bacterial pneumonia, endocarditis, necrotizing fasciitis etc.) many streptococcal species form part of the human microbiota and are important for fermented foods. In fact, *S. thermophilus* is often employed as selected starter culture and it is important for the dairy industry, since it is one of the principal components of many natural cultures used in fermented products such as hard cooked or pasta filata cheeses, yogurt and Cheddar [193]. This species is usually present in high numbers in the first steps of cheese-making and its relationships with BAs, which mainly accumulate during ripening, has been longer neglected. In fact, despite its wide industrial use, there are few papers regarding the decarboxylating potential of this species. Nevertheless, some BA producing strains have been identified and studied in recent years and several screenings on aminobiogenic potential of *Streptococcus* spp. strains have been performed. Ladero et al. [174] studied the BA producing ability of 137 strains of starter and NSLAB belonging to nine species of the genera *Lactobacillus*, *Lactococcus*, *Streptococcus* and *Leuconostoc* (all isolated from artisanal cheeses) in liquid media supplemented with the appropriate precursor amino acid by Ultra-High Performance Liquid Chromatography technique. Moreover, assessing the presence of key genes involved in the biosynthetic pathways of the target BA, they found that two *S. thermophilus* strains possessed *hdc* genes, although they were unable to synthesize histamine in broth. Also strains belonging to *Streptococcus macedonicus* species, isolated from Greek Kasseri cheese, showed tyramine production [194]. Some authors demonstrated that *Streptococcus mutans* expressed an agmatine deiminase system, encoded by the agmatine-inducible *aguBDAC* operon, which was induced in the presence of agmatine and was regulated by carbon catabolite repression. This metabolism was proposed to augment the acid resistance properties and pathogenic potential of *S. mutans*, etiological agent of dental caries and acid tolerance in oral biofilms [195,196].

Elsanhoty and Ramadan [197] reported the presence of *tdc*, *hdc* and *AgDI* genes in a *S. thermophilus* strain. Buňková et al. [48] studied BA production capability of selected technological important LAB belonging to *Lactococcus*, *Lactobacillus* and *Streptococcus* genera. Among these strains, one *S. thermophilus* of 11 was able to produce tyramine. Gezginc et al. [198] investigated the BA production capability of

S. thermophilus isolates in homemade natural yogurt, evidencing the presence of *hdcA* gene in several strains, although it was poorly correlated with histamine production in the decarboxylase medium. Yilmaz and Gökmen [173] investigated tyramine formation during yoghurt fermentation, focusing on interaction between a *S. thermophilus* strain and some *Lactobacillus* species The streptococci cells were able to produce tyramine depending on the fermentation conditions and synergistic interactions between *S. thermophilus* and *L. delbrueckii* subsp. *bulgaricus* were found in terms of BA accumulation.

The decarboxylating potential of the strain *S. thermophilus* NCFB2392 in lysine decarboxylase broth has been reported [199]. This strain was able to accumulate mostly putrescine, cadaverine and agmatine, and, when co-cultered with other BA producer Gram negative strains, had synergistic or antagonistic effect on BA concentrations. In fact, it caused 2-fold lower cadaverine production by *Salmonella* Paratyphi A and stimulated tyramine accumulation of *Escherichia coli*.

A high number of *S. thermophilus* strains have been investigated for tyramine production by La Gioia et al. [49]. Only the strain 1TT45, isolated from Taleggio cheese, demonstrated the capability to accumulate tyramine in broth. For this strain, a tyrosine decarboxylase (*tdc*A) gene was identified, with a nearly identical sequence to a *tdc*A of *L. curvatus*, indicated a horizontal gene transfer event. In the same work *tdc*A expression level and the production of tyramine were evaluated under different conditions during 7 days of incubation in skim milk. High transcript levels were evidenced only at the seventh day in presence of tyrosine, showing that the ability of *S. thermophilus* 1TT45 to form this BA depends on precursor availability in the culture medium, due to the incapability of this species to release peptides and free amino acids from milk proteins when grown in pure culture. On the other hand, the presence in cheeses of highly proteolytic LAB species would likely allow tyramine formation by *S. thermophilus*.

Calles-Enríquez et al. [35] observed two *S. thermophilus* strains able to produce histamine and reported their complete *hdc* gene cluster organization. This cluster began with the *hdcA* gene, was followed by a transporter (*hdcP*) and ended with the *hdcB* gene, located in the chromosome and orientated in the same direction.

The gene order of *hdcAPB* operon is similar to *Staphylococcus capitis* and *Clostridium perfringens*, which, however, lacks *hdcB* [200]. Transcriptional analysis of the *hdc* cluster revealed the maximum expression during the stationary growth phase, with high expression levels correlated with high histamine concentration. In the same work, also some factors affecting histamine biosynthesis and histidine-decarboxylating gene (*hdcA*) expression were studied. In particular, low temperature incubation determined lower levels of histamine in milk than in samples kept at 42 °C. This reduction was attributed to a reduction in the activity of the HDC enzyme itself rather than a reduction in gene expression or the presence of a lower cell number.

The occurrence of a histidine decarboxylase gene (*hdcA*) was demonstrated also in five among 83-screened *S. thermophilus* strains by Rossi et al. [85]. The sequence of the *hdcA* gene and closest flanking regions were determined for the strain PRI60, which produced the highest amounts of histamine. This strain synthesized HDC enzyme in milk even in the absence of histidine and it remained active also in cell-free extracts. Tabanelli et al. [201] continued the study of the histamine potential of PRI60 strain, testing histamine accumulation by cells or crude enzyme preparations with respect to factors related to dairy products, reporting a histamine concentration increase concomitantly with the cell growth. Moreover, HDC was mostly active at pH 4.5 and salt concentration up to 5% (*w/v*) did not affected enzyme activity. These authors evidenced enzyme thermal resistance up to temperature resembling low pasteurization, showing the risk of the presence of histaminogenic *S. thermophilus* strains in products from raw or mildly heat-treated milk. In fact, this strain was able to accumulate histamine in experimental cheeses, both when inoculated as starter or as cell-free crude enzyme preparations, highlighting that histamine formation by *S. thermophilus* in artisanal cheeses must not be overlooked, especially during typical production practices [31].

3.4. Lactococcus

Lactococci are among the most important LAB involved in the dairy industry and some species, including *Lc. lactis* subsp. *lactis*, *Lc. lactis* subsp. *lactis* biovar *diacetylactis* and *Lc. lactis* subsp. *cremoris*, play a critical role in the manufacture of many fermented dairy products [202]. As well as starters, they have been proposed as bioprotective cultures in the food industry and, for these reasons, safety criteria (such as BA production) should be evaluated.

Despite their QPS status (recognized by EFSA) and their generally regarded as safe (GRAS) status (recognized by FDA), some *Lc. lactis* have been reported to have aminobiogenic activity, both in vitro and in real systems. In fact, several strains of *Lc. lactis* subsp. *lactis* and *Lc. lactis* subsp. *cremoris* able to produce putrescine from agmatine via the AgDI pathway have been identified and these species are known to be, together with *E. faecalis*, *E. hirae*, *L. brevis* and *L. curvatus*, the main putrescine producers in dairy products [65,67,174]. The analysis of the *AgDI* cluster and their flanking regions revealed that the capability to produce putrescine via the AgDI pathway could be a specific characteristic that was lost during the adaptation to the milk environment by a process of reductive genome evolution [65]. The AgDI pathway increases the growth of *Lc. lactis* and causes the alkalinization of the culture medium, although it does not seem to be an acid stress resistance mechanism [30]. Linares et al. [203] investigated the role of *aguR* gene in putrescine formation in *Lc. lactis* subsp. *cremoris* CECT 8666, founding that it is essential for putrescine biosynthesis and it is transcribed independently of the polycistronic mRNA encoding the catabolic genes. Moreover, the transmembrane protein encoded by *aguR* can act as a transcription activator of the putrescine biosynthesis operon in response to the agmatine concentration. The same strain was tested in experimental Cabrales-like cheeses with different NaCl concentrations [204]. These authors evidenced that reducing the NaCl concentration of cheese led to increased putrescine accumulation and that NaCl was able to reduce the transcription of the *aguBDAC* operon, even if no effect on the transcription of *aguR* was recorded. The same authors investigated the effect of extracellular pH on putrescine biosynthesis and on the genetic regulation of the AgDI pathway of CECT 8666 strain. They showed increased putrescine biosynthesis at pH 5, when the transcription of the catabolic operon via the activation of the *aguBDAC* promoter P*aguB* was induced and a protection against acidic external conditions was reached through the counteraction of cytoplasm acidification [205].

It has been reported that *Lc. lactis* can produce other BAs in addition to putrescine. Martins Perin et al. [59] evaluated the BA production of bacteriocinogenic lactococci strains isolated from raw goat's milk reporting tyramine and 2-phenylethylamine accumulation capability in some of them. The decarboxylase activity of two aminobiogenic strains of *Lc. lactis* subsp. *cremoris* employed as starters in a model system of Dutch-type cheese was studied during a 90 day ripening period [206]. While in the control samples the amount of BAs was negligible, lactococci accumulated about 500 and 800 mg/kg of tyramine and putrescine, respectively. The putrescine decarboxylase activity observed in the model samples of cheeses with the inoculated strains was consistent with the results by Santos et al. [207].

3.5. Oenococcus and Pediococcus

Pediococcus spp. are often isolated from a large variety of plant materials and are involved in spontaneous fermentation of silage, sauerkraut, beans, cucumbers, olives, and cereals [208]. In addition, pediococci are also associated with fermented sausages and cheese, where selected strains of *Pediococcus pentosaceus* and *Pediococcus acidilactici* are often exploited as commercial starters, to control the development of undesired and pathogenic microbiota given their bacteriocinogenic features [209]. However, they can act also as spoilers in several fermented foods such as beer, wine and cider, causing turbidity, acidic off-tastes, adverse flavors and accumulating undesirable compounds [210].

O. oeni is a wine-associated LAB, considered the dominant species during the malolactic fermentation and possesses remarkable adaptability to harsh physico-chemical conditions. Several *O. oeni* strains have been described as BA producers and therefore many authors considered this species, together with other LAB species such as *L. hilgardii* and *Pediococcus*, as responsible for histamine presence in wine [33,36,43,108,176].

Lonvaud-Funel and Joyeux [44] isolated for the first time a strain of *O. oeni* able to produce histamine via histidine decarboxylase from a wine from the Bordeaux area. Subsequently, Coton et al. [108] purified and characterized this enzyme, concluding that it requires pyridoxal-5-phosphate as cofactor. Other authors have also shown that histamine-producing strains of *O. oeni* are frequent in wine [211] but this feature was strain dependent and in some strains no BA potential has been found [176]. For this reason, the role of *O. oeni* in BA accumulation in wine is still controversial [212]. As a possible explanation for these discrepancies, it has been suggested that the *hdc* genes are located on a large and possibly unstable plasmid and that culture collections will lose the capability to produce histamine in laboratory subcultures because of the loss of this unstable plasmid [36]. In the LAB isolated from wine by Landete et al. [42] mostly of *O. oeni* and *P. parvulus* harbored *hdc* genes and pediococci produced high level of histamine in synthetic medium, showing the highest histaminogenic potential within the tested genus. Nevertheless, in the same conditions *O. oeni* showed lower levels of histamine production, that were even lower when these strains were tested in wine samples. Berbegal et al. [213] showed the presence of both histamine producer and non-producer strains in a Spanish red Ribera del Duero wine and proposed a non histaminogenic strain to be used as starter reducing of 5-fold histamine content in inoculated wine than the non-inoculated control. It has been further demonstrated that, after one year, the barrel-ageing histamine concentrations were 3-fold lower in the inoculated vat than in the non-inoculated one.

Landete et al. [214] studied the influence of enological factors on the *hdc* expression and on HDC activity in *P. parvulus* and *O. oeni*. Gene expression was lowered by glucose, fructose, malic acid, and citric acid, whereas ethanol enhanced the HDC enzyme activity, so that the conditions normally occurring during malolactic fermentation and later on, could favor histamine production. On the other hand, Gardini et al. [177] evaluated the interactive effect of some variables on the BA production of *O. oeni*, demonstrating that high ethanol amounts and low concentration of pyridoxal-5-phosphate reduced their accumulation while higher pH enhanced BA concentrations. In addition, the SO_2 effect on tyramine accumulation depended also on other variables.

In cider, where a microbiological stabilization after malolactic fermentation is not performed, indigenous heterofermentative LAB constitute the predominant microbiota and several species such as oenococci and pediococci (beside to lactobacilli) are able to produce BAs [18,215]. Some pediococci strains isolated from wine and ciders showed the presence of decarboxylase genes, i.e., *AgDI* cluster in *P. parvulus* and *P. pentosaceus* [18,68,188]. In this latter work, also *O. oeni* strains had *AgDI* genes and the authors found a few discrepancies between phenotypic and genotypic data. On the other hand, the identification of an *odc* gene in a putrescine producer *O. oeni* strain has been reported by Marcobal et al. [70]. Later, the *odc* gene was also identified and sequenced in three *O. oeni* wine strains and in two *O. oeni* cider strains [216] and the sequencing of the complete *odc* gene from *O. oeni* and *L. brevis* showed an 83% identity [19].

Low production of cadaverine and tyramine was also found in a *Pediococcus* spp. strain isolated from beer [63] and Izquierdo-Pulido et al. [217] reported *Pediococcus* genus to be mainly responsible for tyramine accumulation in beer.

3.6. Other Genera: Weissella, Carnobacterium, Tetragenococcus, Leuconostoc, Sporolactobacillus

Leuconostocs are LAB associated with plants and decaying plant material, often detected in various fermented vegetable products but also in foods of animal origin [218]. Some species such as *Leuconostoc carnosum*, *Leuconostoc gasicomitatum*, and *Leuconostoc gelidum* have often been associated with food spoilage and some strains have been found to be decarboxylase positive. Although it is known that AgDI pathway has been demonstrated in *Leuconostoc mesenteroides* [18], some strains isolated from wine have been suggested to produce putrescine exclusively from arginine via the arginine deiminase pathway (ADI) pathway, given to the selective effect of the ecological niche on BA biosynthesis pathway [115,219]. Recently, based on current knowledge and QPS/GRAS/dairy (IDF) safety criteria guidelines, the safety of different LAB candidate antifungal bioprotective strains has been evaluated finding a tyramine-producer *Leuc. mesenteroides*. This result confirmed the importance to test decarboxylase activity before considering a candidate strain for use as a bioprotective agent in food products [220]. Dairy strains of leuconostocs have been associated with high levels of BAs in cheese and other dairy products and tyramine and 2-phenylethylamine production in *Leuconostoc* strains isolated from dairy products have been reported [32,221]. Moreno-Arribas et al. [183] found that *Leuc. mesenteroides* may also be responsible for tyramine production in wines and Landete et al. [42] isolated a wine strain belonging to this species able to produce histamine.

This genus can be also implicated in beer BA accumulation. In fact, some authors studied the occurrence of aminobiogenic strains during a craft brewing process, highlighting the presence of *Leuc. mesenteroides* possessing *tdc*, *hdc*, *odc* decarboxylase genes and able to produce tyramine in wort and beer [61]. *Leuc. mesenteroides* ssp. *mesenteroides* isolated from meat, fermented sausages and cheeses was able to form putrescine and cadaverine [56].

Weissella are heterofermentative LAB which occur in a wide range of habitats, i.e., milk, plants and as well as from a variety of fermented foods such as European sourdoughs and Asian and African traditional fermented vegetables. They can be involved in such traditional fermentations and some strains of *Weissella confusa* and *Weissalla cibaria* can produce copious amounts of dextran but strains of certain *Weissella* species are known as opportunistic pathogens involved in human infections [222]. Moreover, some *Weissella* strains have been demonstrated to be able to produce BAs in fermented foods. *Weissella viridescens* isolated from Tofu-misozuke, a traditional Japanese fermented food, resulted tyramine producers and several strains isolated from kimchi belonging to *W. cibaria*, *W. confusa* and *Weissella paramesenteroides* produced multiple BAs, including tyramine and histamine [45,152]. On the other hand, Pereira et al. [71] demonstrated that a *Weissella halotolerans* strain combines an ornithine decarboxylation pathway and an arginine deiminase pathway, leading to the accumulation of putrescine and producing a proton motive force.

Carnobacterium spp. can be found in vacuum or modified atmosphere packed, refrigerated raw or processed meat products and lightly preserved fish products, milk, and certain types of soft cheese [223]. Although *Carnobacterium divergens* and *Carnobacterium maltaromaticum* have been demonstrated to be bacteriocin producers, able to inhibit *Listeria monocytogenes*, some strains appear to display undesirable properties such as amino acid decarboxylation activities. In fact, these two species can produce tyramine while strains belonging to *C. divergens*, *Carnobacterium gallinarum*, *Carnobacterium maltaromaticum* and *C. mobile* can possess ADI pathway [57,104,223]. Curiel et al. [224] studied the BA production capability by LAB and enterobacteria isolated from fresh pork sausages and reported that all the tyramine-producer isolated strains were molecularly identified as *C. divergens*, whose abundance depended from the different packaging conditions. All these strains presented the *tdc* genes. Coton et al. [51] identified the gene encoding a putative tyrosine decarboxylase in *C. divergens*, evidencing the presence of three putative open reading frames (*tyrS*, *tyrDC* and amino acid transporter *PotE*) which showed the strongest homologies with *E. faecium* (94% identity, 98% homology) and *E. faecalis* (85% identity, 92% similarity) and exhibited conserved domains characteristic of the group II (PLP-dependent) decarboxylase family.

Other genera less frequent in fermented foods may also be involved in BA production.

Tetragenococcus is a halophilic facultative aerobic homofermentative coccus, which cannot be readily distinguished from members of the genus *Pediococcus* and which can play a role in halophilic fermentation processes such as the production of soy products, brined anchovies, fish sauce and fermented mustard or can constitute the dominant microbiota in concentrated sugar-concentrated juice [225]. Recently, the safety of 49 *Tetragenococcus halophilus* strains isolated from the Korean sauce doenjang has been assessed. The isolates produced higher tyramine level than reference strains and similar cadaverine, histamine, and putrescine production patterns [72]. A *T. muriaticus* strain isolated from fish sauce produced histamine during the late exponential growth phase, reaching a maximum production of this BA at 5–7% of NaCl, and was able to maintain a histidine decarboxylase activity also in the presence of 20% of salt [46].

Recently, a sporulating LAB, belonging to a novel species of *Sporolactobacillus* genus and isolated from cider must, harbored *tdc* gene, which showed the same organization as already described genes found in other tyramine-producing LAB. Moreover, genes showing the highest identities with mobile elements surrounded the *tdc* operon, suggesting that the tyramine-forming trait was acquired through horizontal gene transfer [58].

4. Conclusions

Biogenic amines can accumulate in high concentrations in fermented foods due to microbial activity and can cause toxic effects in consumers. LAB are considered mainly responsible for BA accumulation in these products and strains belonging to different species and genera, commonly found in fermented foods, have been characterized for their decarboxylase activities.

It is known that this decarboxylase activity provides cell advantages because it allows increasing the environmental pH and leads to the energization of membrane. The genetic clusters responsible for BA production have been described individually and they can show differences, within the same amine, that depend mainly on the species and the strain. Nevertheless, it is interesting to note, that the decarboxylation mechanisms constitute an important ecological tool which can favor strain competitiveness in stressful conditions (i.e., acid stress and nutritional stress) [20,26,27].

Even if differences between the chromosomic decarboxylase clusters are present, some interesting consideration can be drawn. The first is that the presence of these cluster are usually strain and not species dependent and can be regarded as genomic islands, as demonstrated for TDC cluster in *L. brevis* by Coton and Coton [50]. In addition, the genes encoding different decarboxylase pathways in several LAB species (*L. saerimneri*, *L. brevis*, and *O. oeni*) are clustered on the chromosome, acting as a genetic hotspot related to acid stress resistance [19,20,192].

Although the knowledge concerning the origin and factors involved in BA production in fermented foods is well documented, it is difficult to prevent the accumulation of these compounds since the fermentation conditions cannot be easily modified and the aminobiogenic ability is strain dependent. For these reasons, the selection of specific LAB starters lacking the pathways for BA accumulation and able to outgrow autochthonous microbiota under production conditions is essential to obtain high quality food with reduced contents of these toxic compounds. In fact, the inability of a strain to synthesize BAs has to be included as a selective criterion for starter cultures [226]. On the other hand, the metabolic heterogeneity observed in natural starter cultures could open a serious concern about the presence of aminobiogenic LAB strains. This risk could be avoided with the use of defined starters or selected autochthonous strain mixtures, chosen based on the absence of such activity and endowed with taylor-made metabolic and functional features for specific products. Nevertheless, when undefined cultures need to be used, strategies to prevent the presence and growth of aminobiogenic LAB should be actuated. Among them, the use of food microorganisms able to degrade BAs previously synthesized in the food matrix should be taken into consideration.

Author Contributions: Conceptualization, G.T. and F.G.; Literature data collection, G.T., F.B., writing—original draft preparation, G.T., F.B.; writing—review and editing, C.M. and F.G.; supervision, F.G.

Funding: This research received no external funding.

Conflicts of Interest: The authors declare no conflict of interest.

References

1. Pessione, E.; Cirrincione, S. Bioactive molecules released in food by lactic acid bacteria: Encrypted peptides and biogenic amines. *Front. Microbiol.* **2016**, *7*, 876. [CrossRef] [PubMed]

2. Alvarez, M.A.; Moreno-Arribas, M.V. The problem of biogenic amines in fermented foods and the use of potential biogenic amine-degrading microorganisms as a solution. *Trends Food Sci. Technol.* **2014**, *39*, 146–155. [CrossRef]

3. Hungerford, J.M. Scombroid poisoning: A review. *Toxicon* **2010**, *56*, 231–243. [CrossRef] [PubMed]

4. Shalaby, A.R. Significance of biogenic amines to food safety and human health. *Food Res. Int.* **1996**, *29*, 675–690. [CrossRef]

5. Sathyanarayana Rao, T.S.; Yeragani, V.K. Hypertensive crisis and cheese. *Indian J. Psychiatry* **2009**, *51*, 65–66. [CrossRef] [PubMed]

6. Silla Santos, M.H. Biogenic amines: Their importance in foods. *Int. J. Food Microbiol.* **1996**, *29*, 213–231. [CrossRef]

7. McCabe-Sellers, B.; Staggs, C.G.; Bogle, M.L. Tyramine in foods and monoamine oxidase inhibitor drugs: A crossroad where medicine, nutrition, pharmacy, and food industry converge. *J. Food Comp. Anal.* **2006**, *19*, S58–S65. [CrossRef]

8. Knope, K.E.; Sloan-Gardner, T.S.; Stafford, R.J. Histamine fish poisoning in Australia, 2001 to 2013. *Commun. Dis. Intell. Q. Rep.* **2014**, *38*, E285–E293.

9. Marcobal, A.; de Las Rivas, B.; Landete, J.M.; Tabera, L.; Muñoz, R. Tyramine and phenylethylamine biosynthesis by food bacteria. *Crit. Rev. Food Sci. Nutr.* **2012**, *52*, 448–467. [CrossRef]

10. Pegg, A.E. Toxicity of polyamines and their metabolic products. *Chem. Res. Toxicol.* **2013**, *26*, 1782–1800. [CrossRef]

11. European Commission. Commission Regulation (EC) No. 2073/2005 of 15 November 2005 on microbiological criteria for foodstuffs. *Off. J. Eur. Union* **2005**, *50*, 1–26.

12. EFSA. Scientific opinion on risk based control of biogenic amine formation in fermented foods. *EFSA J.* **2011**, *9*, 2393–2486. [CrossRef]

13. Bover-Cid, S.; Latorre-Moratalla, M.L.; Veciana-Nogués, M.T.; Vidal-Carou, M.C. Biogenic amines. In *Encyclopedia of Food Safety*; Motarjemi, Y., Moy, G., Todd, E., Eds.; Academic Press: San Diego, CA, USA, 2014; pp. 381–391. ISBN 978-0-12-378613-5.

14. Halász, A.; Baráth, Á.; Simon-Sarkadi, L.; Holzapfel, W. Biogenic amines and their production by microorganisms in food. *Trends Food Sci. Technol.* **1994**, *5*, 42–49. [CrossRef]

15. Bargossi, E.; Tabanelli, G.; Montanari, C.; Lanciotti, R.; Gatto, V.; Gardini, F.; Torriani, S. Tyrosine decarboxylase activity of enterococci grown in media with different nutritional potential: Tyramine and 2-phenylethylamine accumulation and *tyrDC* gene expression. *Front. Microbiol.* **2015**, *6*, 259. [CrossRef] [PubMed]

16. Marcobal, A.; de las Rivas, B.; Muñoz, R. First genetic characterization of a bacterial b-phenylethylamine biosynthetic enzyme in *Enterococcus faecium* RM58. *FEMS Microbiol. Lett.* **2006**, *258*, 144–149. [CrossRef] [PubMed]

17. Pessione, E.; Pessione, A.; Lamberti, C.; Coïsson, D.J.; Riedel, K.; Mazzoli, R.; Bonetta, S.; Eberl, L.; Giunta, C. First evidence of a membrane-bound, tyramine and beta-phenylethylamine producing, tyrosine decarboxylase in *Enterococcus faecalis*: A two-dimensional electrophoresis proteomic study. *Proteomics* **2009**, *9*, 2695–2710. [CrossRef]

18. Coton, M.; Romano, A.; Spano, G.; Ziegler, K.; Vetrana, C.; Desmarais, C.; Coton, E. Occurrence of biogenic amine-forming lactic acid bacteria in wine and cider. *Food Microbiol.* **2010**, *27*, 1078–1085. [CrossRef]

19. Romano, A.; Trip, H.; Lonvaud-Funel, A.; Lolkema, J.S.; Lucas, P.M. Evidence of two functionally distinct ornithine decarboxylation systems in lactic acid bacteria. *Appl. Environ. Microbiol.* **2012**, *78*, 1953–1961. [CrossRef] [PubMed]

20. Romano, A.; Ladero, V.; Alvarez, M.A.; Lucas, P.M. Putrescine production via the ornithine decarboxylation pathway improves the acid stress survival of *Lactobacillus brevis* and is part of a horizontally transferred acid resistance locus. *Int. J. Food Microbiol.* **2014**, *175*, 14–19. [CrossRef]

21. Arena, M.E.; Manca de Nadra, M.C. Biogenic amine production by *Lactobacillus*. *J. Appl. Microbiol.* **2001**, *90*, 158–162. [CrossRef]

22. Bardócz, S. Polyamines in food and their consequences for food quality and human health. *Trends Food Sci. Technol.* **2005**, *6*, 341–346. [CrossRef]

23. Kalač, P.; Krausová, P. A review of dietary polyamines: Formation, implications for growth and health and occurrence in foods. *Food Chem.* **2005**, *90*, 219–230. [CrossRef]

24. Molenaar, D.; Bosscher, J.S.; Ten Brink, B.; Driessen, A.J.M.; Konings, W.N. Generation of a proton motive force by histidine decarboxylation and electrogenic histidine/histamine antiport in *Lactobacillus buchneri*. *J. Bacteriol.* **1993**, *175*, 2864–2870. [CrossRef]

25. Pessione, A.; Lamberti, C.; Pessione, E. Proteomics as a tool for studying energy metabolism in lactic acid bacteria. *Mol. BioSyst.* **2010**, *6*, 1419–1430. [CrossRef]

26. Pereira, C.I.; Matos, D.; Romão, M.V.S.; Barreto Crespo, M.T. Dual role for the tyrosine decarboxylation pathway in *Enterococcus faecium* E17: Response to an acid challenge and generation of a proton motive force. *Appl. Environ. Microbiol.* **2009**, *75*, 345–352. [CrossRef]

27. Perez, M.; Calles-Enríquez, M.; Nes, I.; Martin, M.C.; Fernández, M.; Ladero, V.; Alvarez, M.A. Tyramine biosynthesis is transcriptionally induced at low pH and improves the fitness of *Enterococcus faecalis* in acidic environments. *Appl. Microbiol. Biotechnol.* **2015**, *99*, 3547–3558. [CrossRef]

28. Konings, W.N. Microbial transport: Adaptations to natural environments. *Antonie Van Leeuwenhoek* **2006**, *90*, 325–342. [CrossRef]

29. Vido, K.; Le Bars, D.; Mistou, M.Y.; Anglade, P.; Gruss, A.; Gaudu, P. Proteome analyses of heme-dependent respiration in *Lactococcus lactis*: Involvement of the proteolytic system. *J. Bacteriol.* **2004**, *186*, 1648–1657. [CrossRef]

30. del Rio, B.; Linares, D.M.; Ladero, V.; Redruello, B.; Fernández, M.; Martin, M.C.; Alvarez, M.A. Putrescine production via the agmatine deiminase pathway increases the growth of *Lactococcus lactis* and causes the alkalinization of the culture medium. *Appl. Microbiol. Biotechnol.* **2015**, *99*, 897–905. [CrossRef]

31. Gardini, F.; Rossi, F.; Rizzotti, L.; Torriani, S.; Grazia, L.; Chiavari, C.; Coloretti, F.; Tabanelli, G. Role of *Streptococcus thermophilus* PRI60 in histamine accumulation in cheese. *Int. Dairy J.* **2012**, *27*, 71–76. [CrossRef]

32. Benkerroum, N. Biogenic amines in dairy products: Origin, incidence, and control means. *Compr. Rev. Food Sci. Food Saf.* **2016**, *15*, 801–826. [CrossRef]

33. Lucas, P.M.; Wolken, W.A.; Claisse, O.; Lolkema, J.S.; Lonvaud-Funel, A. Histamine-producing pathway encoded on an unstable plasmid in *Lactobacillus hilgardii* 0006. *Appl. Environ. Microbiol.* **2005**, *7*, 1417–1424. [CrossRef]

34. Landete, J.M.; de las Rivas, B.; Marcobal, A.; Muñoz, R. Updated molecular knowledge about histamine biosynthesis by bacteria. *Crit. Rev. Food Sci. Nutr.* **2008**, *48*, 697–714. [CrossRef]

35. Calles-Enríquez, M.; Eriksen, B.H.; Andersen, P.S.; Rattray, F.P.; Johansen, A.H.; Fernández, M.; Ladero, V.; Alvarez, M.A. Sequencing and transcriptional analysis of the *Streptococcus thermophilus* histamine biosynthesis gene cluster: Factors that affect differential *hdcA* expression. *Appl. Environ. Microbiol.* **2010**, *76*, 6231–6238. [CrossRef]

36. Lucas, P.M.; Claisse, O.; Lonvaud-Funel, A. High frequency of histamine-producing bacteria in the enological environmental and instability of the histidine decarboxylase production phenotype. *Appl. Environ. Microbiol.* **2008**, *74*, 811–817. [CrossRef]

37. Komprda, T.; Sládková, P.; Petirová, E.; Dohnal, V.; Burdychová, R. Tyrosine- and histidine-decarboxylase positive lactic acid bacteria and enterococci in dry fermented sausages. *Meat Sci.* **2010**, *86*, 870–877. [CrossRef]

38. Ladero, V.; Linares, D.M.; Fernández, M.; Alvarez, M.A. Real time quantitative PCR detection of histamine-producing lactic acid bacteria in cheese: Relation with histamine content. *Food Res. Int.* **2008**, *41*, 1015–1019. [CrossRef]

39. Diaz, M.; del Rio, B.; Sanchez-Llana, E.; Ladero, V.; Redruello, B.; Fernández, M.; Martin, M.C.; Alvarez, M.A. *Lactobacillus parabuchneri* produces histamine in refrigerated cheese at a temperature-dependent rate. *Int. J. Food Sci. Technol.* **2018**, *53*, 2342–2348. [CrossRef]

40. Diaz, M.; del Rio, B.; Ladero, V.; Redruello, B.; Fernández, M.; Martin, M.C.; Alvarez, M.A. Isolation and typification of histamine-producing *Lactobacillus vaginalis* strains from cheese. *Int. J. Food Microbiol.* **2015**, *215*, 117–123. [CrossRef]

41. Thomas, C.M.; Hong, T.; van Pijkeren, J.P.; Hemarajata, P.; Trinh, D.V.; Hu, W.; Britton, R.A.; Kalkum, M.; Versalovic, J. Histamine derived from probiotic *Lactobacillus reuteri* suppresses TNF via modulation of PKA and ERK signaling. *PLoS ONE* **2012**, *7*, e31951. [CrossRef]

42. Landete, J.M.; Ferrer, S.; Pardo, I. Biogenic amine production by lactic acid bacteria, acetic bacteria and yeast isolated from wine. *Food Control* **2007**, *18*, 1569–1574. [CrossRef]

43. Landete, J.M.; Ferrer, S.; Pardo, I. Which lactic acid bacteria are responsible for histamine production in wine? *J. App. Microbiol.* **2005**, *99*, 580–586. [CrossRef] [PubMed]

44. Lonvaud-Funel, A.; Joyeux, A. Histamine production by wine lactic acid bacteria: Isolation of a histamine-producing strain of *Leuconostoc oenos*. *J. Appl. Bacteriol.* **1994**, *77*, 401–407. [CrossRef]

45. Jeong, D.W.; Lee, J.H. Antibiotic resistance, hemolysis and biogenic amine production assessments of *Leuconostoc* and *Weissella* isolates for kimchi starter development. *LWT-Food Sci. Technol.* **2015**, *64*, 1078–1084. [CrossRef]

46. Kimura, B.; Konagaya, Y.; Fujii, T. Histamine formation by *Tetragenococcus muriaticus*, a halophilic lactic acid bacterium isolated from fish sauce. *Int. J. Food Microbiol.* **2001**, *70*, 71–77. [CrossRef]

47. Linares, D.M.; Martin, M.C.; Ladero, V.; Alvarez, M.A.; Fernández, M. Biogenic amines in dairy products. *Crit. Rev. Food Sci. Nutr.* **2011**, *51*, 691–703. [CrossRef] [PubMed]

48. Buňková, L.; Buňka, F.; Hlobilová, M.; Vakátková, Z.; Nováková, D.; Dráb, V. Tyramine production of technological important strains of *Lactobacillus*, *Lactococcus* and *Streptococcus*. *Eur. Food Res. Technol.* **2009**, *229*, 533–538. [CrossRef]

49. La Gioia, F.; Rizzotti, L.; Rossi, F.; Gardini, F.; Tabanelli, G.; Torriani, S. Identification of a tyrosine decarboxylase (*tdcA*) gene in *Streptococcus thermophilus* 1TT45: Analysis of its expression and tyramine production in milk. *Appl. Environ. Microbiol.* **2011**, *77*, 1140–1144. [CrossRef] [PubMed]

50. Coton, E.; Coton, M. Evidence of horizontal transfer as origin of strain to strain variation of the tyramine production trait in *Lactobacillus brevis*. *Food Microbiol.* **2009**, *26*, 52–57. [CrossRef] [PubMed]

51. Coton, M.; Coton, E.; Lucas, P.; Lonvaud, A. Identification of the gene encoding a putative tyrosine decarboxylase of *Carnobacterium divergens* 508. Development of molecular tools for the detection of tyramine-producing bacteria. *Food Microbiol.* **2004**, *21*, 125–130. [CrossRef]

52. Ladero, V.; Linares, D.M.; del Rio, B.; Fernández, M.; Martin, M.C.; Alvarez, M.A. Draft genome sequence of the tyramine producer *Enterococcus durans* strain IPLA 655. *Genome Announc.* **2013**, *1*, e00265-13. [CrossRef] [PubMed]

53. Lucas, P.; Landete, J.; Coton, M.; Coton, E.; Lonvaud-Funel, A. The tyrosine decarboxylase operon of *Lactobacillus brevis* IOEB 9809: Characterization and conservation in tyramine-producing bacteria. *FEMS Microbiol. Lett.* **2003**, *229*, 65–71. [CrossRef]

54. Ladero, V.; Fernández, M.; Calles-Enríquez, M.; Sánchez-Llana, E.; Cañedo, E.; Martin, M.C.; Alvarez, M.A. Is the production of the biogenic amines tyramine and putrescine a species-level trait in enterococci? *Food Microbiol.* **2012**, *30*, 132–138. [CrossRef] [PubMed]

55. Bonnin-Jusserand, M.; Grandvalet, C.; Rieu, A.; Weidmann, S.; Alexandre, H. Tyrosine-containing peptides are precursors of tyramine produced by *Lactobacillus plantarum* strain IR BL0076 isolated from wine. *BMC Microbiol.* **2012**, *12*, 199. [CrossRef] [PubMed]

56. Pircher, A.; Bauer, F.; Paulsen, P. Formation of cadaverine, histamine, putrescine and tyramine by bacteria isolated from meat, fermented sausages and cheeses. *Eur. Food Res. Technol.* **2007**, *226*, 225–231. [CrossRef]

57. Massona, F.; Johansson, G.; Montela, M.C. Tyramine production by a strain of *Carnobacterium divergens* inoculated in meat-fat mixture. *Meat Sci.* **1999**, *52*, 65–69. [CrossRef]

58. Coton, M.; Fernández, M.; Trip, H.; Ladero, V.; Mulder, N.L.; Lolkema, J.S.; Alvarez, M.A.; Coton, E. Characterization of the tyramine-producing pathway in *Sporolactobacillus* sp. P3J. *Microbiology* **2011**, *157*, 1841–1849. [CrossRef]

59. Martins Perin, L.; Belviso, S.; dal Bello, B.; Nero, L.A.; Cocolin, L. Technological properties and biogenic amines production by bacteriocinogenic lactococci and enterococci strains isolated from raw goat's milk. *J. Food Prot.* **2017**, *80*, 151–157. [CrossRef]

60. Guo, Y.Y.; Yang, Y.P.; Peng, Q.; Han, Y. Biogenic amines in wine: A review. *Int. J. Food Sci. Technol.* **2015**, *50*, 1523–1532. [CrossRef]

61. Poveda, J.M.; Ruiz, P.; Seseña, S.; Palop, M.L. Occurrence of biogenic amine-forming lactic acid bacteria during a craft brewing process. *LWT-Food Sci. Technol.* **2017**, *85*, 129–136. [CrossRef]

62. Iacumin, L.; Manzano, M.; Panseri, S.; Chiesa, L.; Comi, G. A new cause of spoilage in goose sausages. *Food Microbiol.* **2016**, *58*, 56–62. [CrossRef] [PubMed]

63. Lorencová, E.; Buňková, L.; Matoulková, D.; Dráb, V.; Pleva, P.; Kubáň, V.; Buňka, F. Production of biogenic amines by lactic acid bacteria and bifidobacteria isolated from dairy products and beer. *Int. J. Food Sci. Technol.* **2012**, *47*, 2086–2091. [CrossRef]

64. Wunderlichová, L.; Buňková, L.; Koutný, M.; Jančová, P.; Buňka, F. Formation, degradation, and detoxification of putrescine by foodborne bacteria: A review. *Compr. Rev. Food Sci. Food Saf.* **2014**, *13*, 1012–1033. [CrossRef]

65. Ladero, V.; Rattray, F.P.; Mayo, B.; Martin, M.C.; Fernández, M.; Alvarez, M.A. Sequencing and transcriptional analysis of the biosynthesis gene cluster of putrescine-producing *Lactococcus lactis*. *Appl. Environ. Microbiol.* **2011**, *77*, 6409–6418. [CrossRef] [PubMed]

66. Marcobal, A.; de las Rivas, B.; Moreno-Arribas, M.V.; Munoz, R. Evidence for horizontal gene transfer as origin of putrescine production in *Oenococcus oeni* RM83. *Appl. Environ. Microbiol.* **2006**, *72*, 7954–7958. [CrossRef]

67. Ladero, V.; Canedo, E.; Perez, M.; Cruz Martin, M.; Fernández, M.; Alvarez, M.A. Multiplex qPCR for the detection and quantification of putrescine-producing lactic acid bacteria in dairy products. *Food Control* **2012**, *27*, 307–313. [CrossRef]

68. Costantini, A.; Pietroniro, R.; Doria, F.; Pessione, E.; Garcia-Moruno, E. Putrescine production from different amino acid precursors by lactic acid bacteria from wine and cider. *Int. J. Food Microbiol.* **2013**, *165*, 11–17. [CrossRef]

69. del Rio, B.; Alvarez-Sieiro, P.; Redruello, B.; Martin, M.C.; Fernandez, M.; Ladero, V.; Alvarez, M.A. *Lactobacillus rossiae* strain isolated from sourdough produces putrescine from arginine. *Sci. Rep.* **2018**, *8*, 3989. [CrossRef]

70. Marcobal, A.; de Las Rivas, B.; Moreno-Arribas, M.V.; Muñoz, R. Identification of the ornithine decarboxylase gene in the putrescine-producer *Oenococcus oeni* BIFI-83. *FEMS Microbiol. Lett.* **2004**, *239*, 213–220. [CrossRef]

71. Pereira, C.I.; San Romão, M.V.; Lolkema, J.S.; Barreto Crespo, M.T. *Weissella halotolerans* W22 combines arginine deiminase and ornithine decarboxylation pathways and converts arginine to putrescine. *J. Appl. Microbiol.* **2009**, *107*, 1894–1902. [CrossRef]

72. Jeong, D.W.; Heo, S.; Le, J.H. Safety assessment of *Tetragenococcus halophilus* isolates from doenjang, a Korean high-salt-fermented soybean paste. *Food Microbiol.* **2017**, *62*, 92–98. [CrossRef] [PubMed]

73. Lorenzo, J.M.; Cachaldora, A.; Fonseca, S.; Gómez, M.; Franco, I.; Carballo, J. Production of biogenic amines "in vitro" in relation to the growth phase by *Enterobacteriaceae* species isolated from traditional sausages. *Meat Sci.* **2010**, *86*, 684–691. [CrossRef]

74. Morii, H.; Kasama, K. Activity of two histidine decarboxylases from *Photobacterium phosphoreum* at different temperatures, pHs, and NaCl concentrations. *J. Food Prot.* **2004**, *67*, 1736–1742. [CrossRef]

75. Ruiz-Capillas, C.; Jiménez-Colmenero, F. Biogenic amines in meat and meat products. *Crit. Rev. Food Sci. Nutr.* **2004**, *44*, 489–499. [CrossRef] [PubMed]

76. Özogul, F.; Özogul, Y. Biogenic amine content and biogenic amine quality indices of sardines (*Sardina pilchardus*) stored in modified atmosphere packaging and vacuum packaging. *Food Chem.* **2006**, *99*, 574–578. [CrossRef]

77. Ladero, V.; Sánchez-Llana, E.; Fernández, M.; Alvarez, M.A. Survival of biogenic amine-producing dairy LAB strains at pasteurisation conditions. *Int. J. Food Sci. Technol.* **2011**, *46*, 516–521. [CrossRef]

78. Gardini, F.; Tofalo, R.; Belletti, N.; Iucci, L.; Suzzi, G.; Torriani, S.; Guerzoni, M.E.; Lanciotti, R. Characterization of yeasts involved in the ripening of Pecorino Crotonese cheese. *Food Microbiol.* **2006**, *23*, 641–648. [CrossRef]

79. Qi, W.; Hou, L.H.; Guo, H.L.; Wang, C.L.; Fan, Z.C.; Liu, J.F.; Cao, X.H. Effect of salt-tolerant yeast of *Candida versatilis* and *Zygosaccharomyces rouxii* on the production of biogenic amines during soy sauce fermentation. *J. Sci. Food Agric.* **2014**, *94*, 1537–1542. [CrossRef]

80. Tristezza, M.; Vetrano, C.; Bleve, G.; Spano, G.; Capozzi, V.; Logrieco, A.; Mita, G.; Grieco, F. Biodiversity and safety aspects of yeast strains characterized from vineyards and spontaneous fermentations in the Apulia Region. *Food Microbiol.* **2013**, *36*, 335–342. [CrossRef] [PubMed]

81. Latorre-Moratalla, M.L.; Bover-Cid, S.; Talon, R.; Garriga, M.; Aymerich, T.; Zanardi, E.; Ianieri, A.; Fraqueza, M.J.; Elias, M.; Drosinos, E.H.; et al. Distribution of aminogenic activity among potential autochthonous starter cultures for dry fermented sausages. *J. Food Prot.* **2010**, *73*, 524–525. [CrossRef]

82. Suzzi, G.; Gardini, F. Biogenic amines in dry fermented sausages: A review. *Int. J. Food Microbiol.* **2003**, *88*, 41–54. [CrossRef]

83. Gardini, F.; Özogul, Y.; Suzzi, G.; Tabanelli, G.; Özogul, F. Technological factors affecting biogenic amine content in foods: A review. *Front. Microbiol.* **2016**, *7*, 1218. [CrossRef]

84. Moreno-Arribas, M.V.; Lonvaud-Funel, A. Tyrosine decarboxylase activity of Lactobacillus brevis IOEB 9809 isolated from wine and L. brevis ATCC 367. *FEMS Microbiol. Lett.* **1999**, *180*, 55–60. [CrossRef]

85. Rossi, F.; Gardini, F.; Rizzotti, L.; La Gioia, F.; Tabanelli, G.; Torriani, S. Quantitative analysis of histidine decarboxylase gene (*hdcA*) transcription and histamine production by *Streptococcus thermophilus* PRI60 under conditions relevant to cheese making. *Appl. Environ. Microbiol.* **2011**, *77*, 2817–2822. [CrossRef]

86. Becker, K.; Southwick, K.; Reardon, J.; Berg, R.; MacCormack, J.N. Histamine poisoning associated with eating tuna burgers. *JAMA* **2001**, *285*, 1327–1330. [CrossRef]

87. Novella-Rodríguez, S.; Veciana-Nogués, M.T.; Izquierdo-Pulido, M.; Vidal-Carou, M.C. Distribution of biogenic amines and polyamines in cheese. *J. Food Sci.* **2003**, *68*, 750–755. [CrossRef]

88. Jairath, G.; Singh, P.K.; Dabur, R.S.; Rani, M.; Chaudhari, M. Biogenic amines in meat and meat products and its public health significance: A review. *J. Food Sci. Technol.* **2015**, *52*, 6835–6846. [CrossRef]

89. Latorre-Moratalla, M.L.; Bover-Cid, S.; Bosch-Fusté, J.; Vidal-Carou, M.C. Influence of technological conditions of sausage fermentation on the aminogenic activity of *L. curvatus* CTC273. *Food Microbiol.* **2012**, *29*, 43–48. [CrossRef]

90. Ruiz-Capillas, C.; Pintado, T.; Jiménez-Colmenero, F. Biogenic amine formation in refrigerated fresh sausage "chorizo" keeps in modified atmosphere. *J. Food Biochem.* **2011**, *36*, 449–457. [CrossRef]

91. Medina-Pradas, E.; Arroyo-López, F.N. Presence of toxic microbial metabolites in table olives. *Front. Microbiol.* **2015**, *6*, 873. [CrossRef]

92. Rabie, M.A.; Siliha, H.; el-Saidy, S.; el-Badawy, A.A.; Malcata, F.X. Reduced biogenic amine contents in sauerkraut via addition of selected lactic acid bacteria. *Food Chem.* **2011**, *129*, 1778–1782. [CrossRef]

93. Prester, L. Biogenic amines in fish, fish products and shellfish: A review. *Food Addit. Contam.* **2011**, *28*, 1547–1560. [CrossRef]

94. Spano, G.; Russo, P.; Lonvaud-Funel, A.; Lucas, P.; Alexandre, H.; Grandvalet, C.; Coton, E.; Coton, M.; Barnavon, L.; Bach, B.; et al. Biogenic amines in fermented foods. *Eur. J. Clin. Nutr.* **2010**, *64*, 64–951. [CrossRef]

95. Gobbetti, M.; De Angelis, M.; Di Cagno, R.; Mancini, L.; Fox, P.F. Pros and cons for using non-starter lactic acid bacteria (NSLAB) as secondary/adjunct starters for cheese ripening. *Trends Food Sci. Technol.* **2015**, *45*, 167–178. [CrossRef]

96. Smid, E.J.; Kleerebezem, M. Production of aroma compounds in lactic fermentations. *Annu. Rev. Food Sci. Technol.* **2014**, *5*, 313–326. [CrossRef]

97. Cocconcelli, P.S.; Fontana, C. Starter cultures for meat fermentation. In *Handbook of Meat Processing*; Toldrà, F., Ed.; Wiley-Blackwell: Ames, IA, USA, 2010; pp. 199–218. ISBN 978-0-81-382089-7.

98. Montanari, C.; Barbieri, F.; Magnani, M.; Grazia, L.; Gardini, F.; Tabanelli, G. Phenotypic diversity of *Lactobacillus sakei* strains. *Front. Microbiol.* **2018**, *9*, 2003. [CrossRef]

99. Sgarbi, E.; Bottari, B.; Gatti, M.; Neviani, E. Investigation of the ability of dairy nonstarter lactic acid bacteria to grow using cell lysates of other lactic acid bacteria as the exclusive source of nutrients. *Int. J. Dairy Technol.* **2014**, *67*, 342–347. [CrossRef]

100. Skeie, S.; Kieronczyka, A.; Næssa, R.M.; Østliea, H. *Lactobacillus* adjuncts in cheese: Their influence on the degradation of citrate and serine during ripening of a washed curd cheese. *Int. Dairy J.* **2008**, *18*, 158–168. [CrossRef]

101. Lucas, P.M.; Blancato, V.S.; Claisse, O.; Magni, C.; Lolkema, J.S.; Lonvaud-Funel, A. Agmatine deiminase pathway genes in *Lactobacillus brevis* are linked to the tyrosine decarboxylation operon in a putative acid resistance locus. *Microbiology* **2007**, *153*, 2221–2230. [CrossRef]

102. Montanari, C.; Kamdem, S.L.S.; Serrazanetti, D.I.; Etoa, F.X.; Guerzoni, M.E. Synthesis of cyclopropane fatty acids in *Lactobacillus helveticus* and *Lactobacillus sanfranciscensis* and their cellular fatty acids changes following short term acid and cold stresses. *Food Microbiol.* **2010**, *27*, 493–502. [CrossRef]

103. Pessione, E. Lactic acid bacteria contribution to gut microbiota complexity: Lights and shadows. *Front. Cell. Infect. Microbiol.* **2012**, *2*, 86. [CrossRef]

104. Connil, N.; Plissoneau, L.; Onno, B.; Pilet, M.F.; Prevost, H.; Dousset, X. Growth of *Carnobacterium divergens* V41 and production of biogenic amines and divercin V41 in sterile cold-smoked salmon extract at varying temperatures, NaCl levels, and glucose concentrations. *J. Food Prot.* **2002**, *65*, 333–338. [CrossRef]

105. Bargossi, E.; Gardini, F.; Gatto, V.; Montanari, C.; Torriani, S.; Tabanelli, G. The capability of tyramine production and correlation between phenotypic and genetic characteristics of *Enterococcus faecium* and *Enterococcus faecalis* strains. *Front. Microbiol.* **2015**, *6*, 1371. [CrossRef]

106. Fernández, M.; Linares, D.M.; Alvarez, M.A. Sequencing of the tyrosine decarboxylase cluster of *Lactococcus lactis* IPLA 655 and the development of a PCR method for detecting tyrosine decarboxylating lactic acid bacteria. *J. Food Prot.* **2004**, *67*, 2521–2529. [CrossRef]

107. Gatto, V.; Tabanelli, G.; Montanari, C.; Prodomi, V.; Bargossi, E.; Torriani, S.; Gardini, F. Tyrosine decarboxylase activity of *Enterococcus mundtii*: New insights into phenotypic and genetic aspects. *Microb. Biotechnol.* **2016**, *9*, 801–813. [CrossRef]

108. Coton, E.; Rollan, G.C.; Lonvaud-Funel, A. Histidine decarboxylase of *Leuconostoc oenos* 9204: Purification, kinetic properties, cloning and nucleotide sequence of the *hdc* gene. *J. Appl. Microbiol.* **1998**, *84*, 143–151. [CrossRef]

109. Coton, E.; Coton, M. Multiplex PCR for colony direct detection of Gram-positive histamine- and tyramine-producing bacteria. *J. Microbiol. Methods* **2005**, *63*, 296–304. [CrossRef]

110. Konagaya, Y.; Kimura, B.; Ishida, M.; Fujii, T. Purification and properties of a histidine decarboxylase from *Tetragenococcus muriaticus*, a halophilic lactic acid bacterium. *J. Appl. Microbiol.* **2002**, *92*, 1136–1142. [CrossRef]

111. Martin, M.C.; Fernández, M.; Linares, D.M.; Alvarez, M.A. Sequencing, characterization and transcriptional analysis of the histidine decarboxylase operon of *Lactobacillus buchneri*. *Microbiology* **2005**, *151*, 1219–1228. [CrossRef]

112. Satomi, M.; Furushita, M.; Oikawa, H.; Yoshikawa-Takahashi, M.; Yano, Y. Analysis of a 30 kbp plasmid encoding histidine decarboxylase gene in *Tetragenococcus halophilus* isolated from fish sauce. *Int. J. Food Microbiol.* **2008**, *126*, 202–209. [CrossRef]

113. Vanderslice, P.; Copeland, W.C.; Robertus, J.D. Cloning and nucleotide sequence of wild type and a mutant histidine decarboxylase from *Lactobacillus* 30a. *J. Biol. Chem.* **1986**, *261*, 15186–15191.

114. Trip, H.; Mulder, N.L.; Rattray, F.P.; Lolkema, J.S. HdcB, a novel enzyme catalysing maturation of pyruvoyl-dependent histidine decarboxylase. *Mol. Microbiol.* **2011**, *79*, 861–871. [CrossRef]

115. Nannelli, F.; Claisse, O.; Gindreau, E.; de Revel, G.; Lonvaud-Funel, A.; Lucas, P.M. Determination of lactic acid bacteria producing biogenic amines in wine by quantitative PCR methods. *Lett. Appl. Microbiol.* **2008**, *47*, 594–599. [CrossRef]

116. Coton, E.; Mulder, N.; Coton, M.; Pochet, S.; Trip, H.; Lolkema, J.S. Origin of the putrescine-producing ability of the coagulase-negative bacterium *Staphylococcus epidermidis* 2015B. *Appl. Environ. Microbiol.* **2010**, *76*, 5570–5576. [CrossRef]

117. Linares, D.M.; Perez, M.; Ladero, V.; del Rio, B.; Redruello, B.; Martin, M.C.; Fernández, M.; Alvarez, M.A. An agmatine-inducible system for the expression of recombinant proteins in *Enterococcus faecalis*. *Microb. Cell Fact.* **2014**, *13*, 169. [CrossRef]

118. Özogul, F.; Hamed, I. The importance of lactic acid bacteria for the prevention of bacterial growth and their biogenic amines formation: A review. *Crit. Rev. Food Sci. Nutr.* **2018**, *58*, 1660–1670. [CrossRef]

119. Coburn, P.S.; Baghdayan, A.S.; Dolan, G.T.; Shankar, N. Horizontal transfer of virulence genes encoded on the *Enterococcus faecalis* pathogenicity island. *Mol. Microbiol.* **2007**, *63*, 530–544. [CrossRef]

120. Davis, I.J.; Roberts, A.P.; Ready, D.; Richards, H.; Wilson, M.; Mullany, P. Linkage of a novel mercury resistance operon with streptomycin resistance on a conjugative plasmid in *Enterococcus faecium*. *Plasmid* **2005**, *54*, 26–38. [CrossRef]

121. Foulquié Moreno, M.; Sarantinopoulos, P.; Tsakalidou, E.; De Vuyst, L. The role and application of enterococci in food and health. *Int. J. Food Microbiol.* **2006**, *106*, 1–24. [CrossRef]

122. Franz, C.M.; Huch, M.; Abriouel, H.; Holzapfel, W.; Gálvez, A. Enterococci as probiotics and their implications in food safety. *Int. J. Food Microbiol.* **2011**, *151*, 125–140. [CrossRef]

123. Giraffa, G. Functionality of enterococci in dairy products. *Int. J. Food Microbiol.* **2003**, *88*, 215–222. [CrossRef]

124. Martino, G.P.; Quintana, I.M.; Espariz, M.; Blancato, V.S.; Gallina Nizo, G.; Esteban, L.; Magni, C. Draft genome sequences of four *Enterococcus faecium* strains isolated from Argentine cheese. *Genome Announc.* **2016**, *4*, e01576-15. [CrossRef]

125. Hanchi, H.; Mottawea, W.; Sebei, K.; Hammami, R. The genus *Enterococcus*: Between probiotic potential and safety concerns-an update. *Front. Microbiol.* **2018**, *9*, 1791. [CrossRef]

126. Bover-Cid, S.; Hugas, M.; Izquierdo-Pulido, M.; Vidal-Carou, M.C. Amino acid decarboxylase activity of bacteria isolated from fermented pork sausages. *Int. J. Food Microbiol.* **2001**, *66*, 185–189. [CrossRef]

127. Capozzi, V.; Ladero, V.; Beneduce, L.; Fernández, M.; Alvarez, M.A.; Benoit, B.; Laurent, B.; Grieco, F.; Spano, G. Isolation and characterization of tyramine-producing *Enterococcus faecium* strains from red wine. *Food Microbiol.* **2011**, *28*, 434–439. [CrossRef]

128. Jiménez, E.; Ladero, V.; Chico, I.; Maldonado-Barragán, A.; López, M.; Martin, V.; Fernández, L.; Fernández, M.; Álvarez, M.A.; Torres, C.; Rodríguez, J.M. Antibiotic resistance, virulence determinants and production of biogenic amines among enterococci from ovine, feline, canine, porcine and human milk. *BMC Microbiol.* **2013**, *13*, 288. [CrossRef]

129. Ladero, V.; Fernández, M.; Alvarez, M.A. Isolation and identification of tyramine-producing enterococci from human fecal samples. *Can. J. Microbiol.* **2009**, *55*, 215–218. [CrossRef]

130. Ladero, V.; Martínez, N.; Cruz Martin, M.; Fernández, M.; Alvarez, M.A. qPCR for quantitative detection of tyramine-producing bacteria in dairy products. *Food Res. Int.* **2010**, *43*, 289–295. [CrossRef]

131. Muñoz-Atienza, E.; Landeta, G.; de Las Rivas, B.; Gómez-Sala, B.; Muñoz, R.; Hernández, P.E.; Cintas, L.M.; Herranz, C. Phenotypic and genetic evaluations of biogenic amine production by lactic acid bacteria isolated from fish and fish products. *Int. J. Food Microbiol.* **2011**, *146*, 212–216. [CrossRef]

132. Sarantinopoulos, P.; Andrighetto, C.; Georgalaki, M.D.; Rea, M.C.; Lombardi, A.; Cogan, T.M.; Kalantzopoulos, G.; Tsakalidou, E. Biochemical properties of enterococci relevant to their technological performance. *Int. Dairy J.* **2001**, *11*, 621–647. [CrossRef]

133. Suzzi, G.; Caruso, M.; Gardini, F.; Lombardi, A.; Vannini, L.; Guerzoni, M.E.; Andrighetto, C.; Lanorte, M.T. A survey of the enterococci isolated from an artisanal Italian goat's cheese (semicotto caprino). *J. Appl. Microbiol.* **2000**, *89*, 267–274. [CrossRef]

134. Bonetta, S.; Bonetta, S.; Carraro, E.; Coïsson, J.D.; Travaglia, F.; Arlorio, M. Detection of biogenic amine producer bacteria in a typical Italian goat cheese. *J. Food Prot.* **2008**, *71*, 205–209. [CrossRef]

135. Fernández, M.; Linares, D.M.; Del Rio, B.; Ladero, V.; Alvarez, M.A. HPLC quantification of biogenic amines in cheeses: Correlation with PCR-detection of tyramine-producing microorganisms. *J. Dairy Res.* **2007**, *74*, 276–282. [CrossRef]

136. Galgano, F.; Suzzi, G.; Favati, F.; Caruso, M.; Martuscelli, M.; Gardini, F.; Salzano, G. Biogenic amines during ripening in 'Semicotto Caprino' cheese: Role of enterococci. *J. Food Sci. Technol.* **2001**, *36*, 153–160. [CrossRef]

137. Joosten, H.M.L.J.; Northolt, M.D. Conditions allowing the formation of biogenic amines in cheese. 2. Decarboxylative properties of some non-starter bacteria. *Neth. Milk Dairy J.* **1987**, *41*, 259–280.

138. Burdychova, R.; Komprda, T. Biogenic amine-forming microbial communities in cheese. *FEMS Microbiol. Lett.* **2007**, *276*, 149–155. [CrossRef]

139. Rea, M.C.; Franz, C.M.A.P.; Holzapfel, W.H.; Cogan, T.M. Development of enterococci and production of tyramine during the manufacture and ripening of cheddar cheese. *Irish J. Agric. Food Res.* **2004**, *43*, 247–258.

140. Kalhotka, L.; Manga, I.; Přichystalová, J.; Hůlová, M.; Vyletělová, M.; Šustová, K. Decarboxylase activity test of the genus *Enterococcus* isolated from goat milk and cheese. *Acta Vet. Brno* **2012**, *81*, 145–151. [CrossRef]

141. Martino, G.P.; Espariz, M.; Gallina Nizo, G.; Esteban, L.; Blancato, V.S.; Magni, C. Safety assessment and functional properties of four enterococci strains isolated from regional Argentinean cheese. *Int. J. Food Microbiol.* **2018**, *277*, 1–9. [CrossRef]

142. Bover-Cid, S.; Holzapfel, W.H. Improved screening procedure for biogenic amine production by lactic acid bacteria. *Int. J. Food Microbiol.* **1999**, *53*, 33–41. [CrossRef]

143. Coloretti, F.; Chiavari, C.; Armaforte, E.; Carri, S.; Castagnetti, G.B. Combined use of starter cultures and preservatives to control production of biogenic amines and improve sensorial profile in low-acid salami. *J. Agric. Food Chem.* **2008**, *56*, 11238–11244. [CrossRef]

144. Hugas, M.; Garriga, M.; Aymerich, M. Functionality of enterococci in meat products. *Int. J. Food Microbiol.* **2003**, *88*, 223–233. [CrossRef]

145. Garriga, M.; Aymerich, T. The microbiology of fermentation and ripening. In *Handbook of Fermented Meat and Poultry*; Toldrá, F., Hui, Y.H., Astiasarán, I., Sebranek, J.G., Talon, R., Eds.; John Wiley & Sons: Chichester, UK, 2014; pp. 107–115. ISBN 978-1-118-52269-1.

146. Latorre-Moratalla, M.L.; Comas-Basté, O.; Bover-Cid, S.; Vidal-Carou, M.C. Tyramine and histamine risk assessment related to consumption of dry fermented sausages by the Spanish population. *Food Chem. Toxicol.* **2017**, *99*, 78–85. [CrossRef]

147. Landeta, G.; Curiel, J.A.; Carrascosa, A.V.; Muñoz, R.; de las Rivas, B. Technological and safety properties of lactic acid bacteria isolated from Spanish dry-cured sausages. *Meat Sci.* **2013**, *95*, 272–280. [CrossRef]

148. Maksimovic, A.Z.; Zunabovic-Pichler, M.; Kos, I.; Mayrhofer, S.; Hulak, N.; Domig, K.J.; Fuka, M.M. Microbiological hazards and potential of spontaneously fermented game meat sausages: A focus on lactic acid bacteria diversity. *LWT-Food Sci. Technol.* **2018**, *89*, 418–426. [CrossRef]

149. Marcobal, A.; de las Rivas, B.; García-Moruno, E.; Muñoz, R. The tyrosine decarboxylation test does not differentiate *Enterococcus faecalis* from *Enterococcus faecium*. *Syst. Appl. Microbiol.* **2004**, *27*, 423–426. [CrossRef]

150. Pérez-Martín, F.; Seseña, S.; Izquierdo-Pulido, M.; Llanos Palop, M. Are Enterococcus populations present during malolactic fermentation of red wine safe? *Food Microbiol.* **2014**, *42*, 95–101. [CrossRef]

151. Jeon, A.R.; Lee, J.H.; Mah, J.H. Biogenic amine formation and bacterial contribution in *Cheonggukjang*, a Korean traditional fermented soybean food. *LWT-Food Sci. Technol.* **2018**, *92*, 282–289. [CrossRef]

152. Takebe, Y.; Takizaki, M.; Tanaka, H.; Ohta, H.; Niidome, T.; Morimura, S. Evaluation of the biogenic amine-production ability of lactic acid bacteria from Tofu-misozuke. *Food Sci. Technol. Res.* **2016**, *22*, 673–678. [CrossRef]

153. Suárez, C.; Espariz, M.; Blancato, V.S.; Magni, C. Expression of the agmatine deiminase pathway in *Enterococcus faecalis* is activated by the *AguR* regulator and repressed by CcpA and PTS^{Man} systems. *PLoS ONE* **2013**, *8*, e76170. [CrossRef]

154. Perez, M.; Victor Ladero, V.; del Rio, B.; Redruello, B.; de Jong, A.; Kuipers, O.; Kok, J.; Martin, M.C.; Fernández, M.; Alvarez, M.A. The relationship among tyrosine decarboxylase and agmatine deiminase pathways in *Enterococcus faecalis*. *Front. Microbiol.* **2017**, *8*, 2107. [CrossRef] [PubMed]

155. Bargossi, E.; Tabanelli, G.; Montanari, C.; Gatto, V.; Chinnici, F.; Gardini, F.; Torriani, S. Growth, biogenic amine production and *tyrDC* transcription of *Enterococcus faecalis* in synthetic medium containing defined amino acid concentrations. *J. Appl. Microbiol.* **2017**, *122*, 1078–1091. [CrossRef] [PubMed]

156. Gardini, F.; Martuscelli, M.; Caruso, M.C.; Galgano, F.; Crudele, M.A.; Favati, F.; Guerzoni, M.E.; Suzzi, G. Effects of pH, temperature and NaCl concentration on the growth kinetics, proteolytic activity and biogenic amine production of *Enterococcus faecalis*. *Int. J. Food Microbiol.* **2001**, *64*, 105–117. [CrossRef]

157. Torriani, S.; Gatto, V.; Sembeni, S.; Tofalo, R.; Suzzi, G.; Belletti, N.; Gardini, F.; Bover-Cid, S. Rapid detection and quantification of tyrosine decarboxylase gene (*tdc*) and its expression in Gram-positive bacteria associated with fermented foods using PCR-based methods. *J. Food Prot.* **2008**, *71*, 93–101. [CrossRef] [PubMed]

158. Fernández, M.; Linares, D.M.; Rodríguez, A.; Alvarez, M.A. Factors affecting tyramine production in *Enterococcus durans* IPLA 655. *Appl. Microbiol. Biotechnol.* **2007**, *73*, 1400–1406. [CrossRef]

159. Linares, D.M.; Fernández, M.; Martín, M.C.; Alvarez, M.A. Tyramine biosynthesis in *Enterococcus durans* is transcriptionally regulated by the extracellular pH and tyrosine concentration. *Microbiol. Biotechnol.* **2009**, *2*, 625–633. [CrossRef] [PubMed]

160. Solheim, M.; Leanti La Rosa, S.; Mathisen, T.; Snipen, L.G.; Nes, I.F.; Anders Brede, D. Transcriptomic and functional analysis of NaCl-induced stress in *Enterococcus faecalis*. *PLoS ONE* **2014**, *9*, e94571. [CrossRef]

161. Liu, F.; Wang, X.; Du, L.; Wang, D.; Zhu, Y.; Geng, Z.; Xu, X.; Xu, W. Effect of NaCl treatments on tyramine biosynthesis of *Enterococcus faecalis*. *J. Food. Prot.* **2015**, *78*, 940–945. [CrossRef]

162. Hugas, M.; Garriga, M.; Aymerich, T.; Monfort, J.M. Biochemical characterization of lactobacilli from dry fermented sausages. *Int. J. Food Microbiol.* **1993**, *18*, 107–113. [CrossRef]

163. Holck, A.; Axelsson, L.; McLeod, A.; Rode, T.M.; Heir, E. Health and safety considerations of fermented sausages. *J. Food Qual.* **2017**, *9753894*. [CrossRef]

164. Pereira, C.I.; Barreto Crespo, M.T.; Romao, M.V.S. Evidence for proteolytic activity and biogenic amines production in *Lactobacillus curvatus* and *L. homohiochii*. *Int. J. Food Microbiol.* **2001**, *68*, 211–216. [CrossRef]

165. Freiding, S.; Gutsche, K.A.; Ehrmann, M.A.; Vogel, R.F. Genetic screening of *Lactobacillus sakei* and *Lactobacillus curvatus* strains for their peptidolytic system and amino acid metabolism, and comparison of their volatilomes in a model system. *Syst. Appl. Microbiol.* **2011**, *34*, 311–320. [CrossRef] [PubMed]

166. Latorre-Moratalla, M.L.; Bover-Cid, S.; Veciana-Nogués, M.T.; Vidal-Carou, M.C. Control of biogenic amines in fermented sausages: Role of starter cultures. *Front. Microbiol.* **2012**, *3*, 169. [CrossRef]

167. Wüthrich, D.; Berthoud, H.; Wechsler, D.; Eugster, E.; Irmler, S.; Bruggmann, R. The histidine decarboxylase gene cluster of *Lactobacillus parabuchneri* was gained by horizontal gene transfer and is mobile within the species. *Front. Microbiol.* **2017**, *8*, 218. [CrossRef] [PubMed]

168. Linares, D.M.; del Rio, B.; Redruello, B.; Fernández, M.; Martin, M.C.; Ladero, V.; Alvarez, M.A. The use of qPCR-based methods to identify and quantify food spoilage microorganisms. In *Novel Food Preservation and Microbial Assessment Techniques*; Boziaris, I.S., Ed.; CRC Press: Boca Raton, FL, USA, 2014; pp. 313–334. ISBN 978-1-46-658075-6.

169. Fröhlich-Wyder, M.T.; Guggisberg, D.; Badertscher, R.; Wechsler, D.; Wittwer, A.; Irmler, S. The effect of *Lactobacillus buchneri* and *Lactobacillus parabuchneri* on the eye formation of semi-hard cheese. *Int. Dairy J.* **2013**, *33*, 120–128. [CrossRef]

170. Bunková, L.; Bunka, F.; Mantlová, G.; Cablová, A.; Sedlácek, I.; Svec, P.; Pachlová, V.; Krácmar, S. The effect of ripening and storage conditions on the distribution of tyramine, putrescine and cadaverine in Edam-cheese. *Food Microbiol.* **2010**, *27*, 880–888. [CrossRef]

171. Komprda, T.; Burdychová, R.; Dohnal, V.; Cwiková, O.; Sládková, P.; Dvorácková, H. Tyramine production in Dutch-type semi-hard cheese from two different producers. *Food Microbiol.* **2008**, *25*, 219–227. [CrossRef] [PubMed]

172. Pachlová, V.; Buňková, L.; Flasarová, R.; Salek, R.N.; Dlabajová, A.; Butor, I.; Buňka, F. Biogenic amine production by nonstarter strains of *Lactobacillus curvatus* and *Lactobacillus paracasei* in the model system of Dutch-type cheese. *LWT-Food Sci. Technol.* **2018**, *97*, 730–735. [CrossRef]

173. Yılmaz, C.; Gökmen, V. Formation of tyramine in yoghurt during fermentation – Interaction between yoghurt starter bacteria and *Lactobacillus plantarum*. *Food Res. Int.* **2017**, *97*, 288–295. [CrossRef] [PubMed]

174. Ladero, V.; Martin, M.C.; Redruello, B.; Mayo, B.; Flórez, A.B.; Fernández, M.; Alvarez, M.A. Genetic and functional analysis of biogenic amine production capacity among starter and non-starter lactic acid bacteria isolated from artisanal cheeses. *Eur. Food Res. Technol.* **2015**, *241*, 377–383. [CrossRef]

175. Linares, D.M.; del Rio, B.; Ladero, V.; Martínez, N.; Fernández, M.; Martin, M.C.; Alvarez, M.A. Factors influencing biogenic amines accumulation in dairy products. *Front. Microbiol.* **2012**, *3*, 180. [CrossRef] [PubMed]

176. Costantini, A.; Cersosimo, M.; Del Prete, V.; Garcia-Moruno, E. Production of biogenic amines by lactic acid bacteria: Screening by PCR, thin-layer chromatography and high-performance liquid chromatography of strains isolated from wine and must. *J. Food Prot.* **2006**, *69*, 391–396. [CrossRef] [PubMed]

177. Gardini, F.; Zaccarelli, A.; Belletti, N.; Faustini, F.; Cavazza, A.; Martuscelli, M.; Mastrocola, D.; Suzzi, G. Factors influencing biogenic amine production by a strain of *Oenococcus oeni* in a model system. *Food Control* **2005**, *16*, 609–616. [CrossRef]

178. Ancín-Azpilicueta, C.; González-Marco, A.; Jiménez-Moreno, N. Current knowledge about the presence of amines in wine. *Crit. Rev. Food Sci. Nutr.* **2008**, *48*, 257–275. [CrossRef] [PubMed]

179. Lerm, E.; Engelbrecht, L.; du Toit, M. Malolactic fermentation: The ABC's of MLF. *S. Afr. J. Enol. Vitic.* **2010**, *31*, 186–212. [CrossRef]

180. Marcobal, A.; Martín-Álvarez, P.J.; Polo, C.; Muñoz, R.; Moreno-Arribas, M.V. Formation of biogenic amines throughout the industrial manufacture of red wine. *J. Food Prot.* **2006**, *69*, 397–404. [CrossRef]

181. Marcobal, A.; Martín-Álvarez, P.J.; Moreno-Arribas, M.V.; Muñoz, R. A multifactorial design for studying factors influencing growth and tyramine production of the lactic acid bacteria *Lactobacillus brevis* CECT 4669 and *Enterococcus faecium* BIFI-58. *Res. Microbiol.* **2006**, *157*, 417–424. [CrossRef]

182. Henríquez-Aedo, K.; Durán, D.; Garcia, A.; Hengst, M.B.; Aranda, M. Identification of biogenic amines-producing lactic acid bacteria isolated from spontaneous malolactic fermentation of chilean red wines. *LWT-Food Sci. Technol.* **2016**, *68*, 183–189. [CrossRef]

183. Moreno-Arribas, M.V.; Polo, M.C.; Jorganes, F.; Muñoz, R. Screening of biogenic amine production by lactic acid bacteria isolated from grape must and wine. *Int. J. Food Microbiol.* **2003**, *84*, 117–123. [CrossRef]

184. Moreno-Arribas, V.; Lonvaud-Funel, A. Purification and characterization of tyrosine decarboxylase of *Lactobacillus brevis* IOEB 9809 isolated from wine. *FEMS Microbiol. Lett.* **2001**, *195*, 103–107. [CrossRef]

185. Arena, M.P.; Romano, A.; Capozzi, V.; Beneduce, L.; Ghariani, M.; Grieco, F.; Lucas, P.; Spano, G. Expression of *Lactobacillus brevis* IOEB 9809 tyrosine decarboxylase and agmatine deiminase genes in wine correlates with substrate availability. *Lett. Appl. Microbiol.* **2001**, *53*, 395–402. [CrossRef] [PubMed]

186. Lucas, P.; Lonvaud-Funel, A. Purification and partial gene sequence of the tyrosine decarboxylase of *Lactobacillus brevis* IOEB 9809. *FEMS Microbiol. Lett.* **2002**, *211*, 85–89. [CrossRef] [PubMed]

187. Geissler, A.J.; Behr, J.; von Kamp, K.; Vogel, R.F. Metabolic strategies of beer spoilage lactic acid bacteria in beer. *Int. J. Food Microbiol.* **2016**, *216*, 60–68. [CrossRef] [PubMed]

188. Ladero, V.; Coton, M.; Fernández, M.; Buron, N.; Martín, M.C.; Guichard, H.; Coton, E.; Alvarez, M.A. Biogenic amines content in Spanish and French natural ciders: Application of qPCR for quantitative detection of biogenic amine-producers. *Food Microbiol.* **2011**, *28*, 554–561. [CrossRef] [PubMed]

189. Corsetti, A.; Settanni, L. Lactobacilli in sourdough fermentation. *Food Res. Int.* **2007**, *40*, 539–558. [CrossRef]

190. Costantini, A.; Vaudano, E.; Del Prete, V.; Danei, M.; Garcia-Moruno, E. Biogenic amine production by contaminating bacteria found in starter preparations used in winemaking. *J. Agric. Food Chem.* **2009**, *57*, 10664–10669. [CrossRef]

191. Rodwell, A.W. The occurrence and distribution of amino-acid decarboxylases within the genus *Lactobacillus.* *J. Gen. Microbiol.* **1953**, *8*, 224–232. [CrossRef]

192. Romano, A.; Trip, H.; Lolkema, J.S.; Lucas, P.M. Three-component lysine/ornithine decarboxylation system in *Lactobacillus saerimneri* 30a. *J. Bacteriol.* **2013**, *195*, 1249–1254. [CrossRef]

193. Delorme, C. Safety assessment of dairy microorganisms: *Streptococcus thermophiles. Int. J. Food Microbiol.* **2008**, *126*, 274–277. [CrossRef]

194. Georgalaki, M.D.; Sarantinopoulos, P.; Ferreira, E.S.; De Vuyst, L.; Kalantzopoulos, G.; Tsakalidou, E. Biochemical properties of *Streptococcus macedonicus* strains isolated from Greek Kasseri cheese. *J. Appl. Microbiol.* **2000**, *88*, 817–825. [CrossRef]

195. Griswold, A.R.; Jameson-Lee, M.; Burne, R.A. Regulation and physiologic significance of the agmatine deiminase system of *Streptococcus mutans* UA159. *J. Bacteriol.* **2006**, *188*, 834–841. [CrossRef] [PubMed]

196. Liu, Y.; Zeng, L.; Burne, R.A. *AguR* is required for induction of the *Streptococcus mutans* agmatine deiminase system by low pH and agmatine. *Appl. Environ. Microbiol.* **2009**, *75*, 2629–2637. [CrossRef] [PubMed]

197. Elsanhoty, R.M.; Ramadan, M.F. Genetic screening of biogenic amines production capacity from some lactic acid bacteria strains. *Food Control* **2016**, *68*, 220–228. [CrossRef]

198. Gezginc, Y.; Akyol, I.; Kuley, E.; Özogul, F. Biogenic amines formation in *Streptococcus thermophilus* isolated from home-made natural yogurt. *Food Chem.* **2013**, *138*, 655–662. [CrossRef] [PubMed]

199. Kuley, E.; Balıkcı, E.; Özoğul, I.; Gökdogan, S.; Ozoğul, F. Stimulation of cadaverine production by foodborne pathogens in the presence of *Lactobacillus, Lactococcus,* and *Streptococcus* spp. *J. Food Sci.* **2012**, *77*, M650–M658. [CrossRef] [PubMed]

200. de las Rivas, B.; Rodríguez, H.; Carrascosa, A.V.; Muñoz, R. Molecular cloning and functional characterization of a histidine decarboxylase from *Staphylococcus capitis. J. Appl. Microbiol.* **2008**, *104*, 194–203. [CrossRef] [PubMed]

201. Tabanelli, G.; Torriani, S.; Rossi, F.; Rizzotti, L.; Gardini, F. Effect of chemico-physical parameters on the histidine decarboxylase (HdcA) enzymatic activity in Streptococcus thermophilus PRI60. *J. Food Sci.* **2012**, *77*, M231–M237. [CrossRef]

202. Fox, P.F.; McSweeney, P.L.H.; Cogan, T.M.; Guinee, T.P. *Cheese: Chemistry, Physics and Microbiology*, 3rd ed.; Volume 1 General aspects; Elsevier Academic Press: London, UK, 2004; ISBN 978-0-12-263652-3.

203. Linares, D.M.; del Rio, B.; Redruello, B.; Ladero, V.; Martin, M.C.; de Jong, A.; Kuipers, O.P.; Fernández, M.; Alvarez, M.A. *AguR*, a transmembrane transcription activator of the putrescine biosynthesis operon in *Lactococcus lactis*, acts in response to the agmatine concentration. *Appl. Environ. Microbiol.* **2015**, *81*, 6145–6157. [CrossRef]

204. del Rio, B.; Redruello, B.; Ladero, V.; Fernández, M.; Martin, M.C.; Alvarez, M.A. Putrescine production by *Lactococcus lactis* subsp. *cremoris* CECT 8666 is reduced by NaCl via a decrease in bacterial growth and the repression of the genes involved in putrescine production. *Int. J. Food Microbiol.* **2016**, *232*, 1–6. [CrossRef]

205. del Rio, B.; Linares, D.; Ladero, V.; Redruello, B.; Fernández, M.; Martin, M.C.; Alvarez, M.A. Putrescine biosynthesis in *Lactococcus lactis* is transcriptionally activated at acidic pH and counteracts acidification of the cytosol. *J. Food Microbiol.* **2016**, *236*, 83–89. [CrossRef]

206. Flasarová, R.; Pachlová, V.; Buňková, L.; Menšíková, A.; Georgová, N.; Dráb, V.; Buňka, F. Biogenic amine production by *Lactococcus lactis* subsp. *cremoris* strains in the model system of Dutch-type cheese. *Food Chem.* **2016**, *194*, 68–75. [CrossRef] [PubMed]

207. Santos, W.C.; Souza, M.R.; Cerqueira, M.M.O.P.; Gloria, M.B.A. Bioactive amines formation in milk by *Lactococcus* in the presence or not of rennet and NaCl at 20 and 32 °C. *Food Chem.* **2003**, *81*, 595–606. [CrossRef]

208. Holzapfel, W.H.; Franz, C.M.A.P.; Ludwig, W.; Back, W.; Dicks, L.M.T. The genera *Pediococcus* and *Tetragenococcus.* In *The Prokaryotes*, 3rd ed.; Dworkin, M., Falkow, S., Rosenberg, E., Schleifer, K.H., Stackebrandt, E., Eds.; Springer: New York, NY, USA, 2006; Volume 4, pp. 229–266. ISBN 978-0-387-30744-2.

209. Hugas, M.; Monfort, J.M. Bacterial starter cultures for meat fermentation. *Food Chem.* **1997**, *59*, 547–554. [CrossRef]

210. Walling, E.; Gindreau, E.; Lonvaud-Funel, A. A putative glucan synthase gene *dps* detected in exopolysaccharide-producing *Pediococcus damnosus* and *Oenococcus oeni* strains isolated from wine and cider. *Int. J. Food Microbiol.* **2005**, *98*, 53–62. [CrossRef] [PubMed]

211. Guerrini, S.; Mangani, S.; Granchi, L.; Vincenzini, M. Biogenic amine production by *Oenococcus oeni*. *Curr. Microbiol.* **2002**, *44*, 374–378. [CrossRef] [PubMed]

212. Garcia-Moruno, E.; Muñoz, R. Does *Oenococcus oeni* produce histamine? *Int. J. Food Microbiol.* **2012**, *157*, 121–129. [CrossRef] [PubMed]

213. Berbegal, C.; Benavent-Gil, Y.; Navascués, E.; Calvo, A.; Albors, C.; Pardo, I.; Ferrer, S. Lowering histamine formation in a red Ribera del Duero wine (Spain) by using an indigenous *O. oeni* strain as a malolactic starter. *Int. J. Food Microbiol.* **2017**, *244*, 11–18. [CrossRef] [PubMed]

214. Landete, J.M.; Pardo, I.; Ferrer, S. Regulation of *hdc* expression and HDC activity by enological factors in lactic acid bacteria. *J. Appl. Microbiol.* **2008**, *105*, 1544–1551. [CrossRef]

215. Garai, G.; Dueñas, M.T.; Irastorza, A.; Moreno-Arribas, M.V. Biogenic amine production by lactic acid bacteria isolated from cider. *Lett. Appl. Microbiol.* **2007**, *45*, 473–478. [CrossRef]

216. Bonnin-Jusserand, M.; Grandvalet, C.; David, V.; Alexandre, H. Molecular cloning, heterologous expression, and characterization of Ornithine decarboxylase from *Oenococcus oeni*. *J. Food Prot.* **2011**, *74*, 1309–1314. [CrossRef]

217. Izquierdo-Pulido, M.; Mariné-Font, A.; Vidal-Carou, M.C. Effect of tyrosine on tyramine formation during beer fermentation. *Food Chem.* **2000**, *70*, 329–332. [CrossRef]

218. Huys, G.; Leisner, J.; Björkroth, J. The lesser LAB gods: *Pediococcus, Leuconostoc, Weissella, Carnobacterium*, and affiliated genera. In *Lactic Acid Bacteria: Microbiological and Functional Aspects*, 4th ed.; Lahtinen, S., Ouwehand, A.C., Salminen, S., von Wright, A., Eds.; CRC Press: Boca Raton, FL, USA, 2011; pp. 93–121. ISBN 978-1-43-983677-4.

219. Liu, S.; Pritchard, G.G.; Hardman, M.J.; Pilone, G.J. Occurrence of arginine deiminase pathway enzymes in arginine catabolism by wine lactic acid bacteria. *Appl. Environ. Microbiol.* **1995**, *61*, 310–316. [PubMed]

220. Coton, M.; Lebreton, M.; Marcia Leyva Salas, M.L.; Garnier, L.; Navarri, M.; Pawtowski, A.; Le Bla, G.; Valence, F.; Coton, E.; Mounier, J. Biogenic amine and antibiotic resistance profiles determined for lactic acid bacteria and a propionibacterium prior to use as antifungal bioprotective cultures. *Int. Dairy J.* **2018**, *85*, 21–26. [CrossRef]

221. González del Llano, D.; Cuesta, P.; Rodríguez, A. Biogenic amine production by wild lactococal and leuconostoc strains. *Lett. Appl. Microbiol.* **1998**, *26*, 270–274. [CrossRef]

222. Fusco, V.; Quero, G.M.; Cho, G.S.; Kabisch, J.; Meske, D.; Neve, H.; Bockelmann, W.; Franz, C.M.A.P. The genus *Weissella*: Taxonomy, ecology and biotechnological potential. *Front. Microbiol.* **2015**, *6*, 155. [CrossRef] [PubMed]

223. Leisner, J.J.; Laursen, B.G.; Prévost, H.; Drider, D.; Dalgaard, P. *Carnobacterium*: Positive and negative effects in the environment and in foods. *FEMS Microbiol. Rev.* **2007**, *31*, 592–613. [CrossRef] [PubMed]

224. Curiel, J.A.; Ruiz-Capillas, C.; de las Rivas, B.; Carrascosa, A.V.; Jiménez-Colmenero, F.; Muñoz, R. Production of biogenic amines by lactic acid bacteria and enterobacteria isolated from fresh pork sausages packaged in different atmospheres and kept under refrigeration. *Meat Sci.* **2011**, *88*, 368–373. [CrossRef]

225. Justé, A.; Lievens, B.; Frans, I.; Marsh, T.L.; Klingeberg, M.; Michiels, C.W.; Willems, K.A. Genetic and physiological diversity of *Tetragenococcus halophilus* strains isolated from sugar- and salt-rich environments. *Microbiology* **2008**, *154*, 2600–2610. [CrossRef]

226. Torriani, S.; Felis, G.E.; Fracchetti, F. Selection criteria and tools for malolactic starters development: An update. *Ann. Microbiol.* **2001**, *61*, 33–39. [CrossRef]

Article

Influence of Iodine Feeding on Microbiological and Physico-Chemical Characteristics and Biogenic Amines Content in a Raw Ewes' Milk Cheese

Maria Schirone *, Rosanna Tofalo, Giorgia Perpetuini, Anna Chiara Manetta, Paola Di Gianvito, Fabrizia Tittarelli, Noemi Battistelli, Aldo Corsetti, Giovanna Suzzi and Giuseppe Martino *

Faculty of Bioscience and Technology for Food, Agriculture and Environment, University of Teramo, Via R. Balzarini, 1, 64100 Teramo, Italy; rtofalo@unite.it (R.T.); giorgia.perpetuini@gmail.com (G.P.); acmanetta@unite.it (A.C.M.); digianvito.paola@gmail.com (P.D.G.); ftittarelli@unite.it (F.T.); noemi.battistelli@gmail.com (N.B.) acorsetti@unite.it (A.C.); gsuzzi@unite.it (G.S.)
* Correspondence: mschirone@unite.it (M.S.); gmartino@unite.it (G.M.);
  Tel.: +39-086-126-6911 (M.S.); +39-086-126-6950 (G.M.)

Received: 27 May 2018; Accepted: 6 July 2018; Published: 7 July 2018

Abstract: Iodine is an essential trace element involved in the regulation of thyroid metabolism and antioxidant status in humans and animals. The aim of this study was to evaluate the effect of ewes' dietary iodine supplementation on biogenic amines content as well as microbiological and physico-chemical characteristics in a raw milk cheese at different ripening times (milk, curd, and 2, 7, 15, 30, 60, and 90 days). Two cheese-making trials were carried out using milk from ewes fed with unifeed (Cheese A) or with the same concentrate enriched with iodine (Cheese B). The results indicated that the counts of principal microbial groups and physico-chemical characteristics were quite similar in both cheeses at day 90. Cheese B was characterized by a higher content of biogenic amines and propionic acid. Propionic bacteria were found in both cheeses mainly in Trial B in agreement with the higher content of propionic acid detected.

Keywords: raw milk cheese; biogenic amines; iodine feed; physico-chemical composition

---

1. Introduction

Milk and dairy products represent the second most important source of iodine in the European Union or in the United States [1] particularly for infants and children. Iodine deficit in the diet causes various thyroid dysfunctions and infant mortality [2]; iodine has a recommended daily intake of 150 µg for both adolescents and adults [3,4]. The concentration of this element in milk and dairy products has been reported in different papers and it can vary in terms of animal feed, the season (the higher concentration is in winter), and exposure to iodophors [2]. Changes in animal feeding have been proposed as one of the most promising approaches to modify iodine content in milk [5]. Some studies [6,7] have been carried out to evaluate the effects of dietary iodine supplementation in dairy goats and cows on milk iodine content and milk production traits. Nudda et al. [6] reported that the iodine supplementation in dairy goat diets doubled the milk iodine content when compared with the control group, even if no evident effect was observed in the gross composition of milk. On the contrary, Weiss et al. [7] found that iodine concentration increased in serum but not in milk after supplementation of this element in diets of dairy cows. In fact, very little information is available about the effects of iodine addition on ewes' milk and milk-based product composition, nor about the response of dairy product microbiota.

Pecorino Incanestrato di Castel del Monte (ICM) is an artisanal semi-hard pasta filata cheese obtained starting from ewes' raw milk without the addition of starter cultures. ICM is produced in the Abruzzo region (Central Italy) and is included in the list of typical products (PAT—Prodotti Agroalimentari Tradizionali). As other raw milk cheeses, the characteristics of the final product are influenced by several parameters such as raw milk microbiota, microorganisms deriving from equipment and from the dairy environments, and outside and inside grazing animal feeding systems [8].

In this study, the effect of dietary iodine supplementation in dairy ewes on biogenic amine (BA) content as well as microbiological and physico-chemical characteristics in ICM cheese was evaluated.

2. Materials and Methods

2.1. Cheese-Making Procedure

Cheese samples were manufactured in a small factory, located in the production area of ICM (L'Aquila, Abruzzo Region, Italy), from raw whole ewes' milk of one or two daily milking without the addition of natural or commercial starter cultures. The milk was filtered and heated at 35–40 °C for 15–25 min and coagulated with lamb rennet at 38 °C, according to routine manufacture. Afterwards, the curd was broken and fit into special baskets, the so-called *fiscelle*. The product was salted and ripened up to 3 months. The final products weighted about 2 kg. Two different cheese-making trials were carried out in triplicate using milk (100 L). In a completely randomized block design, 2 groups of 15 Sopravissana ewes were assigned to 2 diets. In the first group, ewes were fed with unifeed (hay and concentrate) (Cheese A), while in the second group ewes were fed with unifeed enriched with iodine at a final concentration of 10 mg/kg (Cheese B). This concentration of iodine was selected according to Regulation EC No. 1459/2005 [9]. The cheese yield was about 24% in both cheese-making trials. Analyses were performed in triplicate on milk, curd, and cheese samples at different ripening times: 2, 7, 15, 30, 60, and 90 days.

2.2. Microbiological Analyses

Milk and cheese samples (10 mL or g) were diluted in 90 mL of a sodium citrate (2% w/v) solution and homogenized with a Stomacher Lab-Blender 400 (Steward Medical, London, UK) for 1 min. Serial dilutions in sterile peptone water (0.1% w/v) were plated in triplicate on different media to enumerate the following microorganisms: mesophilic lactobacilli, lactococci, aerobic mesophilic bacteria (AMB), yeasts, *Enterobacteriaceae*, enterococci, and coagulase-negative staphylococci (CNS), according to Schirone et al. [10]. The presence of *Escherichia coli* O157:H7, *Salmonella* spp., and *Listeria monocytogenes* was determined according to standard methods reported in ISO [11–13].

For the detection of propionibacteria, a semi-quantitative approach was applied as described previously [14,15]. DNA was extracted using PowerSoil DNA Isolation Kit (MoBio Laboratories) according to manufacturer's protocol starting from 5 g of cheese as previously described [14]. PB1 (5′-AGTGGCGAAGGCGGTTCTCTGGA-3′) and PB2 (5′-TGGGGTCGAGTTGCAGACCCCAAT-3′) primer set was used. PCR amplification program consisted of denaturation at 94 °C for 4 min, 40 cycles of denaturation at 94 °C for 30 s, annealing at 70 °C for 15 s, and extension at 72 °C for 1 min followed by a final extension at 72 °C for 5 min.

2.3. Gross Physico-Chemical Composition

A radial slice of each cheese was randomly taken and used for physico-chemical assays. The rind of each slice was carefully removed, and the rind-less material was fully shredded. pH, water activity (a_w), dry matter, total protein, fat, and ash content were determined according to Schirone et al. [9]. Iodine concentration was evaluated using a commercial kit according to manufacture instructions (Celltech, Turin, Italy) in milk and cheese samples.

Organic acids (mg/g) were determined as reported by Tofalo et al. [15] and Bouzas et al. [16] using an HPLC 200 series (Perkin Elmer, Monza, Italy) connected to a UV VIS detector at 210 nm. ROA Organic Acid H$^+$ column (Phenomenex, Bologna, Italy) was used for the analyses.

All determinations were performed isocratically with a flow rate of 0.7 mL/min at 65 °C using H_2SO_4 solution 0.009 N as mobile phase.

The nitrogen fractions determined were water-soluble nitrogen (WSN, expressed in %N) [17], trichloroacetic acid-soluble nitrogen (12% TCA-SN, expressed as %N) [18], and amino acid nitrogen (AAN, expressed as mg leucine/g) [19].

2.4. BA Determination

Determination of BA (mg/kg) was carried out as described by Schirone et al. [20]. In brief, 10 g of cheese samples were extracted and derivatized with dansyl chloride (Fluka Chimica, Milan, Italy). The chromatographic system consisted of an HPLC Waters Alliance (Waters SpA, Vimodrone, Italy), equipped with a Waters 2695 separation module connected to a Waters 2996 photodiode array detector. The separation of the analytes was carried out using a Waters Spherisorb C18 S3ODS-2 column (3 μm particle size, 150 mm × 4.6 mm Inner Diameter) equipped with a Waters Spherisorb S5ODS2 guard column. A linear gradient made up of acetonitrile and ultrapure water was applied: acetonitrile 57% (v/v) for 5 min; acetonitrile 80% (v/v) for 4 min, acetonitrile 90% (v/v) for 5 min. The peaks were detected at 254 nm.

2.5. Statistical Analyses

Statistical analyses were performed using the software STATISTICA for Windows (STAT. version 8.0, StatSoft Inc., Tulsa, OK, USA). Collected data were subjected to two-way analysis of variance (ANOVA) to detect significant differences. The principal component analysis (PCA) was performed on physico-chemical and microbiological data after auto-scaling.

3. Results

3.1. Microbial Analyses

Microbial counts are shown in Table 1. Overall, mesophilic lactobacilli, lattococci, AMB, enterococci, and yeasts showed a significant increase during the first days ripening. This was partly due to both microbial growth during coagulation and the physical retention of microorganisms in curds. The count of AMB obtained from the milk was higher in Cheese A (6.9 log CFU/mL) than in Cheese B (5.5 log CFU/mL) and then increased up to 8.4 log and 8.7 log CFU/g at 90 days of ripening, respectively. These counts are common in cheeses produced from raw milk, and they agree with those obtained in different cheese varieties such as Montasio [21] or Cebreiro [22,23].

Lactic acid bacteria (LAB) dominated in ICM cheeses during all ripening. In Cheese B, a higher number of lactococci and mesophilic lactobacilli was observed than in the Cheese A during the first stages of ripening, while at the end of ripening both cheeses showed similar counts, more than 8 log CFU/g. In general, LAB dominated the adventitious microbiota prevailing in all cheeses. Overall, in the early phase of manufacture, non-starter lactic acid bacteria (NSLAB) were present at very low values, whereas during ripening they increase from approximately 2.0 to 6.0 log CFU/g in ripened cheese [24]. Enterococci counts during ripening resulted to be quite different in ICM Cheeses A and B. In Trial A, their number increased from 2.8 log CFU/mL in milk up to a maximum value of 6.5 log CFU/g at 2 days and then decreased at 5.5 log CFU/g at the end of ripening. In Trial B, the counts started from 2.5 log CFU/g in milk, increased up to 6.3 log CFU/g after 15 days, and then decreased at 3.8 log CFU/g at 90 days of ripening. Enterococci represent the major part of curd microbiota and in some cases, they are the predominant microorganisms in the fully ripened product, constituting about the 41% of the LAB population [25]. In particular, enterococci have been recognized as an essential part of the natural microbial population of many dairy products, where they can sometimes prevail over lactobacilli and lactococci [22,26,27]. High levels of enterococci observed in other cheeses have been suggested to have a relevant role during the whole ripening process, because of their proteolytic and lipolytic activities that contribute to aroma compounds production (C4 metabolites such as diacetyl acetoin or 2,3-butanediol) [28,29].

Table 1. Evolution of principal microbial groups during the ripening in the two different trials expressed as log CFU/g.

Microbial Groups	Trial	Milk	Curd	2 Days	7 Days	15 Days	30 Days	60 Days	90 Days
Mesophilic lactobacilli	A	5.2 (0.02) *	6.6 (0.006) *	7.2 (0.01) *	7.8 (0.002) *	8.3 (0.02) *	7.2 (0.009) *	8.4 (0.006) *	8.7 (0.001) *
	B	6.8 (0.003) *	7.9 (0.001) *	8.6 (0.004) *	8.5 (0.004) *	8.8 (0.002) *	8.5 (0.002) *	8.1 (0.09) *	8.7 (0.001) *
Lactococci	A	5.4 (0.006) *	6.1 (0.013) *	7.5 (0.01) *	7.6 (0.01) *	8.1 (0.02) *	7.9 (0.004) *	8.7 (0.05) *	8.5 (0.004) *
	B	6.8 (0.001) *	8.4 (0.007) *	8.6 (0.005) *	8.7 (0.001) *	8.8 (0.001) *	8.8 (0.002) *	8.5 (0.05) *	8.5 (0.007) *
AMB [a]	A	6.9 (0.001) *	6.7 (0.001) *	7.9 (0.001) *	7.4 (0.007) *	8.6 (0.004) *	8.9 (0.001) *	8.4 (0.001) *	8.4 (0.006) *
	B	5.5 (0.01) *	7.0 (0.014) *	8.7 (0.002) *	8.8 (0.004) *	8.9 (0.001) *	8.9 (0.001) *	8.9 (0.09) *	8.7 (0.006) *
Yeasts	A	-	4.3 (0.005) *	3.5 (0.007) *	4.6 (0.002) *	4.5 (0.006) *	4.4 (0.005) *	4.5 (0.004)	4.4 (0.007) *
	B	-	3.5 (0.003) *	4.7 (0.004) *	4.4 (0.004) *	5.2 (0.009) *	5.1 (0.009) *	5.3 (0.4)	5.1 (0.003) *
Enterobacteriaceae	A	3.4 (0.009) *	5.4 (0.007) *	4.7 (0.003) *	4.3 (0.009) *	4.5 (0.007) *	2.6 (0.005) *	2.4 (0.002) *	2.4 (0.004)
	B	3.4 (0.001) *	5.4 (0.005) *	5.8 (0.002) *	5.8 (0.001) *	5.8 (0.001) *	3.8 (0.002) *	3.9 (0.003) *	<1
Enterococci	A	2.8 (0.001) *	4.7 (0.004) *	6.5 (0.004) *	5.3 (0.002) *	4.5 (0.004) *	4.3 (0.004) *	5.8 (0.003) *	5.5 (0.003) *
	B	2.5 (0.006) *	5.7 (0.002) *	5.6 (0.005) *	6.0 (0.001) *	6.3 (0.002) *	6.1 (0.001) *	5.1 (0.07) *	3.8 (0.001) *
CNS [b]	A	4.9 (0.002) *	6.4 (0.003) *	5.4 (0.003) *	6.5 (0.001) *	6.5 (0.003) *	6.7 (0.001) *	3.1 (0.003) *	3.4 (0.01) *
	B	6.3 (0.005) *	6.4 (0.003) *	7.3 (0.007) *	6.8 (0.001) *	5.7 (0.002) *	5.6 (0.003) *	5 (1) *	4.3 (0.01) *

The data are reported as mean (S.D.); samples for each microbial group at the same ripening time marked with * showed statistically significant differences ($p < 0.05$). [a] aerobic mesophilic bacteria, [b] coagulase-negative staphylococci.

As regards *Enterobacteriaceae*, they are associated to the natural microbiota of many dairy products, and together with coliforms are considered indicators of the microbiological quality of cheese. These microorganisms were present in milk of both cheeses and after 15 days of ripening, ranging from about 4.5 to 5.8 log CFU/g for Cheeses A and B respectively, whereas they were not enumerable (<10 CFU/g) in Cheese B at 90 days of ripening. *Enterobacteriaceae* are generally considered as microorganisms with a high decarboxylase activity, particularly in relation to the production of cadaverine and putrescine [30] and are common in many traditional cheeses of Mediterranean area [31]. The counts of CNS were higher in Milk B (6.3 log CFU/mL) than in Milk A (4.9 log CFU/mL). These microorganisms increased during the first days of ripening and decreased at 90 days with values of 3.4 and 4.3 log CFU/g in Cheeses A and B, respectively.

Yeasts, absent in milk in both trials, were present in curd and reached values of 4.4 and 5.1 log CFU/g in Cheeses A and B, respectively, at the end of ripening. Similar data have been reported in other raw milk cheeses such as Pecorino di Farindola [32,33].

Pathogens such as *Salmonella* spp., *L. monocytogenes*, and *E. coli* O157:H7 resulted absent in all the examined samples.

3.2. Gross Physico-Chemical Composition

The physico-chemical parameters and organic acids content for the two cheese-making procedures at 90 days of ripening are reported in Table 2. After 2 days of ripening, the pH values were 5.75 and 5.44 in Cheeses A and B, respectively (data not shown), and they then slightly decreased at the end of ripening. These differences, generally attributed to the metabolic activity of different species and strains of LAB, are typical of low acidified cheese produced with ewes' raw milk [10]. The mean a_w values decreased as ripening progressed and at day 90 they were similar in both cheeses (about 0.97). A higher percentage of fat was observed in Cheese B (51.69% w/w) than in Cheese A (47.03% w/w) at the end of ripening, whereas proteins were present in the amount in Cheese A (44.12% w/w) than in Cheese B (40.92% w/w), even if no statistical differences were observed ($p < 0.05$). The average values of iodine concentration were 86.1 and 481.3 µg/100 mL in Milks A and B, respectively. At day 90, the iodine concentration was 128.7 µg/100 g in Cheeses A and 375.9 µg/100 g in Cheese B. The iodine amount in milk has been reported to reflect the dietary iodine content, and it is an indicator of the iodine status of the animal [34]. The iodine concentration in milk is directly proportional to the iodine levels in feedstuffs. Moreover, the season of milk production and fat content of milk can significantly affect its rate [35]. Manca et al. [36] found that iodine supplementation did not influence the goat milk fatty acid profile, except for some short-chain fatty acids. Milk fat and protein content did not vary between two groups of dairy sheep fed with iodine supplementation in diets at different concentrations [37].

Lactic acid was the most abundant organic acid with values of about 35 mg/g in both Cheeses A and B (Table 2). Similar concentrations of citric, acetic, and succinic acids were detected in both cheeses with values of about 0.5, 0.9, and 0.2 mg/g, respectively. Propionic acid was present in higher concentration in Cheese B. Therefore, to verify the origin of this organic acid, a genus-specific PCR was used. *Propionibacterium freudenreichii* was the only propionic bacteria detected, as demonstrated by the presence of a specific fragment of 850 bp. It was present in all samples, the only exception being Cheese A at 7 days of ripening (data not shown). Band intensities were correlated to propionic bacteria abundance. In Cheese A, the *P. freudenreichii* presence ranged from 10 to 10^3 CFU/g—in Cheese B, from 10^3 to 10^6 CFU/g. Similar results were found by Tofalo et al. [15] in a traditional Abruzzo cheese where the presence of *P. freudenreichii* has been shown to play an important role in its sensorial characteristic and aromatic quality conferring an intense flavor.

Assessment of proteolysis in Cheeses A and B, through the determination of WSN, 12% TCA-SN and AAN over three months of ripening, is reported in Figure 1. The WSN value was 9%N in Curd A and about 12%N in Curd B. During the first weeks of ripening, there were no statistically significant differences between the examined cheeses in the level of WSN and the concentrations increased with a more intense rise in Cheese B, reaching a final rate of 14.5%N at day 90. The effect of feeding system on nitrogen fractions was more marked starting from 30 days of ripening, probably due to the impact of the milk as a source of microbial enzymes. The amount of 12% TCA-SN also increased progressively in both cheeses, but it was stronger always in Cheese B. Starter LAB and non-starter LAB (NSLAB) proteinases are principally responsible for the formation of 12% TCA-SN [38], that contains small peptides (2–20 residues) and free amino acids [39]. The average content of AAN showed a general similar evolution in both cheeses, but Trial A showed a slower proteolytic activity than that in Trial B. However, the final values obtained were similar: 8.42 and 8.60 mg leucine/g, respectively, for Cheeses A and B at 90 days of ripening.

Table 2. Physico-chemical characteristics and organic acids content (mg/g) in cheeses at the end of ripening.

Parameters	Cheese A	Cheese B
Physico-chemical		
pH	5.55 ± 0.20	5.39 ± 0.10
a_w	0.98 ± 0.01	0.97 ± 0.01
% Dry matter	67.68 ± 4.80	66.15 ± 2.23
% Fat [1]	47.03 ± 3.13	51.69 ± 1.73
% Protein [1]	44.12 ± 6.68	40.92 ± 7.54
% Ash [1]	8.85 ± 0.26	7.39 ± 0.47
Organic acids		
Citric acid	0.5 ± 0.1	0.40 ± 0.07
Succinic acid	0.10 ± 0.05	0.21 ± 0.03
Lactic acid	35 ± 4	36 ± 4
Acetic acid	0.80 ± 0.05	0.90 ± 0.06
Propionic acid	0.04 ± 0.03	0.13 ± 0.06

Data are expressed as mean $\pm$ S.D.; [1] these parameters (fat, protein, and ash) are expressed in dry matter; no statistically significant differences were observed ($p < 0.05$).

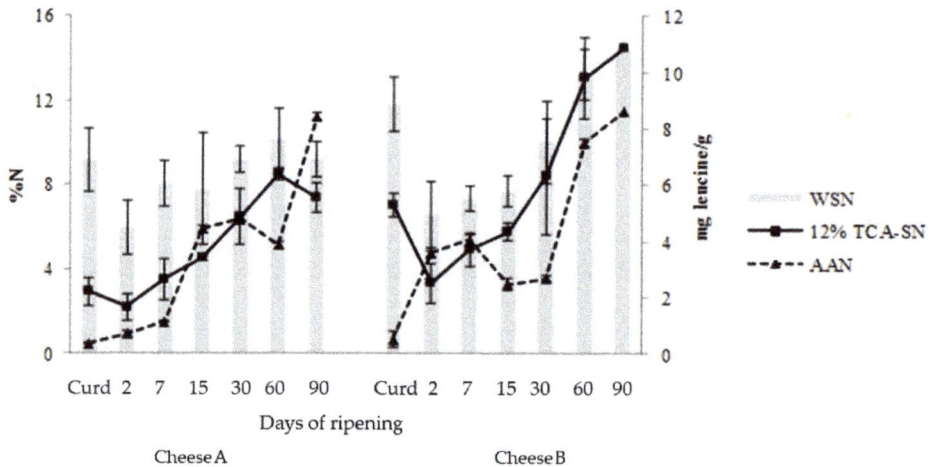

Figure 1. Evolution of nitrogen fractions during ripening in Cheeses A and B. WSN: water-soluble nitrogen; TCA-SN: trichloroacetic acid-soluble nitrogen; AAN: amino acid nitrogen.

3.3. BA Content

The high content of BA in cheese is well documented [20,40]. The accumulation of BA has been mainly ascribed to the activity of NSLAB, even if an indirect role of LAB proteolytic activity could be hypothesized providing the precursor amino acids used for BA synthesis. Moreover, some factors, such as environmental hygienic conditions, decarboxylase microorganisms, and temperature and ripening of cheese can contribute to the qualitative and quantitative BA profiles [31]. The accumulation of BA at high concentrations and the presence of BA-producing microorganisms cannot be avoided in raw milk cheeses as well as in fermented foods and beverages. Total BA content was found to be similar up to 60 days in both cheeses with an average content of about 400 mg/kg (Figure 2A,B). In raw milk A and B, only putrescine was detected at low concentrations (about 2 mg/kg). At day 60, the main amine was putrescine, followed by cadaverine, tyramine, and histamine. In Cheese B, a significant increase was observed in the total BA content at day 90 (760.7 mg/kg); in Cheese A, a decrease was detected at that time (244.30 mg/kg). The reduction was particularly evident for histamine (5.80 mg/kg) and cadaverine that disappeared. This fact could be explained by the presence of some BA-degrading strains, as reported by other authors [41–43]. Recently, Alvarez et al. [41] reported that a significant alternative to reduce BA content in fermented foods (such as cheese, wine, and sausages) was the use of BA-degrading strains, isolated from different origins. Fresno et al. [42] suggested that the addition of two strains (*Lactobacillus casei* 4a and 5b) were able to reduce BA contents in a Cabrales-like mini-cheese manufacturing model, although the exact mechanism via which this occurs remains unknown. In order to identify the pathways involved in the catabolism of these compounds, Ladero et al. [43] reported the draft genome of the *L. casei* 5b strain isolated from cheese. The use of BA-degrading strains could be particularly useful during cheeses manufacturing from raw milk in which a specific non-starter microbiota is essential for the organoleptic characteristic of the final product.

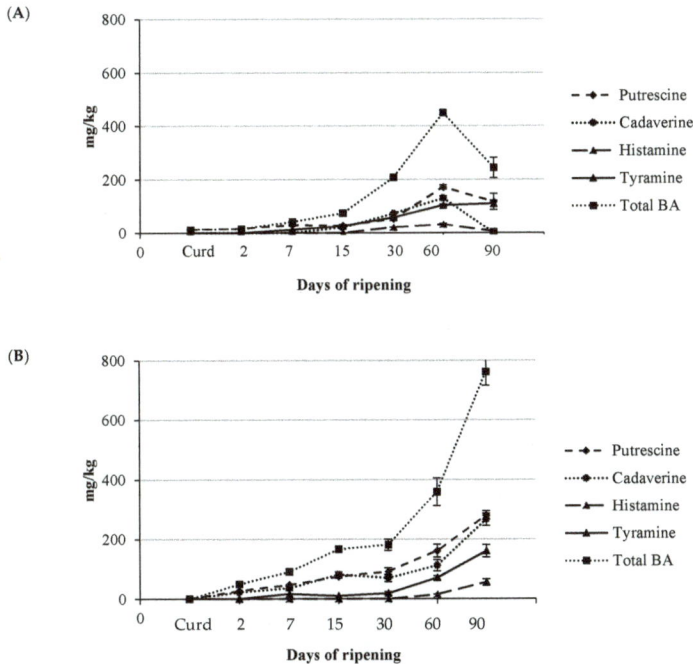

Figure 2. Biogenic amines content (mg/kg) during the ripening in Cheese A (**A**) and Cheese B (**B**). BA, biogenic amine.

In order to understand the variability between the two different cheeses, PCA was carried out using as variables physico-chemical and microbiological data. The PCA results were shown in Figure 3, the score plot (A) and the loading plot (B). The two principal components (PCs) captured 60.64% of total variance in the first two dimensions with 43.30% and 17.34% explained by Factors 1 and 2, respectively. In the score plot (A), both Cheeses A and B at days 60 and 90 of ripening clustered together in the positive section of PC 1 and were closely related to the high values of dry matter, fat, protein, and ash content as well as propionic and lactic acids; meanwhile, the negative counterpart of PC 1 was mainly associated to the succinic acid and grouped the samples of Cheese A in the first month of ripening. The different ripening times of Cheese B were discriminated over the first PC, based above all on the counts of microorganisms.

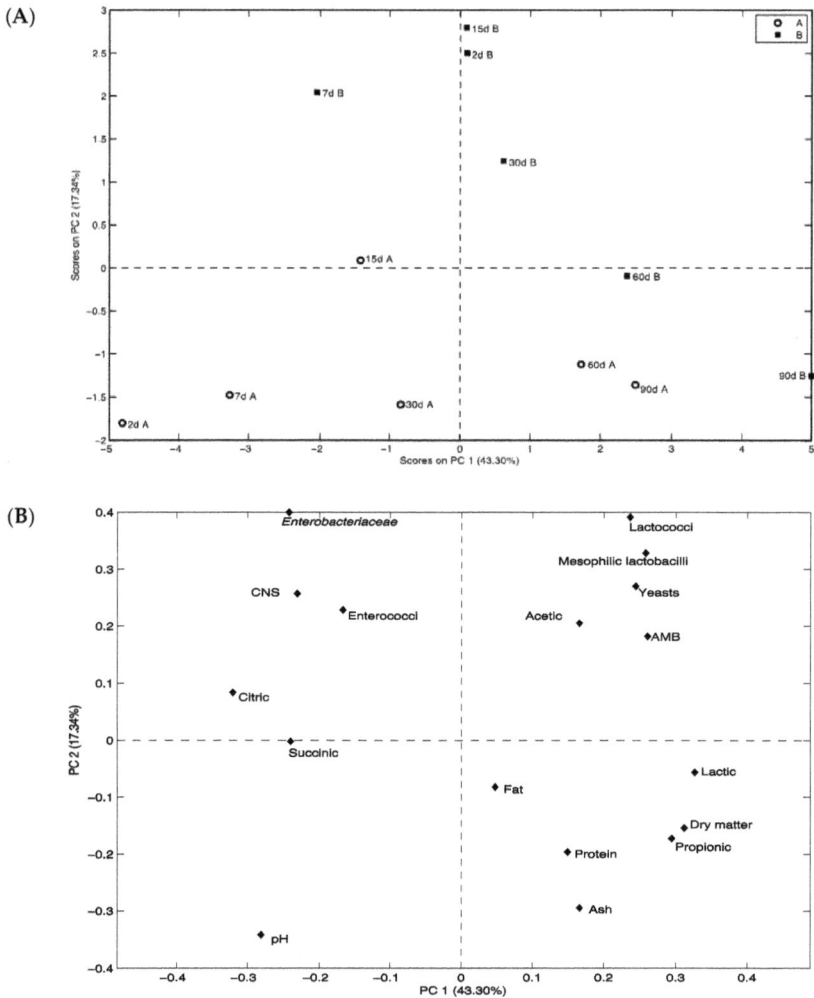

Figure 3. Score plot (**A**) and loading plot (**B**) of the first and second principal components (PCs) after PC analysis encompassing microbiological and physico-chemical parameters.

4. Conclusions

The overall management system of the farm was the same, so it is possible to hypothesize that iodine influenced the main features of Pecorino Incanestrato di Castel del Monte cheese. The physico-chemical and microbiological data highlighted a relevant effect of dietary iodine supplementation on ewes' raw milk and cheese microbiota. Even if the counts of principal microbial groups were quite similar in both cheeses, differences were found in some biochemical activities of microorganisms such as proteolytic/peptidasic activities or total BA content at day 90 of ripening. The findings call for a deep study on the selective effect of iodine on microbial populations in raw milk cheeses.

Author Contributions: Conceptualization: G.M. and R.T. Methodology: A.C.M., P.D.G., and N.B. Software: N.B., and G.P. Formal Analysis: A.C.M., P.D.G., and G.P. Investigation: A.C. and F.T. Resources: G.M. Writing—Original Draft Preparation: M.S., R.T., and G.S. Writing—Review & Editing: F.T., A.C., G.S. Supervision: R.T.

Funding: This work is part of the project named "Cararatterizzazione e miglioramento degli indici salutistici e sicurezza alimentare delle produzioni ovine tipiche abruzzesi a marchio di origine" (IprOv) funded by the Abruzzo Region within the Rural Development Programme 2007/2013.

Acknowledgments: The authors thank Dr. Giuseppe Fasoli for molecular analysis.

Conflicts of Interest: The authors declare no conflict of interest.

References

1. Bader, N.; Möller, U.; Leiterer, M.; Franke, K.; Jahreis, G. Pilot study: Tendency of increasing iodine content in human milk and cow's milk. *Exp. Clin. Endocrinol. Diabetes* **2005**, *113*, 8–12. [CrossRef] [PubMed]
2. Gaucheron, F. Milk minerals, trace elements, and macroelements. In *Milk and Dairy Products in Human Nutrition: Production, Composition and Health*; Park, Y.W., Haenlein, G.F.W., Eds.; Wiley & Sons Ltd.: Hoboken, NJ, USA, 2013; pp. 172–199. ISBN 9781118534168.
3. Haldimann, M.; Alt, A.; Blank, A.; Blondeau, K. Iodine content of food groups. *J. Food Compos. Anal.* **2005**, *18*, 461–471. [CrossRef]
4. World Health Organization (WHO). *Assessment of Iodine Deficiency Disorders and Monitoring Their Elimination*, 3rd ed.; World Health Organization (WHO): Geneva, Switzerland, 2007; ISBN 9789241595827.
5. Knowles, S.O.; Grace, N.D.; Knight, T.W.; McNabba, W.C.; Lee, J. Reasons and means for manipulating the micronutrient composition of milk from grazing dairy cattle. *Anim. Feed Sci. Technol.* **2006**, *131*, 154–167. [CrossRef]
6. Nudda, A.; Battacone, G.; Decandia, M.; Acciaro, M.; Aghini-Lombardi, F.; Frigeri, M.; Pulina, G. The effect of dietary iodine supplementation in dairy goats on milk production traits and milk iodine content. *J. Dairy Sci.* **2009**, *92*, 5133–5138. [CrossRef] [PubMed]
7. Weiss, W.P.; Wyatt, D.J.; Kleinschmit, D.H.; Socha, M.T. Effect of including canola meal and supplemental iodine in diets of dairy cows on short-term changes in iodine concentrations in milk. *J. Dairy Sci.* **2015**, *98*, 4841–4849. [CrossRef] [PubMed]
8. Giello, M.; La Storia, A.; Masucci, F.; Di Francia, A.; Ercolini, D.; Villani, F. Dynamics of bacterial communities during manufacture and ripening of traditional Caciocavallo of Castelfranco cheese in relation to cows' feeding. *Food Microbiol.* **2017**, *63*, 170–177. [CrossRef] [PubMed]
9. European Commission (EC). No. 1459/2005 of 8 September 2005 amending the conditions for the authorization of a number of feed additives belonging to the group of trace elements. *Off. J. Eur. Union* **2005**, *L233*, 8–10.
10. Schirone, M.; Tofalo, R.; Mazzone, G.; Corsetti, A.; Suzzi, G. Biogenic amine content and microbiological profile of Pecorino di Farindola cheese. *Food Microbiol.* **2011**, *28*, 128–136. [CrossRef] [PubMed]
11. International Organization for Standardization (ISO). *Microbiology of Food and Animal Feeding Stuffs—Horizontal Method for the Detection and Enumeration of Escherichia coli O157:H7*; ISO 16654; International Organization for Standardization (ISO): Geneva, Switzerland, 2001.
12. International Organization for Standardization (ISO). *Microbiology of Food and Animal Feeding Stuffs—Horizontal Method for the Detection of Salmonella spp.*; ISO 6579; International Organization for Standardization (ISO): Geneva, Switzerland, 2002.

13. International Organization for Standardization (ISO). *Microbiology of Food and Animal Feeding Stuffs—Horizontal Method for the Detection and Enumeration of Listeria monocytogenes—Part 1: Enumeration Method*; ISO 11290-1; International Organization for Standardization (ISO): Geneva, Switzerland, 2005.

14. Rossi, F.; Torriani, S.; Dellaglio, F. Genus and species-specific PCR-based detection of dairy propionibacteria in environmental samples using primers targeted to the 16S rDNA. *Appl. Environ. Microbiol.* **1999**, *65*, 4241–4244. [PubMed]

15. Tofalo, R.; Schirone, M.; Fasoli, G.; Perpetuini, G.; Patrignani, F.; Manetta, A.C.; Lanciotti, R.; Corsetti, A.; Martino, G.; Suzzi, G. Influence of pig rennet on proteolysis, organic acids content and microbiota of Pecorino di Farindola, a traditional Italian ewe's raw milk cheese. *Food Chem.* **2015**, *175*, 121–127. [CrossRef] [PubMed]

16. Bouzas, J.; Kantt, C.A.; Bodyfelt, F.; Torres, J.A. Simultaneous determination of sugars and organic acids in Cheddar cheese by high-performance liquid chromatography. *J. Food Sci.* **1991**, *56*, 276–278. [CrossRef]

17. Kuchroo, C.N.; Fox, P.F. Soluble nitrogen in Cheddar cheese: Comparison of extraction procedures. *Milchwissenschaft* **1982**, *37*, 331–335.

18. Polychroniadou, A.; Michaelidou, A.; Paschaloudis, N. Effect of time, temperature and extraction method on the trichloroacetic acid-soluble nitrogen of cheese. *Int. Dairy J.* **1999**, *9*, 559–568. [CrossRef]

19. Folkertsma, B.; Fox, P.F. Use of Cd-ninhydrin reagent to assess proteolysis in cheese during ripening. *J. Dairy Res.* **1992**, *59*, 217–224. [CrossRef]

20. Schirone, M.; Tofalo, R.; Fasoli, G.; Perpetuini, G.; Corsetti, A.; Manetta, A.C.; Ciarrocchi, A.; Suzzi, G. High content of biogenic amines in Pecorino cheeses. *Food Microbiol.* **2013**, *34*, 137–144. [CrossRef] [PubMed]

21. Manzano, M.; Sarais, I.; Stecchini, M.L.; Rondinini, G. Etiology of gas defects in Montasio cheese. *Ann. Microbiol.* **1990**, *40*, 255–259.

22. Centeno, J.A.; Menéndez, S.; Rodriguez-Otero, J.L. Main microbial flora present as natural starters in Cebreiro raw cow's-milk cheese (Northwest Spain). *Int. J. Food Microbiol.* **1996**, *33*, 307–313. [CrossRef]

23. Quinto, E.; Franco, C.; Rodriguez-Otero, J.L.; Fente, C.; Cepeda, A. Microbiological quality of Cebreiro cheese from Northwest Spain. *J. Food Saf.* **1994**, *14*, 1–8. [CrossRef]

24. Berthier, F.; Beuvier, E.; Dasen, A.; Grappin, R. Origin and diversity of mesophilic lactobacilli in Comté cheese, as revealed by PCR with repetitive and species-specific primers. *Int. Dairy J.* **2001**, *11*, 293–305. [CrossRef]

25. Samelis, J.; Lianou, A.; Kakouri, A.; Delbes, C.; Rogelj, I.; Bogovič-Matijašić, B.O.J.A.N.A.; Montel, M.C. Changes in the microbial composition of raw milk induced by thermization treatments applied prior to traditional Greek hard cheese processing. *J. Food Prot.* **2009**, *72*, 783–790. [CrossRef] [PubMed]

26. Poullet, B.; Huertas, M.; Sánchez, A.; Cáceres, P.; Larriba, G. Main lactic acid bacteria isolated during ripening of Casar de Cáceres cheese. *J. Dairy Res.* **1993**, *60*, 123–127. [CrossRef]

27. Suzzi, G.; Caruso, M.; Gardini, F.; Lombardi, A.; Vannini, L.; Guerzoni, M.E.; Andrighetto, C.; Lanorte, M.T. A survey of enterococci isolated from an artisanal Italian goat's cheese (semicotto caprino). *J. Appl. Microbiol.* **2000**, *89*, 267–274. [CrossRef] [PubMed]

28. Giraffa, G. Functionality of enterococci in dairy products. *Int. J. Food Microbiol.* **2003**, *88*, 215–222. [CrossRef]

29. Martino, G.P.; Quintana, I.M.; Espariz, M.; Blancato, V.S.; Magni, C. Aroma compounds generation in citrate metabolism of *Enterococcus faecium*: Genetic characterization of type I citrate gene cluster. *Int. J. Food Microbiol.* **2016**, *218*, 27–37. [CrossRef] [PubMed]

30. Suzzi, G.; Gardini, F. Biogenic amines in dry fermented sausages: A review. *Int. J. Food Microbiol.* **2003**, *88*, 41–54. [CrossRef]

31. Schirone, M.; Tofalo, R.; Visciano, P.; Corsetti, A.; Suzzi, G. Biogenic amines in Italian Pecorino cheese. *Front. Microbiol.* **2012**, *3*, 171. [CrossRef] [PubMed]

32. Tofalo, R.; Fasoli, G.; Schirone, M.; Perpetuini, G.; Pepe, A.; Corsetti, A.; Suzzi, G. The predominance, biodiversity and biotechnological properties of *Kluyveromyces marxianus* in the production of Pecorino di Farindola cheese. *Int. J. Food Microbiol.* **2014**, *187*, 41–49. [CrossRef] [PubMed]

33. Coloretti, F.; Chiavari, C.; Luise, D.; Tofalo, R.; Fasoli, G.; Suzzi, G.; Grazia, L. Detection and identification of yeasts in natural whey starter for Parmigiano Reggiano cheese-making. *Int. Dairy J.* **2017**, *66*, 13–17. [CrossRef]

34. Moschini, M.; Battaglia, M.; Beone, G.M.; Piva, G.; Masoero, F. Iodine and selenium carry over in milk and cheese in dairy cows: Effect of diet supplementation and milk yield. *Animal* **2010**, *4*, 147–155. [CrossRef] [PubMed]

35. O'Kane, S.M.; Pourshadidi, L.K.; Mulhern, M.S.; Weir, R.R.; Hill, S.; O'Reilly, J.; Kmiotek, D.; Deitrich, C.; Mackle, E.M.; Fitzgerald, E.; et al. The effect of processing and seasonality on the iodine and selenium concentration of cow's milk produced in Northern Ireland (NI): Implications for population dietary intake. *Nutrients* **2018**, *10*, 287. [CrossRef] [PubMed]

36. Manca, M.G.; Nudda, A.; Rubattu, R.; Boe, R.; Pulina, G. Fatty acid profile of milk fat in goat supplemented with iodized salt. *Ital. J. Anim. Sci.* **2009**, *8* (Suppl. 2), 460.

37. Secchiari, P.; Mele, M.; Frigeri, M.; Serra, A.; Conte, G.; Pollicardo, A.; Formisano, G.; Alghini-Lombardi, F. Iodine supplementation in dairy ewes nutrition: Effect on milk production traits and milk iodine content. *Ital. J. Anim. Sci.* **2009**, *8* (Suppl. 2), 463.

38. Fox, P.F.; Law, J.; McSweeney, P.L.H.; Wallace, J. Biochemistry of cheese ripening. In *Cheese: Chemistry, Physics and Microbiology. General Aspects*, 3rd ed.; Fox, P.F., McSweeney, P.L.H., Cogan, T.M., Guinee, T.P., Eds.; Chapman & Hall: London, UK, 1993; pp. 389–438. ISBN 97800805000942.

39. Sousa-Gallagher, M.J.; Ardo, Y.; McSweeney, P.L.H. Advances in study of proteolysis during cheese ripening. *Int. Dairy J.* **2001**, *11*, 327–345. [CrossRef]

40. Spano, G.; Russo, P.; Lonvaud-Funel, A.; Lucas, P.; Alexandre, H.; Grandvalet, C.; Coton, E.; Coton, M.; Barnavon, L.; Bach, B.; et al. Biogenic amines in fermented foods. *Eur. J. Clin. Nutr.* **2010**, *64*, S95–S100. [CrossRef] [PubMed]

41. Alvarez, M.A.; Moreno-Arribas, M.V. The problem of biogenic amines in fermented foods and the use of potential biogenic amine-degrading microorganisms as a solution. *Trends Food Sci. Technol.* **2014**, *39*, 146–155. [CrossRef]

42. Herrero-Fresno, A.; Martínez, N.; Sánchez-Llana, E.; Díaz, M.; Fernández, M.; Martin, M.C.; Ladero, V.; Alvarez, M.A. *Lactobacillus casei* strains isolated from cheese reduce biogenic amine accumulation in an experimental model. *Int. J. Food Microbiol.* **2012**, *157*, 297–304. [CrossRef] [PubMed]

43. Ladero, V.; Herrero-Fresno, A.; Martinez, N.; del Río, B.; Linares, D.M.; Fernández, M.; Martin, M.C.; Alvarez, M.A. Genome sequence analysis of the biogenic amine-degrading strain *Lactobacillus casei* 5b. *Genome Announc.* **2014**, *2*, 1–2. [CrossRef] [PubMed]

foods

MDPI

Article

High Hydrostatic Pressure as a Tool to Reduce Formation of Biogenic Amines in Artisanal Spanish Cheeses

Diana Espinosa-Pesqueira, Maria Manuela Hernández-Herrero and Artur X. Roig-Sagués *

CIRTTA-Departament de Ciència Animal i dels Aliments, Universitat Autònoma de Barcelona, Travessera dels Turons S/N, 08193 Barcelona, Spain; diespe@gmail.com (D.E.-P.); manuela.hernandez@uab.cat (M.M.H.-H.)
* Correspondence: arturxavier.roig@uab.cat; Tel.: +34-935-812-582

Received: 28 July 2018; Accepted: 24 August 2018; Published: 30 August 2018

Abstract: Two artisanal varieties of cheese made in Spain, one made of ewes' raw milk and the other of goats' raw milk were selected to evaluate the effect of a high hydrostatic pressure (HHP) treatment at 400 MPa during 10 min at 2 °C on the formation of biogenic amines (BA). These conditions were applied at the beginning of the ripening (before the 5th day; HHP1) and in the case of ewes' milk cheeses also after 15th days (HHP15). BA formation was greatly influenced by HHP treatments in both types of cheese. HHP1 treatments significantly reduced the amounts of BA after ripening, being tyramine and putrescine the most affected BA in goats' milk cheeses and tyramine and cadaverine in ewes' milk cheeses. The BA reduction in the HHP1 samples could be explained by the significant decrease in microbiological counts, especially in the LAB, enterococci and enterobacteria groups at the beginning of ripening. The proteolysis in these samples was also affected reducing the amount of free amino acids. Although proteolysis in ewes' milk cheeses HHP15 was similar than in control samples a reduction of BA was observed probably because the decrease caused on microbial counts.

Keywords: biogenic amines; cheese; high hydrostatic pressure

1. Introduction

Biogenic amines (BA) are basic nitrogenous compounds formed in different foodstuffs due to the microbial decarboxylation of amino acids. The kind and the amount of BA formed depends on the decarboxylase capability of the bacterial strains present, the availability of substrate amino acids and the physicochemical properties of the matrix [1]. Some aromatic BA, such as histamine (HIS), tyramine (TY), β-phenylethylamine (PHE) and tryptamine (TR), have psychoactive and vasoactive properties that may cause food poisoning when present in foodstuffs, while the diamines putrescine (PU) and cadaverine (CA) can boost the toxic action of aromatic BA [2,3]. Outbreaks caused by HIS are frequent, although not always correctly diagnosed or declared. It causes intoxication—called "histamine food poisoning" or "scombroid food poisoning"—since it is frequently associated with consumption of scombroid fish, especially tuna fish and mackerel [4]. After fish, cheese is the most commonly implicated food product associated with histamine poisoning and outbreaks have been associated with cheeses made from both raw and pasteurized milk [5]. The HIS concentrations in cheeses that were implicated in outbreaks ranged between 850 to 1870 mg/kg [6]. Tyramine is another BA associated with food-borne poisoning, being one of the causative agents of the so called "cheese reaction" [7], causing migraine, headache and, in extreme cases, hypertensive breakdown [5,8,9]. The seriousness of any BA food-borne poisoning depends on the ingested dose and the sensitiveness of the consumer (genetic or acquired) [3,10].

During cheese manufacture several factors may contribute to the accumulation of toxic amounts of BA. Good manufacturing practices to minimize the occurrence of BA-producing microorganisms in raw materials, pasteurization of milk or addition of BA-non-producing starter cultures have been suggested as BA risk mitigation options. Although it has been described that milk pasteurization reduces the level of decarboxylase-positive bacteria, later contamination of milk and curd during cheese manufacturing by decarboxylating bacteria and their subsequent growth and metabolic activity during cheese ripening usually results in BA build up [10–13].

High hydrostatic pressure (HHP) is non-thermal processing method used to extend shelf-life of foods. This induces morphological changes and inhibition of enzymes and genetic mechanisms of microorganism [14]. HHP offers the advantage that can be applied after cheese manipulation is over and no further contamination of curd is expected. The effect of HHP treatments has been evaluated in different kind of cheeses made from cows' milk [15], ewes' milk [16] and goats' milk [17,18], being able to eliminate most of the pathogenic bacteria associated with this product but it may also be useful to eliminate decarboxylating bacteria and avoid BA formation during cheese ripening. Calzada et al. [19] evaluated HHP treatments up to 600 MPa at different ripening stages to control the excessive proteolysis and BA formation on blue veined cheese, observing a lower concentration of TY. In another study [20], it was observed that HHP treatments significantly reduced the BA build up when applied on days 21 and 35 of ripening in "Torta del Casar" type cheese. This kind of cheese is made of raw milk and vegetable rennet, which causes a strong proteolytic activity and leads to extensive caseins breakdown in cheese matrix. Nevertheless, much less information was found about the HHP effect on the BA formation in other type of cheeses made of pressed paste.

The main objective of this survey was to evaluate HHP processing to reduce BA formation in two varieties of artisanal Spanish cheese made of raw milk from ewe and goat. Treatments were applied at different stages of ripening and the consequences of these treatments on the BA formation and proteolytic activity were evaluated during the ripening.

2. Materials and Methods

2.1. Cheese Manufacturing

Two types of artisanal ripened cheeses elaborated in Spain were studied in this work, both made with starter culture, enzymatic curd and pressed paste. The first one was produced from goats' raw milk in the region of Catalonia, northeast of Spain, and the second was made from ewes' raw milk in Castilla y León, central Spain. Three independent batches of each type of cheese were produced following the usual manufacturing procedures used by the manufacturers.

2.2. High-Hydrostatic Pressure (HHP) Treatments

HHP treatments were performed at 400 MPa for 10 min at a temperature of 2 °C using an Alstom HHP equipment (Alstom, Nantes, France) with a 2 L pressure chamber. A mixture of alcohol and water (1:9) was used into the chamber. The pressurization rate and depressurization time were 268 MPa/min and 55 s, respectively. Before processing, goats' milk cheese samples were separated in two batches: samples not HHP treated (Control samples) and samples HHP treated before the 5th day of ripening (HHP1). In the case of ewes' milk cheese samples a third batch for samples that were treated after 15 days of ripening (HHP15) was included. In all cases a portion of about 8.0 cm diameter was obtained and shaped to adequate it to the diameter of the cylinder of the HHP equipment and then vacuum packaged. After the HHP treatments, samples were deprived of the plastic bag and kept into a ripening chamber at 14 °C and 88% of relative humidity to continue with the ripening process for 60 days. Analyses of cheeses were performed at the 5th, 15th, 30th, 45th and 60th day of ripening.

2.3. Microbiological Analysis of Cheeses

Ten grams of each sample were homogenized in 90 mL of sterile Buffered Peptone Water (Oxoid, Basingstoke, Hampshire, UK) with a paddle blender (BagMixer, Interscience, France). Counts of *Lactococcus* spp. were made on M-17 agar (Oxoid) supplemented with a bacteriological grade lactose solution (5 g/L, Oxoid) and incubated at 30 °C for 48 h; Lactobacilli were determined on Man Rogose Sharpe agar (MRS, Oxoid) and incubated at 30 °C for 48 h; Enterococci were enumerated using KF *Streptococcus* Agar (Oxoid) supplemented with 2,3,5-trifeniltetrazolium chloride solution 1% (Oxoid) and incubated at 37 °C for 48 h; enumeration of *Enterobacticeae* was performed on Violet Red Bile Glucose Agar (VRBG, Oxoid) and counts of *Escherichia coli* was made on the chromogenic selective media Coli ID (BioMérieux, Marcy l'Etoile, France) and incubated at 30 °C for 24 h; *Staphylococcus aureus* counts were determined on Bair Parker Agar supplemented with rabbit plasma fibrinogen (BP-RPF agar, BioMérieux, Marcy L'Etoile, France) and incubated at 37 °C for 24–48 h.

2.4. Assessment of Proteolysis Activity on Cheeses

Water Soluble Extracts (WSE) of cheese were prepared according to the method described by Kuchroo and Fox [21]. From the WSE adjusted at pH 4.6, Water-Soluble Nitrogen (*WSN*) fraction was obtained and determined by the Dumas combustion method [22]. The nitrogen content of the *WSN* fraction was expressed as a percentage of Total Nitrogen (*TN*), described as the Ripening Index (*RI*), according to the formula:

$$RI = (WSN/TN) \times 100 \tag{1}$$

The measurement of the amino group content was determined on WSE by the Trinitrobenzensulphonic Acid method (TNBS) according to the procedure described by Hernández-Herrero et al. [23]. Results were expressed as mg of L-Leucine per g of cheese. The total Free Amino Acids (FAA) content was determined on WSE by the cadmium-ninhydrin method described by Folkertsmaa and Fox [24] and the results expressed as mg of L-Leucine per g of cheese.

2.5. Determination of Biogenic Amines in Cheese

The RP-HPLC method described by Eerola et al. [25] and modified by Roig Sagués et al. [12] was used to determine the BA content in cheese samples. The BA determined were β-phenylethylamine (PHE), tryptamine (TR), putrescine (PU), cadaverine (CA), histamine (HIS), tyrosine (TY), spermidine (SD) and spermine (SM).

2.6. Statistical Analysis

Analysis of variance (ANOVA) was performed on all data from each batch and treatment of goats' and ewes' milk cheeses at different ripening stages. Comparisons of mean values of physicochemical, microbiological and BA were followed by Duncan test with significance level set on $p < 0.05$. Comparisons of mean values of proteolysis indexes were performed using the Student-Newman-Keuls test with the significance level set at $p < 0.05$. All tests were performed with the SPSS for windows (v.15.01) program (SPSS Inc., Chicago, IL, USA).

3. Results and Discussion

3.1. Effect of HHP Treatment on Microbial Counts

Tables 1 and 2 shows the growth of the microbial groups counts along the ripening process in the goats' and ewes' cheeses, respectively, in both control and HHP processed samples.

Table 1. Changes in microbial population (mean values ± standard deviation as $\log_{10}$ CFU/g) of the goats' raw milk cheese samples with and without high hydrostatic pressure (HHP) treatment during ripening.

Microbial Group	Day of Ripening	Control	HHP1
Lactococci	5	9.55 ± 0.27 [a,A]	7.30 ± 0.22 [B]
	15	8.75 ± 0.34 [b,A]	6.76 ±0.49 [B]
	30	8.47 ± 0.26 [b,A]	7.03 ± 0.64 [B]
	45	8.26 ± 0.20 [b,A]	7.19 ± 0.25 [B]
	60	7.54 ± 0.51 [c,A]	7.05 ± 0.17 [A]
Lactobacilli	5	6.54 ± 0.31 [a,A]	3.13 ± 0.26 [a,B]
	15	8.36 ± 0.42 [b,A]	5.23 ± 1.54 [b,B]
	30	8.28 ± 0.15 [b,A]	6.31 ± 0.33 [c,B]
	45	8.18 ± 0.19 [b,A]	6.80 ± 0.50 [c,B]
	60	7.89 ± 0.45 [b,A]	7.03 ± 0.22 [c,A]
Enterococci	5	6.23 ± 0.24 [a,A]	4.44 ± 0.44 [ab,B]
	15	6.10 ± 0.70 [a,A]	3.99 ± 0.24 [ab,B]
	30	6.17 ± 0.47 [a,A]	4.60 ± 0.96 [b,B]
	45	6.17 ± 0.48 [a,A]	3.45 ± 0.34 [a,B]
	60	5.32 ± 0.64 [a,A]	3.85 ± 0.59 [ab,B]
Enterobacteria	5	6.33 ± 0.26 [a,A]	1.92 ± 0.54 [a,B]
	15	3.15 ± 2.75 [b,A]	2.49 ± 0.36 [a,A]
	30	4.54 ± 0.15 [b,A]	0.86 ± 1.48 [a,B]
	45	3.58 ± 0.40 [b,A]	ND [b,B]
	60	2.94 ± 0.54 [b,A]	ND [b,B]
E. coli	5	2.56 ± 0.62 [b,A]	ND [B]
	15	0.90 ± 0.78 [ab,A]	ND [A]
	30	1.60 ± 1.96 [ab,A]	0.83 ± 1.44 [A]
	45	0.77 ± 0.68 [ab,A]	ND [A]
	60	ND [a,A]	ND [A]
S. aureus	5	2.64 ± 0.15 [a,A]	0.43 ± 0.75 [a,B]
	15	1.06 ± 0.96 [b,A]	ND [A]
	30	1.18 ± 0.31 [b,A]	ND [B]
	45	0.77 ± 1.33 [b,A]	0.33 ± 0.58 [B]
	60	ND [b,A]	ND [A]

ND: not detected (<10 CFU/g); means with different superscript small letter differ significant ($p < 0.05$) in the same column for the same parameter and HHP treatment; means with different superscript capital letter differ significant ($p < 0.05$) in the same row for the same parameter and day of ripening; HHP1: HHP treated the 5th day of the ripening.

Table 2. Changes in microbial population (mean values ± standard deviation as $\log_{10}$ CFU/g) of the ewes' raw milk cheese samples with and without HHP treatment during ripening.

Microbial Group	Day of Ripening	Control	HHP1	HHP15
Lactococci	5	9.34 ± 0.33 [a,A]	7.80 ± 0.31 [a,B]	
	15	9.16 ± 0.20 [a,A]	6.87 ± 0.39 [b,B]	7.25 ± 0.22 [a,B]
	30	9.21 ± 0.18 [a,A]	8.04 ± 0.74 [b,B]	7.67 ± 0.14 [ab,B]
	45	8.74 ± 0.20 [ab,A]	8.07 ± 0.61 [b,B]	7.92 ± 0.39 [ab,B]
	60	8.38 ± 0.16 [b,A]	8.32 ± 0.76 [b,A]	7.57 ± 0.32 [b,B]
Lactobacilli	5	5.57 ± 0.35 [a,A]	4.54 ± 0.64 [a,B]	
	15	7.52 ± 1.07 [b,A]	5.60 ± 0.37 [b,B]	6.19 ± 0.50 [a,B]
	30	8.64 ± 0.41 [c,A]	6.61 ± 0.82 [c,B]	6.78 ± 1.00 [ab,B]
	45	8.68 ± 0.12 [c,A]	8.09 ± 0.55 [d,A]	7.82 ± 0.27 [bc,A]
	60	8.15 ± 0.39 [c,A]	8.12 ± 0.51 [d,A]	7.53 ± 0.15 [c,B]

Table 2. *Cont.*

Microbial Group	Day of Ripening	Control	HHP1	HHP15
Enterococci	5	5.59 ± 0.74 [A]	3.16 ± 1.05 [A]	
	15	6.18 ± 2.14 [AB]	3.50 ± 1.13 [C]	3.90 ± 1.11 [B]
	30	6.16 ± 1.18 [A]	1.76 ± 1.53 [B]	3.96 ± 1.32 [A]
	45	4.63 ± 1.04 [A]	2.76 ± 2.26 [B]	3.27 ± 0.48 [A]
	60	4.23 ± 1.40 [A]	2.58 ± 1.93 [B]	3.88 ± 0.93 [A]
Enterobacteria	5	4.65 ± 0.23 [a,A]	ND [B]	
	15	4.46 ± 0.30 [a,A]	0.29 ± 0.58 [B]	ND ± ND [B]
	30	4.42 ± 1.92 [a,A]	ND [B]	1.22 ± 1.11 [B]
	45	2.87 ± 1.21 [b,A]	ND [B]	0.57 ± 0.98 [B]
	60	2.35 ± 0.96 [b,A]	ND [B]	ND [B]
E. coli	5	1.16 ± 1.01	ND	
	15	1.15 ± 1.37	ND	ND
	30	0.65 ± 0.79	ND	ND
	45	0.64 ± 0.73	ND	0.78 ± 1.35
	60	0.73 ± 0.91	ND	ND
S. aureus	5	2.35 ± 0.58 [a]	1.07 ± 1.35	
	15	1.74 ± 1.19 [ab]	0.50 ± 0.58	0.33 ± 0.58
	30	1.17 ± 0.35 [ab]	0.29 ± 0.58	0.33 ± 0.58
	45	0.61 ± 0.71 [bc]	ND	ND
	60	0.25 ± 0.50 [c]	ND	ND

ND: Not detected (<10 CFU/g); means with different superscript small letter differ significant ($p < 0.05$) in the same column for the same parameter and HHP treatment; means with different superscript capital letter differ significant ($p < 0.05$) in the same row for the same parameter and day of ripening; HHP1: HHP treated the 5th day of the ripening; HHP15: HHP treated after 15 days of ripening.

In control samples (non HHP-treated) of both types of cheese, Lactoccocci was the main microbial group at the beginning of the ripening, showing counts above 8 $\log_{10}$ CFU/g, probably because strains of *Lactococcus lactis* were added as the starter culture, while lactobacilli counts were significantly lower at the beginning of the ripening, although their counts rose as the ripening progressed, achieving similar counts to lactococci. Enterococci counts remained steady at a level about 6 $\log_{10}$ CFU/g at the beginning of the process, showing a slight decrease around the 15th and the 30th day in the goats' and ewes' cheeses, respectively. *S. aureus*, enterobacteria and *E. coli* counts decreased during the ripening until becoming undetectable in most of the samples. These results are in agreement with those reported by other authors in other varieties of cured cheeses [26–30].

HHP treatments reduced significantly the Lactic Acid Bacteria (LAB) counts. In HHP1 treated samples, lactoccocci counts decreased 2.2 $\log_{10}$ CFU/g and 1.5 $\log_{10}$ CFU/g in the goats' and ewes' milk cheeses, respectively. In the case of lactobacilli mean reductions were 3.41 $\log_{10}$ CFU/g and 1.03 $\log_{10}$ CFU/g. However, these counts recovered along the ripening process and no statistical differences were observed at the end of the work with respect to control samples. Ewes' milk cheeses HHP15 showed a reduction of about 1.91 $\log_{10}$ CFU/g and 1.33 $\log_{10}$ CFU/g for lactoccocci and lactobacilli counts, respectively. In that case the subsequent recovery of these counts was not so clear and they did not achieve the same counts than control samples. Novella-Rodríguez et al. [31] also observed that starter counts were reduced about 2 $\log_{10}$ CFU/g in goats' milk cheeses as a consequence of HHP, although a subsequent increase was found during ripening. Rynne et al. [32] reported significant reductions of about 1.5 $\log_{10}$ CFU/g in starter and non-starter LAB in cheddar cheese treated at 400 MPa on the first day of ripening. In ovine milk ripened cheeses treated on day 1 and 15 at 300 MPa and 400 MPa proved to cause similar reductions although recovery of LAB counts was observed only in samples treated 1st day. HHP treatments also affected significantly ($p < 0.05$) the counts of enteroccocci in both type of cheeses in either HHP1 and HHP15 samples, being unable to recover the initial counts in any case. Arqués et al. [27] reported a reduction of 2.68 $\log_{10}$ CFU/g in enteroccocci counts when a treatment of 400 MPa for 10 min at 10 °C was applied on the 2nd day of ripening to "La Serena" cheeses, remaining constant during the rest of the ripening. Although high counts of enteroccoci have been associated with the unhygienic processing of cheese, their presence is also considered important for the development of the typical aroma and flavour of traditional Mediterranean cheeses. Their counts may range from 10^4 to 10^6 CFU/g in curds and 10^5 to 10^7 CFU/g in ripened cheeses [33]. In the case of *S. aureus* HHP1 treatments also caused significant reductions in both kind of cheeses and HHP15 in ewes' milk cheeses, becoming undetectable in most cases at the last day of the work. Nevertheless, the initial counts were already quite low. Similar results were reported in "La Serena" cheeses treated at 400 MPa on the 2nd day of ripening [27] and in Cheddar cheeses and their slurries treated at 400 MPa for 20 min at 20 °C [34]. *S. aureus* has been described as one of the most HHP resistant non-sporulated bacteria. López-Pedemonte et al. [35] described that at least 500 MPa HHP treatments would be necessary to achieve reductions of about 6 $\log_{10}$ CFU/g.

HHP treatments showed to be very effective reducing Gram-negative bacteria except for HHP1 in goats' milk cheese samples, where a slight growth was noted on the 15th day probably due to a possible recovery of the sub lethal injured cells after the HHP treatment. However, cheese ripening conditions, with low pH, increasing salt concentration and presence of LAB, made difficult this recovery to consolidate and no positive counts were detected later at the 30th and 45th days of ripening. Juan et al. [28] also described reductions above 3 $\log_{10}$ CFU/g of enterobacteria in ewes' milk cheeses after a 400 MPa treatment applied the 1st and the 15th of ripening. Initial counts of *E. coli* were significantly lower than enterobacteria. O'Reilly et al. [34], Capellas et al. [36] and De Lamo-Castellvi et al. [37] have previously pointed out the greatest sensitivity of *E. coli* to HHP in cheeses with reductions above 6 $\log_{10}$ CFU/g after HHP treatments at 400 MPa.

3.2. Effect of HHP on the Proteolysis of Cheeses

The proteolytic activity parameters measured in the goats' and ewes' cheeses are presented in Tables 3 and 4, respectively.

Table 3. Changes on proteolysis index (mean values ± standard deviation) during the ripening of goats' raw milk cheese samples with and without HHP treatment (TNBS: amino group content determined by the Trinitrobenzensulphonic Acid method, RI: Ripening Index, FFA: free Amino Acid content; TNBS and FAA expressed in mg L-Leucine/g).

Proteolysis Index	Day of Ripening	Control	HHP1
TNBS	5	10.40 ± 2.24 [a]	9.41 ± 1.80 [a]
	15	14.90 ± 1.21 [ab]	15.33 ± 2.15 [ab]
	30	22.54 ± 2.50 [b]	21.14 ± 4.59 [b]
	45	44.89 ± 1.19 [c,B]	32.65 ± 5.81 [c,A]
	60	53.66 ± 2.14 [d,B]	39.96 ± 4.52 [c,A]
RI	5	10.99 ± 3.05 [a]	11.09 ± 2.60 [a]
	15	22.97 ± 1.23 [b]	22.97 ± 1.23 [b]
	30	21.88 ± 0.60 [b]	24.39 ± 3.15 [bc]
	45	22.14 ± 1.78 [b]	25.71 ± 2.15 [bc]
	60	25.07 ± 0.66 [b]	28.03 ± 0.35 [c]
FAA	5	2.08 ± 0.57 [a]	1.48 ± 0.72 [a]
	15	4.85 ± 1.42 [a]	3.71 ± 1.06 [ab]
	30	10.05 ± 2.06 [b,B]	5.90 ± 1.59 [bc,A]
	45	15.07 ± 3.52 [c,B]	8.40 ± 1.43 [cd,A]
	60	15.71 ± 3.06 [c,B]	10.40 ± 1.29 [d,A]

Means with different superscript small letter differ significant ($p < 0.05$) in the same column for the same parameter and HHP treatment; means with different superscript capital letter differ significant ($p < 0.05$) in the same row for the same parameter and day of ripening.

Significant differences between control and HHP1 samples in TNBS were observed from the 45th to the 60th days of ripening in goats' milk cheeses, while the differences in FAA content were mainly noted from day 30 to 60. In both parameters control samples showed the highest values. The ratio of RI displayed a different trend where an increment of around two times was observed during the first 15 days period in control and HHP-treated samples but after this point the proteolysis rate became slower. In the ewes' milk cheeses, an increment on the three-proteolysis index evaluated was observed during the ripening period (Table 4). In the ewes' milk cheeses HHP15 samples presented slightly higher values than control samples, especially in TNBS, although no significant differences were observed at the end of the ripening. Although the HHP1 samples showed an intense proteolysis in the first 30 days of ripening, after this time a decrease on the rate was noticed, obtaining a considerable reduction on the three proteolysis index values at the 60th day, reflecting that pressure treatment caused a deceleration in the rate of proteolysis (Table 4).

Table 4. Evolution of proteolysis index (mean values ± standard deviation) during the ripening of ewes' raw milk cheese samples with and without HHP treatment.

Proteolysis Index	Day of Ripening	Control	HHP1	HHP15
TNBS	5	7.66 ± 1.65 [a]	7.92 ± 0.44 [a]	
	15	13.87 ± 1.37 [ab]	13.36 ± 0.82 [a]	15.03 ± 0.40 [a]
	30	22.13 ± 8.22 [b,AB]	26.33 ± 17.48 [b,BC]	34.31 ± 6.11 [b,C]
	45	34.93 ± 2.60 [c,BC]	28.00 ± 2.84 [b,AB]	44.79 ± 4.93 [bc,C]
	60	47.40 ± 2.74 [d,BC]	37.59 ± 6.86 [b,A]	52.50 ± 2.21 [c,C]
RI	5	9.70 ± 3.55 [a]	10.44 ± 4.39 [a]	
	15	19.38 ± 4.24 [b]	18.51 ± 4.23 [b]	19.15 ± 2.57 [a]
	30	24.96 ± 3.64 [bc]	23.35 ± 2.38 [b]	25.39 ± 3.57 [ab]
	45	29.59 ± 2.97 [c,B]	25.09 ± 2.95 [b,A]	30.54 ± 2.46 [b,B]
	60	31.92 ± 6.63 [c,B]	26.38 ± 3.77 [b,A]	31.62 ± 2.21 [b,B]
FAA	5	2.36 ± 0.04 [a]	2.18 ± 0.22 [a]	
	15	5.01 ± 1.27 [a]	3.66 ± 1.79 [ab]	5.59 ± 2.94 [a]
	30	11.66 ± 2.76 [b,B]	6.85 ± 3.07 [ab,AB]	11.22 ± 1.11 [b,B]
	45	11.39 ± 2.69 [b,B]	6.55 ± 1.10 [ab,AB]	14.86 ± 4.89 [bc,C]
	60	16.56 ± 3.51 [c,B]	8.63 ± 1.71 [b,A]	16.03 ± 5.14 [c,B]

Means with different superscript small letter differ significant (*p* < 0.05) in the same column for the same parameter and HHP treatment; means with different superscript capital letter differ significant (*p* < 0.05) in the same row for the same parameter and day of ripening; TNBS: amino group content determined by the Trinitrobenzensulphonic Acid method, RI: Ripening Index, FFA: free Amino Acid content; TNBS and FAA expressed in mg L-Leucine/g.

RI values indicated that proteolysis was more intense during the first 15 days of ripening in the goats' and ewes' cheeses in control and HHP treated samples. This proteolytic activity was probably caused by milk and rennet proteinases, being not so clear the role of microbial proteinases. On fact, *WSN* is produced mainly by the rennet and to a lesser extent by plasmin or cellular proteinases, whereas starter peptidases are primarily responsible for the formation of small peptides and free amino acids [38]. Messens et al. [39] observed that chymosin and plasmin activity in "Gouda" cheese was not influenced by pressure (from 50 to 400 MPa for 20–100 min). Similarly, Malone et al. [40] in a study about HHP effects (100–800 MPa, 5 min, 25 °C) on the activity of proteolytic and glycolytic enzymes, observed that plasmin was insensible to pressure treatments and the chymosin activity was unaffected by treatments up to 400 MPa, decreasing by 50% after an 800 MPa treatment. On the other hand, Juan et al. [28] found in ewes' milk cheese that the chymosin activity decreases depending on the age of the cheese and the pressure applied (above 400 MPa, 1-day old cheeses), whereas plasmin activity was not significantly affected by HHP treatments (200–500 MPa, for 10 min) applied on the 1st and 15th day of ripening.

On ewes' milk cheeses HHP15 samples showed slightly higher values on the three-proteolysis index evaluated than control samples during the ripening period, although no significant differences were observed at the end of the ripening, reflecting that this treatment did not significantly affect the proteolysis process. However, the HHP application during the initial stages of the ripening in ovine and caprine milk cheeses led to a decrease of the proteolysis rate. Similar results were obtained by Juan et al. [16] who in pressurized ovine milk cheeses on the 15th day of ripening obtained similar *WSN/TN* values than control samples, at the end of the ripening but higher than those obtained in samples with HHP-treatment on the 1st day. In contrast, some works reported after application of HHP treatments not differences on proteolysis indexes during ripening in Gouda [39] and cheddar cheeses [32] or an increase of proteolysis in goat milk cheese [41,42]. Starter bacteria are one of the primary sources of ripening intracellular enzymes (proteinases and peptidases). Cellular lysis is required for their release in the cheese matrix [43], being one of the main factors that influence the rate of secondary proteolysis [38]. HHP-induced cell lysis is pressure and strain-dependent [44] and possibly ripening time-HHP dependent. While Messens et al. [39] indicated that, in Gouda cheese, the possible lysis of the starter bacterial cells resulting from the damage suffered at 400 MPa did not increase proteolysis because endocellular enzymes were inactivated by the pressure, O'Reilly et al. [45] pointed out that ripening enzymes in cheddar cheese would probably begin to denature after the application of pressure treatments between 350–400 MPa and Juan et al. [16] and Calzada et al. [30] noticed that HHP-treatments above 400 MPa delay the proteolysis in ewes' and cows' milk, respectively. On the other hand, Saldo et al. [17,42] suggested that in goats' milk cheeses, the release of starter enzymes probably caused an increase in proteolysis two weeks after HHP treatment at 400 MPa for 5 min was applied.

3.3. Effect of HHP on the Formation of BA on Cheeses

Tables 5 and 6 show the values of the total BA content of the goats' and ewes' milk cheeses after HHP treatments. In general, the sum of all BA, including polyamines, showed a constant and significant increase in control samples during ripening from the 5th (59.36 mg/kg) to the 60th day (1156 mg/kg) of ripening in the goats' milk cheese and from the 6th (85.08 mg/kg) to the 60th day (622.39 mg/kg) in ewes' milk cheese samples. This is in agreement with the results reported by other authors [46–51]. The application of pressure during the first stages of ripening (HHP1) caused an initial reduction of BA formation from the 15th day, observing that at the end of the ripening these samples displayed concentrations around 75% lower than control samples. HHP15 ewes' milk cheeses also resulted in a decrease of the BA content with respect to control samples, although this reduction was less pronounced (38% lower than control ones). The BA reduction in the HHP1 ewes' and goats' milk cheeses could be explained because of the significant decrease on microbiological counts observed one day after the treatment (specially of LAB, enteroccocci and enterobacteria) and the lower proteolysis presented in these samples with respect to the control ones, showing a reduction of about 34% and 49% of FAA content, respectively, at the end of the ripening. HHP15 samples also showed lower microbial counts when compared with the untreated cheeses but this did not affect the proteolysis and consequently the release of amino acids. Novella-Rodriguez et al. [52] found that the total BA content found in goats' milk cheeses HHP treated at 400 MPa during 5 min was similar than the untreated cheeses, although TY content was significantly reduced in HHP samples. Ruiz-Capillas et al. [53] observed that HHP treatments at 350 MPa for 15 min used to treat "Chorizo" slices caused a significant decrease in BA content (TY, PU, CA and SM), being the reduction of these amines coincidental with the decrease in microbial counts, especially of LAB.

Table 5. Monoamine and diamines content (mean values ± standard deviation expressed as mg/kg in dry basis) formed during the ripening of goats' raw milk cheese samples with and without HHP treatment.

BA	Day of Ripening	Control	HHP1
	5	8.99 ± 7.94 [a]	23.12 ± 5.93 [a]
	15	27.76 ± 15.00 [ab]	10.81 ± 9.37 [a]
TR	30	95.96 ± 34.38 [c]	82.10 ± 26.33 [b]
	45	68.39 ± 20.42 [bc]	70.11 ± 40.01 [b]
	60	63.69 ± 27.90 [bc]	89.08 ± 49.98 [b]
	5	1.40 ± 0.65 [a]	0.26 ± 0.45 [a]
	15	14.62 ± 1.42 [b]	17.96 ± 4.07 [b]
PHE	30	17.27 ± 5.94 [b]	17.16 ± 4.29 [b]
	45	20.75 ± 4.20 [ab]	14.80 ± 0.45 [ab]
	60	31.13 ± 14.19 [b]	25.56 ± 7.47 [b]
	5	4.07 ± 0.51 [a]	3.29 ± 0.42
	15	136.56 ± 21.57 [ab]	59.09 ± 7.34
PU	30	225.16 ± 22.31 [b]	67.09 ± 8.50
	45	463.88 ± 60.73 [c,B]	42.69 ± 33.89 [A]
	60	476.41 ± 126.21 [c,B]	79.80 ± 19.51 [A]
	5	30.20 ± 16.19 [a]	24.07 ± 9.04
	15	29.63 ± 4.35 [a]	26.20 ± 3.13
CA	30	50.14 ± 10.97 [ab]	35.29 ± 12.70
	45	69.53 ± 16.34 [b]	36.22 ± 17.00
	60	70.45 ± 27.21 [b]	44.22 ± 15.61
	5	1.27 ± 0.56 [a]	1.00 ± 0.87
	15	3.02 ± 0.43 [a]	2.44 ± 0.34
HIS	30	6.38 ± 3.31 [a]	6.27 ± 3.16
	45	18.04 ± 9.36 [b,B]	6.51 ± 1.75 [A]
	60	15.41 ± 7.05 [b,B]	4.85 ± 2.20 [A]
	5	10.04 ± 6.80 [a]	6.11 ± 6.95
	15	130.51 ± 42.98 [ab]	18.96 ± 1.07
TY	30	234.74 ± 69.16 [b,B]	15.59 ± 3.56 [A]
	45	443.87 ± 105.10 [c,B]	16.17 ± 0.68 [A]
	60	491.89 ± 67.45 [c,B]	28.93 ± 5.91 [A]

Means with different superscript small letter differ significant ($p < 0.05$) in the same column for the same parameter and HHP treatment; means with different superscript capital letter differ significant ($p < 0.05$) in the same row for the same parameter and day of ripening. TR: tryptamine, PHE: β-phenylethylamine, PU: putrescine, CA: cadaverine, HIS: histamine, TY: tyramine.

Table 6. Monoamine and diamines content (mean values ± standard deviation expressed as mg/kg in dry basis) formed during the ripening of ewes' raw milk cheese samples with and without HHP treatment.

BA	Day of Ripening	Control	HHP1	HHP15
	5	1.66 ± 3.31 [a]	4.92 ± 5.86	-
	15	4.51 ± 3.32 [a]	10.03 ± 7.29	8.76 ± 6.98
TR	30	9.87 ± 4.43 [ab]	9.83 ± 4.66	6.25 ± 0.80
	45	9.51 ± 3.85 [ab]	12.49 ± 6.76	12.32 ± 8.43
	60	15.73 ± 2.15 [b]	11.06 ± 6.78	11.66 ± 6.50
	5	1.02 ± 1.23 [a]	0.40 ± 0.48	-
	15	2.78 ± 1.44 [a]	2.39 ± 0.56	4.39 ± 1.84
PHE	30	5.67 ± 5.25 [a]	3.04 ± 1.01	3.71 ± 1.45
	45	12.69 ± 5.08 [ab]	4.37 ± 1.52	5.78 ± 1.53
	60	12.74 ± 2.62 [b]	4.43 ± 2.37	13.31 ± 15.37

Table 6. *Cont.*

BA	Day of Ripening	Control	HHP1	HHP15
PU	5	3.62 ± 0.40 [a]	5.20 ± 2.41	-
	15	42.42 ± 23.00 [ab]	11.25 ± 9.80	15.14 ± 11.99
	30	65.89 ± 30.90 [bc,B]	5.45 ± 1.38 [A]	22.33 ± 14.49 [AB]
	45	87.89 ± 76.10 [c,B]	7.81 ± 3.03 [A]	22.92 ± 17.51 [A]
	60	74.89 ± 54.74 [bc,B]	5.24 ± 0.76 [A]	26.08 ± 6.26 [A]
CA	5	62.13 ± 33.00 [a]	55.55 ± 28.54	-
	15	159.07 ± 37.23 [b,B]	48.75 ± 44.16 [A]	87.50 ± 70.43 [AB]
	30	141.63 ± 79.34 [ab,B]	48.40 ± 16.94 [A]	80.65 ± 76.20 [AB]
	45	129.82 ± 67.03 [ab,B]	41.91 ± 22.89 [A]	72.47 ± 29.78 [AB]
	60	105.90 ± 21.87 [ab,B]	28.51 ± 17.32 [A]	80.28 ± 71.97 [AB]
HIS	5	4.81 ± 4.54 [a]	5.33 ± 3.79	-
	15	39.98 ± 11.42 [b]	12.71 ± 10.97	29.14 ± 22.84 [a]
	30	74.34 ± 29.42 [c,B]	6.38 ± 2.77 [A]	47.52 ± 31.34 [ab,B]
	45	81.66 ± 46.30 [c,B]	5.94 ± 1.25 [A]	58.99 ± 3.14 [b,B]
	60	91.02 ± 5.73 [c,C]	7.07 ± 4.30 [A]	57.65 ± 13.16 [ab,B]
TY	5	4.37 ± 0.71 [a]	3.15 ± 1.94 [a]	-
	15	123.72 ± 40.62 [ab,B]	13.79 ± 5.28 [b,A]	115.79 ± 55.44 [B]
	30	222.32 ± 84.26 [bc,B]	22.16 ± 10.34 [b,A]	162.64 ± 36.21 [A]
	45	268.02 ± 130.91 [c,B]	80.49 ± 126.16 [ab,A]	147.73 ± 30.57 [AB]
	60	277.30 ± 114.08 [c,B]	32.69 ± 16.54 [b,A]	147.62 ± 26.64 [AB]

Means with different superscript small letter differ significant ($p < 0.05$) in the same column for the same parameter and HHP treatment; means with different superscript capital letter differ significant ($p < 0.05$) in the same row for the same parameter and day of ripening. TR: tryptamine, PHE: β-phenylethylamine, PU: putrescine, CA: cadaverine, HIS: histamine, TY: tyramine.

TY and PU were the main BA formed in untreated goats' milk cheeses, showing concentrations of about 492 and 476 mg/kg, respectively, at the end of the ripening. Whereas in ewes' milk control samples the predominant BA were TY and CA with 277 and 106 mg/kg, respectively. Several authors reported, in variable ranges, TY (88.6–445 mg/kg), HIS (not detected–697 mg/kg), PU (74.15–446.5 mg/kg) and CA (44–269.77 mg/kg) as the most abundant BA in goats' and ewes' milk ripened cheeses [26,46,48,49,54–56]. These variable contents depended on the type of cheese, length of the ripening period, the manufacturing process and the type of microorganisms present (starters and non-starter bacteria with decarboxylase activity). Post-ripening processing (cutting, slicing and grating) also has an important influence on the presence of decarboxylating bacteria in cheese and the formation of BA, such as HIS, may be greater than in entire cheeses [57]. TY levels in pressurized cheeses were lower than those presented in control cheeses. This aromatic amine increased slowly until the 15th day of ripening on goats' milk HHP1 samples and remained constant throughout the rest of the ripening. At the end of the ripening the concentrations were about 93% lower than in control samples. Likewise, ewes' milk HHP1 samples reached concentrations of 33 mg/kg at the 60th day, being 88% lower than control samples. The application of HHP treatment on the 15th day of ripening resulted in a slight decrease of TY levels in ewes' milk cheeses and the final content was not significantly different than control samples at the 60th day of ripening. The reduction of the TY levels in pressurized goats' and ewes' milk cheese samples coincided mainly with the decrease in LAB and enteroccocci counts and with the amount of FAA. Novella-Rodriguez et al. [52] observed HHP-treated goats' milk cheeses showed a significant lower TY content than untreated samples, attributing this behaviour to the reducing effect of HHP on the microbiological counts, especially on the non- starter LAB, although levels of TY in this kind of cheeses were similar. High "in vitro" capability to form TY has been described in different species of *Enterococcus* spp. and LAB isolated from cheese samples [12,58–61].

PU amounts in control goats' milk cheeses were almost the same than TY but HHP1-treatments caused a decrease on the first 15 days, remaining almost stable throughout the rest of the ripening. PU levels were about 83% lower in the HHP1 samples than in control samples at the 60th day. The pressure application in ewes' milk cheeses also affected the PU amounts formed, showing that HHP1 treatments limited the production of this diamine around 93% compared with control cheeses at the end of the ripening. HHP15 showed to be less efficient reducing the formation of PU. With respect to CA, untreated and HHP-treated goats' milk cheeses displayed amounts below 100 mg/kg without appreciate significant differences between them. In control ewes' milk cheeses this diamine increased mainly during the first 15 days, while in HHPI and HHP15 cheeses remained without significant changes throughout the ripening. The amounts of TR and PHE increased during the ripening in control goats' milk cheese samples without significant differences in relation to HHP1-treated samples, while low amounts were detected in ewes' cheeses remaining practically constant during ripening and without showing significant differences between treatments. Formation of PU and CA is usually associated with Gram negative bacteria, although some strains of *Enterococcus* spp. also have shown this capability "in vitro" [61]. Some LAB, such as *Lactococcus lactis*, are also able to form PU via the agmatine deiminase [62]. The application of the HHP treatments in the early stages of maturation causes an important reduction of both the Gram-negative microbiota, which practically disappeared and the enterococci that could not recover their initial counts during the rest of the ripening, which would explain that both PU and CA, together with TY, are the AB that present the greatest reduction. When treatments are applied after 15 days of maturation, the decarboxylating capacity of these microorganisms in the early stages of maturation is still present and consequently there would be a lesser effect on the formation of these AB. LAB are also affected by the HHP treatments, although later they are able to recover their initial counts, which is important to develop the cheese's own characteristics, although they may also be responsible for the residual decarboxylating activity [30].

In goats' raw milk control cheeses the concentration of HIS was very low at the beginning of ripening but increased later reaching its maximum after 45 days (18 mg/kg). No significant changes were observed until the 60th day. In HHP1-treated samples HIS showed a similar behaviour but in this case, differences were found after the 45th day displaying levels 68% lower than control samples at the 60th day of ripening. HHP1 ewes' milk cheese samples showed a reduction of 92% of HIS when compared with control samples. This treatment was more efficient than the HHP15. Diverse authors reported low HIS amounts in cheese (below 100 mg/kg), relating the production of this BA with some LAB [12,31,54,56,63,64]. Some works have also described *Enterobacteriaceae* strains able to decarboxylate histidine in diverse foodstuffs [12,58,65–69]. Enterococci have also been related with histamine formation in cheeses [48,59,70].

In cheeses made from raw goat milk, low levels of polyamines were found at the beginning of ripening, increasing their concentration very slightly during ripening until the 60th day (Table 7). However, they showed an increase in their amount when HHP1- treatment was applied. Higher levels of SD were observed in ewes' milk cheeses showing maximum amounts the 30th day of ripening (about 64 mg/kg in control cheeses) followed by a decrease at the end of the ripening. However, HHP1 treatment caused a significant increase of this polyamine at the 30th day, reaching maximum levels of about 122 mg/kg. SM was detected at constant amounts throughout the ripening in both control and HHP treated samples. SD has been reported as the main polyamine in some cheeses [10,46,55]. Novella-Rodríguez et al. [26] found in HHP treated goat cheeses an increase of polyamines, especially SD but no data were reported to elucidate the cause of this phenomenon. Polyamines are described as natural amines of non-microbial origin present generally at a lower concentration than other BA of bacterial origin. No toxic effects have been attributed to them although some authors have mentioned their importance for the intestine cell growth and proliferation in childhood [3,71].

Table 7. Polyamine content (mean values ± standard deviation expressed as mg/kg in dry basis) during the ripening of goats' and ewes' raw milk cheese samples with and without HHP treatment. (SD: spermidine, SM: spermine).

	Day of Ripening	Control	HHP1	HHP15
		Goats' milk chesses		
SD	5	1.13 ± 0.51 [a]	0.32 ± 0.50 [a]	-
	15	1.92 ± 0.18 [aA]	4.34 ± 0.21 [b,B]	-
	30	2.09 ± 0.24 [abA]	3.70 ± 0.89 [b,B]	-
	45	4.32 ± 1.75 [c]	3.28 ± 1.30 [b]	-
	60	3.92 ± 1.49 [bcA]	6.53 ± 1.99 [c,B]	-
SM	5	2.25 ± 2.07 [a]	1.47 ± 1.48 [a]	-
	15	4.35 ± 1.25 [ab,A]	7.80 ± 0.60 [bc,B]	-
	30	3.03 ± 0.75 [ab,A]	6.83 ± 0.46 [b,B]	-
	45	5.57 ± 3.91 [b,B]	2.16 ± 0.30 [a,A]	-
	60	3.90 ± 0.75 [ab,A]	10.46 ± 2.18 [c,B]	-
		Ewes' milk cheeses		
SD	5	0.73 ± 0.65 [a]	0.51 ± 0.90 [a]	-
	15	55.94 ± 29.00 [ab]	90.23 ± 67.99 [bc]	73.15 ± 38.57 [b]
	30	64.22 ± 8.22 [b,A]	122.25 ± 48.16 [c,B]	65.61 ± 10.97 [b,A]
	45	14.74 ± 14.76 [ab]	52.04 ± 36.91 [ab]	34.49 ± 19.56 [ab]
	60	30.14 ± 23.04 [ab]	19.49 ± 33.57 [a]	17.36 ± 9.35 [a]
SM	5	6.76 ± 5.17	5.94 ± 4.03 [a]	
	15	14.13 ± 12.84	18.69 ± 14.57 [a]	8.00 ± 6.99 [a]
	30	25.18 ± 18.41	28.39 ± 7.52 [b]	39.50 ± 32.16 [b]
	45	17.57 ± 9.32	13.13 ± 7.26 [ab]	20.08 ± 21.91 [ab]
	60	14.66 ± 17.02	15.27 ± 17.76 [a]	26.92 ± 17.93 [ab]

Means with different superscript small letter differ significant ($p < 0.05$) in the same column for the same parameter and HHP treatment; means with different superscript capital letter differ significant ($p < 0.05$) in the same row for the same parameter and day of ripening.

4. Conclusions

The effectiveness of HHP treatments depends on different factors, such as the type of cheese, the stage of ripening, the HHP processing conditions applied, the kind and number of microorganisms present. In this work, the use of HHP applied at the initial phases of ripening affected significantly the microorganisms responsible of forming BA, and, in consequence, reduced its content, especially of TY and PU in goats' milk cheeses and TY and CA in ewes' milk cheeses, assuring the safety of this product for the most BA sensitive consumers. As previously mentioned, HIS and TY are the BA that most frequently have been related with food-borne outbreaks, being suggested threshold values in cheese between 200–400 mg/kg of HIS [6,72,73] and between 100–800 mg kg of TY [73,74]. TY dose of 6 mg [6] or above 20 mg of HIS [75] have been suggested as toxic to very sensitive individuals. If we consider the consumption of 30 g of cheese as a serving the goats' and ewes' cheeses analysed in this work would provide around 15 and 8.3 mg of TY, respectively, that can be dangerous for the most sensible consumers but HHP1-treated cheeses would provide with less than 1 mg of TY considering the same size of serving.

Author Contributions: Conceived and designed the experiments: D.E.-P., M.M.H.-H. and A.X.R.-S.; performed the experiments: D.E.-P.; analysed the data: D.E.-P. and M.M.H.-H.; wrote the paper: D.E.-P. and A.X.R.-S.

Conflicts of Interest: The authors declare no conflict of interest.

References

1. Naila, A.; Flint, S.; Fletcher, G.; Bremer, P.; Meerdink, G. Control of biogenic amines in food—Existing and emerging approaches. *J. Food Sci.* **2010**, *75*, 139–150. [CrossRef] [PubMed]
2. Fernández, E.J.Q.; Ventura, M.T.M.; Sagués, A.X.R.; Jerez, J.J.R.; Herrero, M.M.H. Aminas biogenas en queso: Riesgo toxicológico y factores que influyen en su formación. *Alimentaria* **1998**, *294*, 59–66.
3. Mariné Font, A. *Les Amines Biògenes en els Aliments: Història i Recerca en el Marc de les Ciències de L'alimentació*; Institut d'Estudis Catalans, Secció de Ciències Biològiques: Barcelona, Spain, 2005; ISBN 8472837882.
4. Sumner, S.S.; Taylor, S.L. Detection Method for Histamine-Producing, Dairy-Related Bacteria using Diamine Oxidase and Leucocrystal Violet. *J. Food Prot.* **1989**, *52*, 105–108. [CrossRef]
5. Stratton, J.E.; Hutkins, R.W.; Taylor, S.L. Biogenic Amines in Cheese and other Fermented Foods: A Review. *J. Food Prot.* **1991**, *54*, 460–470. [CrossRef]
6. Daniel Collins, J.; Noerrung, B.; Budka, H.; Andreoletti, O.; Buncic, S.; Griffin, J.; Hald, T.; Havelaar, A.; Hope, J.; Klein, G.; et al. Scientific Opinion on risk based control of biogenic amine formation in fermented foods. *EFSA J.* **2011**, *9*, 2393. [CrossRef]
7. Shalaby, A.R. Significance of biogenic amines to food safety and human health. *Food Res. Int.* **1996**, *29*, 675–690. [CrossRef]
8. Taylor, S.L.; Eitenmiller, R.R. Histamine food poisoning: Toxicology and clinical aspects. *CRC Crit. Rev. Toxicol.* **1986**, *17*, 91–128. [CrossRef] [PubMed]
9. Nordisk Ministerråd; Nordisk Råd. *Present Status of Biogenic Amines in Foods in Nordic Countries*; Nordisk Ministerråd: Copenhagen, Denmark, 2002; ISBN 9289307730.
10. Novella-Rodriguez, S.; Veciana-Nogues, M.T.; Izquierdo-Pulido, M.; Vidal-Carou, M.C. Distribution of Biogenic Amines and Polyamines in Cheese. *J. Food Sci.* **2003**, *68*, 750–756. [CrossRef]
11. Joosten, H.M.L.J. Conditions allowing the formation of biogenic amines in cheese. *Neth Milk Dairy J.* **1988**, *4*, 329–357.
12. Roig-Sagués, A.; Molina, A.; Hernández-Herrero, M. Histamine and tyramine-forming microorganisms in Spanish traditional cheeses. *Eur. Food Res. Technol.* **2002**, *215*, 96–100. [CrossRef]
13. Kalac, P.; Abreu Gloria, M.B. Biogenic amine in cheeses, wines, beers and sauerkraut. In *Biological Aspects of Biogenic Amines, Poliamines and Conjugates*; Dandrifosse, G., Ed.; Transworld Research Network: Kerala, 2009; pp. 267–285, ISBN 9788178952499.
14. Hoover, D.G.; Metrick, C.; Papineau, A.M.; Farkas, D.F.; Knorr, D. Biological effects of high hydrostatic pressure on food microorganisms. *Food Technol.-Chic.* **1989**, *43*, 99–107.
15. Evert-Arriagada, K.; Hernández-Herrero, M.M.; Juan, B.; Guamis, B.; Trujillo, A.J. Effect of high pressure on fresh cheese shelf-life. *J. Food Eng.* **2012**, *110*, 248–253. [CrossRef]
16. Juan, B.; Ferragut, V.; Guamis, B.; Buffa, M.; Trujillo, A.J. Proteolysis of a high pressure-treated ewe's milk cheese. *Milchwissenschaft* **2004**, *59*, 616–619.
17. Saldo, J.; McSweeney, P.L.H.; Sendra, E.; Kelly, A.L.; Guamis, B. Proteolysis in caprine milk cheese treated by high pressure to accelerate cheese ripening. *Int. Dairy J.* **2002**, *12*, 35–44. [CrossRef]
18. Delgado, F.J.; González-Crespo, J.; Cava, R.; Ramírez, R. Changes in microbiology, proteolysis, texture and sensory characteristics of raw goat milk cheeses treated by high-pressure at different stages of maturation. *LWT—Food Sci. Technol.* **2012**, *48*, 268–275. [CrossRef]
19. Calzada, J.; Del Olmo, A.; Picon, A.; Gaya, P.; Nuñez, M. Proteolysis and biogenic amine buildup in high-pressure treated ovine milk blue-veined cheese. *J. Dairy Sci.* **2013**, *96*, 4816–4829. [CrossRef] [PubMed]
20. Calzada, J.; del Olmo, A.; Picón, A.; Gaya, P.; Nuñez, M. Reducing Biogenic-Amine-Producing Bacteria, Decarboxylase Activity, and Biogenic Amines in Raw Milk Cheese by High-Pressure Treatments. *Appl. Environ. Microbiol.* **2013**, *79*, 1277–1283. [CrossRef] [PubMed]
21. Kuchroo, C.N.; Fox, P.F. Soluble nitrogen in Cheddar cheese: Comparison of extraction procedures. *Milchwissenschaft* **1982**, *37*, 331–335.
22. *IDF (International Dairy Federation) Standard 185: 2002(E), I; 14891:2002(E) Milk and milk products—Determination of nitrogen content—Routine method using combustion according to the Dumas principle*; IDF: Brussels, Belgium, 2002.

23. Hernández-Herrero, M.M.; Roig-Sagués, A.X.; López-Sabater, E.I.; Rodríguez-Jerez, J.J.; Mora-Ventura, M.T. Protein hydrolysis and proteinase activity during the ripening of salted anchovy (*Engraulis encrasicholus* L.). A microassay method for determining the protein hydrolysis. *J. Agric. Food Chem.* **1999**, *47*, 3319–3324. [CrossRef] [PubMed]

24. Folkertsma, B.; Fox, P.F. Use of the Cd-ninhydrin reagent to assess proteolysis in cheese during ripening. *J. Dairy Res.* **1992**, *59*, 217–224. [CrossRef]

25. Eerola, S.; Hinkkanen, R.; Lindfors, E.; Hirvi, T. Liquid chromatographic determination of biogenic amines in dry sausages. *J. AOAC Int.* **1993**, *76*, 575–577. [PubMed]

26. Novella-Rodríguez, S.; Veciana-Nogués, M.T.; Roig-Sagués, A.X.; Trujillo-Mesa, A.J.; Vidal-Carou, M.C. Comparison of biogenic amine profile in cheeses manufactured from fresh and stored (4 degrees C, 48 hours) raw goat's milk. *J. Food Prot.* **2004**, *67*, 110–116. [CrossRef] [PubMed]

27. Arqués, J.L.; Garde, S.; Gaya, P.; Medina, M.; Nuñez, M. Short Communication: Inactivation of Microbial Contaminants in Raw Milk La Serena Cheese by High-Pressure Treatments. *J. Dairy Sci.* **2006**, *89*, 888–891. [CrossRef]

28. Juan, B.; Ferragut, V.; Buffa, M.; Guamis, B.; Trujillo, A.J. Effects of High Pressure on proteolytic enzymes in cheese: Relationship with the proteolysis of ewe milk cheese. *J. Dairy Sci.* **2007**, *90*, 2113–2125. [CrossRef] [PubMed]

29. Cabezas, L.; Sánchez, I.; Poveda, J.M.; Seseña, S.; Palop, M.L. Comparison of microflora, chemical and sensory characteristics of artisanal Manchego cheeses from two dairies. *Food Control* **2007**, *18*, 11–17. [CrossRef]

30. Calzada, J.; del Olmo, A.; Picon, A.; Gaya, P.; Nuñez, M. Effect of High-Pressure Processing on the Microbiology, Proteolysis, Biogenic Amines and Flavour of Cheese Made from Unpasteurized Milk. *Food Bioprocess Technol.* **2015**, *8*, 319–332. [CrossRef]

31. Novella-Rodriguez, S.; Veciana-Nogues, M.T.; Trujillo-Mesa, A.J.; Vidal-Carou, M.C. Profile of Biogenic Amines in Goat Cheese Made from Pasteurized and Pressurized Milks. *J. Food Sci.* **2002**, *67*, 2940–2944. [CrossRef]

32. Rynne, N.M.; Beresford, T.P.; Guinee, T.P.; Sheehan, E.; Delahunty, C.M.; Kelly, A.L. Effect of high-pressure treatment of 1 day-old full-fat Cheddar cheese on subsequent quality and ripening. *Innov. Food Sci. Emerg. Technol.* **2008**, *9*, 429–440. [CrossRef]

33. Foulquié Moreno, M.R.; Sarantinopoulos, P.; Tsakalidou, E.; De Vuyst, L. The role and application of enterococci in food and health. *Int. J. Food Microbiol.* **2006**, *106*, 1–24. [CrossRef] [PubMed]

34. O'Reilly, C.E.; O'Connor, P.M.; Kelly, A.L.; Beresford, T.P.; Murphy, P.M. Use of hydrostatic pressure for inactivation of microbial contaminants in cheese. *Appl. Environ. Microbiol.* **2000**, *66*, 4890–4896. [CrossRef] [PubMed]

35. López-Pedemonte, T.; Roig-Sagués, A.X.; Lamo, S.D.; Gervilla, R.; Guamis, B. High hydrostatic pressure treatment applied to model cheeses made from cow's milk inoculated with Staphylococcus aureus. *Food Control* **2007**, *18*, 441–447. [CrossRef]

36. Capellas, M.; Mor-Mur, M.; Sendra, E.; Pla, R.; Guamis, B. Populations of aerobic mesophils and inoculated E. coli during storage of fresh goat's milk cheese treated with high pressure. *J. Food Prot.* **1996**, *59*, 582–587. [CrossRef]

37. De Lamo-Castellví, S.; Capellas, M.; Roig-Sagués, A.X.; López-Pedemonte, T.; Hernández-Herrero, M.M.; Guamis, B. Fate of Escherichia coli strains inoculated in model cheese elaborated with or without starter and treated by high hydrostatic pressure. *J. Food Prot.* **2006**, *69*, 2856–2864. [CrossRef] [PubMed]

38. Fox, P.F. Proteolysis During Cheese Manufacture and Ripening. *J. Dairy Sci.* **1989**, *72*, 1379–1400. [CrossRef]

39. Messens, W.; Estepar-Garcia, J.; Dewettinck, K.; Huyghebaert, A. Proteolysis of high-pressure-treated Gouda cheese. *Int. Dairy J.* **1999**, *9*, 775–782. [CrossRef]

40. Malone, A.S.; Wick, C.; Shellhammer, T.H.; Courtney, P.D. High Pressure Effects on Proteolytic and Glycolytic Enzymes Involved in Cheese Manufacturing. *J. Dairy Sci.* **2003**, *86*, 1139–1146. [CrossRef]

41. Trujillo, A.; Buffa, M.; Casals, I.; Fernández, P.; Guamis, B. Proteolysis in goat cheese made from raw, pasteurized or pressure-treated milk. *Innov. Food Sci. Emerg. Technol.* **2002**, *3*, 309–319. [CrossRef]

42. Saldo, J.; Sendra, E.; Guamis, B. High Hydrostatic Pressure for Accelerating Ripening of Goat's Milk Cheese: Proteolysis and Texture. *J. Food Sci.* **2000**, *65*, 636–640. [CrossRef]

43. Thomas, T.D.; Pritchard, G.G. Proteolytic enzymes of dairy starter cultures. *FEMS Microbiol. Lett.* **1987**, *46*, 245–268. [CrossRef]

44. Malone, A.S.; Shellhammer, T.H.; Courtney, P.D. Effects of high pressure on the viability, morphology, lysis, and cell wall hydrolase activity of Lactococcus lactis subsp. cremoris. *Appl. Environ. Microbiol.* **2002**, *68*, 4357–4363. [CrossRef] [PubMed]

45. O'Reilly, C.E.; Kelly, A.L.; Oliveira, J.C.; Murphy, P.M.; Auty, M.A.; Beresford, T.P. Effect of varying high-pressure treatment conditions on acceleration of ripening of cheddar cheese. *Innov. Food Sci. Emerg. Technol.* **2003**, *4*, 277–284. [CrossRef]

46. Ordóñez, A.I.; Ibáñez, F.C.; Torre, P.; Barcina, Y. Formation of biogenic amines in Idiazabal ewe's-milk cheese: Effect of ripening, pasteurization, and starter. *J. Food Prot.* **1997**, *60*, 1371–1375. [CrossRef]

47. Gardini, F.; Martuscelli, M.; Caruso, M.C.; Galgano, F.; Crudele, M.A.; Favati, F.; Guerzoni, M.E.; Suzzi, G. Effects of pH, temperature and NaCl concentration on the growth kinetics, proteolytic activity and biogenic amine production of Enterococcus faecalis. *Int. J. Food Microbiol.* **2001**, *64*, 105–117. [CrossRef]

48. Galgano, F.; Suzzi, G.; Favati, F.; Caruso, M.; Martuscelli, M.; Gardini, F.; Salzano, G. Biogenic amines during ripening in 'Semicotto Caprino' cheese: Role of enterococci. *Int. J. Food Sci. Technol.* **2001**, *36*, 153–160. [CrossRef]

49. Martuscelli, M.; Gardini, F.; Torriani, S.; Mastrocola, D.; Serio, A.; Chaves-López, C.; Schirone, M.; Suzzi, G. Production of biogenic amines during the ripening of Pecorino Abruzzese cheese. *Int. Dairy J.* **2005**, *15*, 571–578. [CrossRef]

50. Combarros-Fuertes, P.; Fernández, D.; Arenas, R.; Diezhandino, I.; Tornadijo, M.E.; Fresno, J.M. Biogenic amines in Zamorano cheese: factors involved in their accumulation. *J. Sci. Food Agric.* **2016**, *96*, 295–305. [CrossRef] [PubMed]

51. Poveda, J.M.; Molina, G.M.; Gómez-Alonso, S. Variability of biogenic amine and free amino acid concentrations in regionally produced goat milk cheeses. *J. Food Compost. Anal.* **2016**, *51*, 85–92. [CrossRef]

52. Novella-Rodríguez, S.; Veciana-Nogués, M.T.; Saldo, J.; Vidal-Carou, M.C. Effects of high hydrostatic pressure treatments on biogenic amine contents in goat cheeses during ripening. *J. Agric. Food Chem.* **2002**, *50*, 7288–7292. [CrossRef] [PubMed]

53. Ruiz-Capillas, C.; Jiménez Colmenero, F.; Carrascosa, A.V.; Muñoz, R. Biogenic amine production in Spanish dry-cured "chorizo" sausage treated with high-pressure and kept in chilled storage. *Meat Sci.* **2007**, *77*, 365–371. [CrossRef] [PubMed]

54. Valsamaki, K.; Michaelidou, A.; Polychroniadou, A. Biogenic amine production in Feta cheese. *Food Chem.* **2000**, *71*, 259–266. [CrossRef]

55. Pinho, O.; Ferreira, I.M.P.L.V.O.; Mendes, E.; Oliveira, B.M.; Ferreira, M. Effect of temperature on evolution of free amino acid and biogenic amine contents during storage of Azeitão cheese. *Food Chem.* **2001**, *75*, 287–291. [CrossRef]

56. Pintado, A.I.E.; Pinho, O.; Ferreira, I.M.P.L.V.O.; Pintado, M.M.E.; Gomes, A.M.P.; Malcata, F.X. Microbiological, biochemical and biogenic amine profiles of Terrincho cheese manufactured in several dairy farms. *Int. Dairy J.* **2008**, *18*, 631–640. [CrossRef]

57. Ladero, V.; Fernandez, M.; Alvarez, M.A. Effect of post-ripening processing on the histamine and histamine-producing bacteria contents of different cheeses. *Int. Dairy J.* **2009**, *19*, 759–762. [CrossRef]

58. Pircher, A.; Bauer, F.; Paulsen, P. Formation of cadaverine, histamine, putrescine and tyramine by bacteria isolated from meat, fermented sausages and cheeses. *Eur. Food Res. Technol.* **2007**, *226*, 225–231. [CrossRef]

59. Leuschner, R.G.K.; Kurihara, R.; Hammes, W.P. Formation of biogenic amines by proteolytic enterococci during cheese ripening. *J. Sci. Food Agric.* **1999**, *79*, 1141–1144. [CrossRef]

60. Perin, L.M.; Belviso, S.; Dal Bello, B.; Nero, L.A.; Cocolin, L. Technological properties and biogenic amines production by bacteriocinogenic lactococci and enterococci strains isolated from raw goat's milk. *J. Food Protect.* **2017**, *80*, 151–157. [CrossRef] [PubMed]

61. Torracca, B.; Pedonese, F.; Turchi, B.; Fratini, F.; Nuvoloni, R. Qualitative and quantitative evaluation of biogenic amines in vitro production by bacteria isolated from ewes' milk cheeses. *Eur. Food Res. Technol.* **2018**, *244*, 721–728. [CrossRef]

62. del Rio, B.; Redruello, B.; Ladero, V.; Fernandez, M.; Cruz Martin, M.; Alvarez, M.A. Putrescine production by *Lactococcus lactis* subsp. cremoris CECT 8666 is reduced by NaCl via a decrease in bacterial growth and the repression of the genes involved in putrescine production. *Int. J. Food Microbiol.* **2016**, *232*, 1–6. [CrossRef] [PubMed]

63. Joosten, H.M.L.J.; Northolt, M.D. Detection, growth, and amine-producing capacity of lactobacilli in cheese. *Appl. Environ. Microbiol.* **1989**, *55*, 2356–2359. [PubMed]

64. Burdychova, R.; Komprda, T. Biogenic amine-forming microbial communities in cheese. *FEMS Microbiol. Lett.* **2007**, *276*, 149–155. [CrossRef] [PubMed]

65. Roig-Sagues, A.X.; Hernandez-Herrero, M.; Lopez-Sabater, E.I.; Rodriguez-Jerez, J.J.; Mora-Ventura, M.T. Histidine decarboxylase activity of bacteria isolated from raw and ripened Salchichon, a Spanish cured sausage. *J. Food Prot.* **1996**, *59*, 516–520. [CrossRef]

66. Roig-Sagués, A.X.; Hernàndez-Herrero, M.M.; López-Sabater, E.I.; Rodríguez-Jerez, J.J.; Mora-Ventura, M.T. Evaluation of three decarboxylating agar media to detect histamine and tyramine-producing bacteria in ripened sausages. *Lett. Appl. Microbiol.* **1997**, *25*, 309–312. [CrossRef] [PubMed]

67. Silla Santos, M.H. Amino acid decarboxylase capability of microorganisms isolated in Spanish fermented meat products. *Int. J. Food Microbiol.* **1998**, *39*, 227–230. [CrossRef]

68. Hernández-Herrero, M.M.; Roig-Sagués, A.X.; Rodríguez-Jerez, J.J.; Mora-Ventura, T.M. Halotolerant and Halophilic Histamine-Forming Bacteria Isolated during the Ripening of Salted Anchovies (*Engraulis encrasicholus*). *J. Food Prot.* **1999**, *62*, 509–514. [CrossRef] [PubMed]

69. Özogul, F.; Özogul, Y. The ability of biogenic amines and ammonia production by single bacterial cultures. *Eur. Food Res. Technol.* **2007**, *225*, 385–394. [CrossRef]

70. Tham, W.; Karp, G.; Danielsson-Tham, M.L. Histamine formation by enterococci in goat cheese. *Int. J. Food Microbiol.* **1990**, *11*, 225–229. [CrossRef]

71. Bardócz, S. Polyamines in food and their consequences for food quality and human health. *Trends Food Sci. Technol.* **1995**, *6*, 341–346. [CrossRef]

72. Rauscher-Gabernig, E.; Grossgut, R.; Bauer, F.; Paulsen, P. Assessment of alimentary histamine exposure of consumers in Austria and development of tolerable levels in typical foods. *Food Control* **2009**, *20*, 423–429. [CrossRef]

73. Benkerroum, N. Biogenic Amines in Dairy Products: Origin, Incidence, and Control Means. *Compr. Rev. Food Sci. Food Saf.* **2016**, *15*, 801–826. [CrossRef]

74. Karovicova, J.; Kohajdova, Z. Biogenic Amines in Food. *ChemInform* **2005**, *36*, 70–79. [CrossRef]

75. Vind, S.; Søndergaard, I.; Poulsen, L.K.; Svendsen, U.G.; Weeke, B. Comparison of methods for intestinal histamine application: Histamine in enterosoluble capsules or via a duodeno-jenunal tube. *Allergy* **1991**, *46*, 191–195. [CrossRef] [PubMed]

Review

Bacterial Production and Control of Biogenic Amines in Asian Fermented Soybean Foods

Jae-Hyung Mah *, Young Kyoung Park, Young Hun Jin, Jun-Hee Lee and Han-Joon Hwang

Department of Food and Biotechnology, Korea University, 2511 Sejong-ro, Sejong 30019, Korea; eskimo@korea.ac.kr (Y.K.P.); younghoonjin3090@korea.ac.kr (Y.H.J.); bory92@korea.ac.kr (J.-H.L.); hjhwang@korea.ac.kr (H.-J.H.)
* Correspondence: nextbio@korea.ac.kr; Tel.: +82-44-860-1431

Received: 27 January 2019; Accepted: 19 February 2019; Published: 25 February 2019

Abstract: Fermented soybean foods possess significant health-promoting effects and are consumed worldwide, especially within Asia, but less attention has been paid to the safety of the foods. Since fermented soybean foods contain abundant amino acids and biogenic amine-producing microorganisms, it is necessary to understand the presence of biogenic amines in the foods. The amounts of biogenic amines in most products have been reported to be within safe levels. Conversely, certain products contain vasoactive biogenic amines greater than toxic levels. Nonetheless, government legislation regulating biogenic amines in fermented soybean foods is not found throughout the world. Therefore, it is necessary to provide strategies to reduce biogenic amine formation in the foods. Alongside numerous existing intervention methods, the use of *Bacillus* starter cultures capable of degrading and/or incapable of producing biogenic amines has been proposed as a guaranteed way to reduce biogenic amines in fermented soybean foods, considering that *Bacillus* species have been known as fermenting microorganisms responsible for biogenic amine formation in the foods. Molecular genetic studies of *Bacillus* genes involved in the formation and degradation of biogenic amines would be helpful in selecting starter cultures. This review summarizes the presence and control strategies of biogenic amines in fermented soybean foods.

Keywords: food safety; biogenic amines; fermented soybean foods; intervention methods; control; starter culture; *Bacillus* spp.

1. Introduction

Microbial fermentation is one of the oldest and most practical technologies used in food processing and preservation. However, fermentation of protein-rich raw materials such as fish, meat, and soybean commonly provides abundant precursor amino acids of biogenic amines. Even though most fermented foods have been found to be beneficial to human health, biogenic amines produced through fermentation and/or contamination of protein-rich raw materials by amino acid-decarboxylating microorganisms may cause intoxication symptoms in human unless they are detoxified by human intestinal amine oxidases, viz., detoxification system [1,2]. Thus, the presence of biogenic amines in fermented foods (and non-fermented foods as well) has become one of the most important food safety issues.

According to old documents, the cultivation and use of soybeans, dating back to B.C., were launched in Manchuria on the north side of the Korean Peninsula and have spread to other regions of the world. Hence, a variety of fermented soybean foods have been developed and consumed in north-east Asian countries around the Korean Peninsula, and consequently humans in this region have steadily taken the fermented foods for a long period of time from hundreds to thousands of years, depending on the types of fermented soybean foods consumed [3]. Presently, fermented soybean foods are of public interest and consumed more frequently even in western leading countries because the fermented foods, particularly fermented soybean pastes, not only have been believed by many people, but also have been scientifically proven by researchers to have health-promoting and -protective effects [4]. However, much less attention has been paid to the safety issues of fermented soybean foods [5].

Fermented soybean foods, including various types of fermented soybean pastes and soy sauces, are commonly made from whole soybeans containing abundant amino acids through microbial fermentation. If the fermenting (or sometimes contaminating) microorganisms are significantly capable of decarboxylating amino acids, the resultant fermented soybean foods may contain unignorable amounts of biogenic amines. Indeed, the presence of biogenic amines seems to be quite frequent and inevitable in fermented soybean foods. Therefore, the present review provides information on the presence, bacterial production, and control strategies of biogenic amines in fermented soybean foods, especially focusing on fermented soybean pastes usually considered as heathy foods.

2. A Brief on Biogenic Amines

Biogenic amines are defined as harmful nitrogenous compounds produced mainly by bacterial decarboxylation of amino acids in various foods. The bacterial decarboxylation of amino acids to biogenic amines have been well illustrated in literature and can be found elsewhere [6–8]. Biogenic amines are also endogenous and indispensable components of living cells, and consequently most food materials, including fruit, vegetables, and grains, contain different levels of biogenic amines depending on their variety, maturity and cultivation condition [7]. Usual intake of dietary biogenic amines generally causes no adverse reactions because human intestinal amine oxidases, such as monoamine oxidase (MAO), diamine oxidase (DAO) and polyamine oxidase (PAO), quickly metabolize and detoxify the biogenic amines. If the capacity of amine-metabolizing enzymes is over-saturated and/or the metabolic activity is impaired by specific inhibitors, vasoactive biogenic amines, including histamine, tyramine and β-phenylethylamine, may cause food intoxication and in turn be considered to be toxic substances in humans [1,2]. Furthermore, the toxicity of biogenic amines can be enhanced by putrefactive biogenic amines such as putrescine and cadaverine [9]. The most common symptoms of biogenic amine intoxication in human are nausea, respiratory distress, hot flushes, sweating, heart palpitation, headache, a bright red rash, oral burning, and hypo- or hypertension [10]. Figure 1 schematically illustrates the detoxification and toxicological risks of biogenic amines.

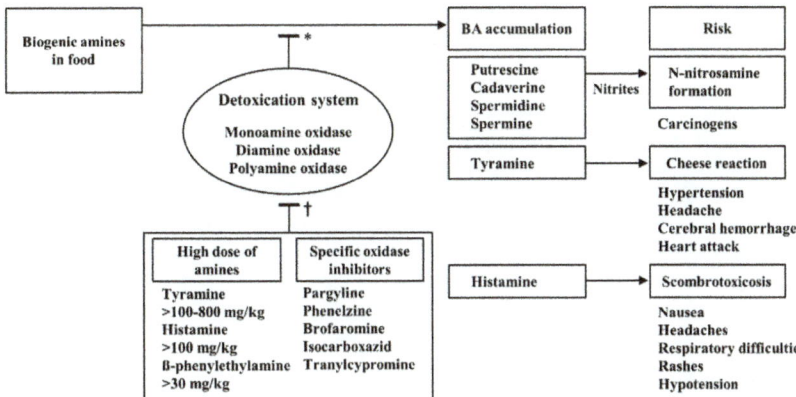

Figure 1. Detoxification and toxicological risks of biogenic amines. *: Metabolic inactivation of biogenic amines through oxidative deamination by oxidases. †: Incapacitation of intestinal detoxication system through saturation by biogenic amines or inhibition by antidepressant medications. BA: Biogenic amines.

The biosynthesis, toxicity and physiological effects have been well reviewed in recent articles [11,12], and will not be summarized here. In addition, it is worth mentioning that in particular, vulnerable people who are immune compromised, such as the elderly, children, and infants, may exhibit intolerance to even low levels of biogenic amines and suffer more severe symptoms [13]. The maximum tolerance levels of vasoactive biogenic amines (mostly histamine in fish and fish products) have established and proposed by government agencies or individual researchers as described below (refer to Table 1), but may need to be further studies and subdivided, considering the vulnerable people.

3. Legal Limits and Toxic Levels of Biogenic Amines in Foods

Early in 1980, the U.S. Food and Drug Administration (FDA) first established regulations for tuna and mahimahi that consider 200 mg histamine/kg as an indication of prior mishandling and 500 mg histamine/kg as an indication of a potential health hazard [14]. Early in 1990, the European Economic Community (EEC) also established regulation for fish species of the *Scombridae* and *Clupeidae* families and fixed a three-class plan for maximum allowable levels of histamine in fresh fish ($n = 9$; $c = 2$; $m = 100$ ppm; $M = 200$ ppm) and enzymatically ripened fish products ($n = 9$; $c = 2$; $m = 200$ ppm; $M = 400$ ppm) where n is the number of units to be analyzed from each lot, m and M are the histamine tolerances, and c is the number of units allowed to contain a histamine level higher than m but lower than M [15]. In 1996, Shalaby [16] suggested the guidelines for histamine content of fish as follows: <50 mg/kg (safe for consumption), 50–200 mg/kg (possibly toxic), 200–1000 mg/kg (probably toxic), >1000 mg/kg (toxic and unsafe for human consumption) based on the review of the regulations and other literature. In the meantime, values of 100–800 mg/kg of tyramine and 30 mg/kg of β-phenylethylamine were reported to be toxic doses in food, respectively, and 100 mg histamine per kg of food and 2 mg histamine per liter of alcoholic beverage were suggested as upper limits for human consumption [6]. The upper limits and toxic doses (stated right above) suggested by Brink et al. [6] have been steadily used by numerous investigators as threshold values to assess human health risks derived from exposure to vasoactive biogenic amines in foods because there have been no other reports describing the guidelines for respective vasoactive biogenic amines in general foods, except for histamine (particularly in fish; not applicable to other foods).

At present, histamine is the only biogenic amine for which the U.S. FDA has set a guidance level, i.e., 50 mg/kg of histamine in the edible portion of fish [17], whereas the European Commission (EC) has established regulatory limits of 100 mg/kg for histamine in fish species and 400 mg/kg for histamine in fish sauce produced by fermentation of fishery products [18]. In the meantime, the European Food Safety Authority [19] reported that although a dose of 50 mg histamine is the no-observed-adverse-effect level (NOAEL), healthy individuals do not experience symptoms unless they ingest a larger amount of histamine than NOAEL. Then, the Food and Agriculture Organization/World Health Organization [20] announced 200 mg histamine/kg as the maximum allowable level for consumption of fish and fish products. According to the Codex standard [21], 200 mg/kg of histamine in fish and fish products and 400 mg/kg of histamine in fish sauce are set as the hygiene and handling indicator levels in the corresponding products, respectively. In addition, the governments of several countries in Asia and Oceania have lately established regulatory limits for histamine in fish and fish products [22–24]. Legal limits and toxic levels set by government agencies or individual researchers for biogenic amines in food products are listed in Table 1. Although several food scientists have referred the suggestion of Brink et al. [6], as described above, there have not been government regulations on maximum allowable levels of biogenic amines, other than histamine, in history. Besides, any government legislation or guidelines on the contents of biogenic amines in fermented soybean foods are not found throughout the world.

Table 1. Legal limits and toxic levels set by agencies for biogenic amines in food products.

Agency	Food	Toxicity Classification	Biogenic Amines (mg/kg)[1]			Governing Entity	Ref.
			PHE	HIS	TYR		
Government	Fish[2] and fish products	Defect action level		50		United States	[17]
		Toxicity level		500		United States	[17]
		Maximum allowable level		200		Australia and New Zealand	[22]
		Maximum allowable level		200		Korea	[23]
		Maximum allowable level		400		China	[24]
	Fish[3] and fish products	Maximum allowable level		200		China	[24]
International organization	Fresh fish[4]	Defect action level		100		Europe	[15]
		Maximum allowable level		200		Europe	
	Enzymatically ripened fish products[4]	Defect action level		200		Europe	
		Maximum allowable level		400		Europe	
	Fish[2]	Regulatory limit		100			[18]
	Fish sauce[5]	Regulatory limit		400			[18]
	Fish[2] and fish products	Maximum allowable level		200			[20]
		Decomposition indicator		100			
		Hygiene and handling indicator		200			[21]
	Fish sauce[6]	Hygiene and handling indicator		400			
Independent research	General foods	Toxicity threshold	30	100	100–800		[6]
	Fish[2]	Safe for consumption Possibly toxic Probably toxic Toxic and unsafe for human consumption		<50 50–200 200–1000 >1000			[16]

[1] PHE: β-phenylethylamine, HIS: histamine, TYR: tyramine; [2] *Scombridae, Clupeidae, Engraulidae, Pomatomide, Scombresosidae* and other fish species well known for high histamine content; [3] fish species without high histamine content; [4] *Scombridae* and *Clupeidae* families only; [5] produced by fermentation of fishery products; [6] prepared from fresh fish.

4. Fermented Soybean Foods and Vasoactive Biogenic Amines

Fermented soybean foods have not only been commonly consumed as they are, but have also been frequently used in a variety of processed products, which make them become a necessity in the household in Asian cultures. Moreover, fermented soybean food products have recently gained popularity, crossing from Asian communities to mainstream markets, in many western countries due to the healthy functions of the foods [3,4]. Aside from soy sauces, the most popular fermented soybean foods produced mainly by bacterial fermentation (sometimes with molds) are Natto, Miso (Japanese fermented soybean pastes), Cheonggukjang, Doenjang, Gochujang (Korean fermented soybean pastes), Chunjang, Doubanjiang, Douchi (Chinese fermented soybean pastes) and Tempeh (an Indonesian fermented soybean paste). Some other soybean foods such as Sufu (a Chinese fermented Tofu) and Tauco (an Indonesian fermented yellow soybeans) prepared by mold fermentation are also available in local area (but were excluded from this review due to great differences in the microorganisms involved in fermentation processes as well as little data available in literature). The safety issues of traditional fermented soybean foods have heretofore been overlooked because humans have consistently taken the foods at least for centuries or millennia. However, considering that fermented soybean foods contain not only abundant dietary amino acid precursors of biogenic amines, as mentioned at the beginning of this article, but also significant biogenic amine-producing microorganisms, mainly bacterial species, it is critically important to assess the levels of biogenic amines in the foods.

Based on a critical review of published data (refer to Table 2) [25–37], it seems that the amounts of biogenic amines in most fermented soybean food products are usually within the safe levels for human consumption. It is noteworthy, however, that some specimens of the fermented soybean food products, including both fermented soybean pastes and soy sauces, have been reported to contain vasoactive biogenic amines greater than toxic dose of each amine. For instance, β-phenylethylamine has been detected at concentrations up to 185.6 mg/kg and 239.0 mg/kg in Doubanjiang and

Douchi, respectively [26,36], which are approximately 6-8 times higher than toxic dose of this amine (30 mg/kg) suggested by Brink et al. [6]. In another report, β-phenylethylamine was determined to be 8704.6 mg/kg in a Doenjang sample [34], but which is unreliably larger than those in other articles in which maximum β-phenylethylamine concentrations of 529.2 mg/kg and 544.0 mg/kg have been reported [28,32]; this report was thus excluded from further review. In the meantime, histamine has been detected at concentrations up to 952.0 mg/kg and 808.0 mg/kg in Doenjang and Douchi, respectively [28,36], whereas tyramine has been detected up to 1430.7 mg/kg and 2539.0 mg/kg in Doenjang and Cheonggukjang, respectively [28,33]. The maximum concentrations of histamine and tyramine reported are approximately 8–10 times higher than upper limit of histamine (100 mg/kg) and 14–25 times (on lower toxic dose basis; 2–3 times, upper dose basis) higher than toxic dose of tyramine (100–800 mg/kg), respectively, suggested by Brink et al. [6]. Like the fermented soybean pastes described above, some specimens of soy sauces have been reported to contain high levels of vasoactive biogenic amines, including β-phenylethylamine (up to maximum 121.6 mg/kg), histamine (398.8 mg/kg) and tyramine (794.3 mg/kg), which are much greater than toxic doses of respective amines [28]. As a counter-example, there is a report in which the amounts of respective vasoactive biogenic amines were very low or not detected in all samples (i.e., three batches) of commercial Natto, Miso, Tempeh, and soy sauce products; however, this report seems to insufficiently brief the presence of biogenic amines in the products because samples (batches) of only a single brand for each type of product were available in local stores [4]. The contents of biogenic amines in different types of fermented soybean food products reported in literature have been reviewed once in a book chapter in 2011 [38], and those in the representative fermented soybean food products reviewed herein are compiled in Table 2. After all, it seems likely that there may occasionally be a risk of food poisoning associated with eating fermented soybean pastes, especially when the pastes contain significant amounts of vasoactive biogenic amines, because some types of the pastes, for instance, Natto, Tempeh and sometimes Cheonggukjang, are taken not only as side dishes, but also main dishes. In the case of soy sauces, the risk to consumers may not be so great, considering the small quantity of intake per serve [29].

Table 2. Biogenic amine content in fermented soybean food products.

Fermented Soybean Products	N[1]	Biogenic Amines (mg/kg)[2]								Ref.
		TRP	PHE	PUT	CAD	HIS	TYR	SPD	SPM	
Cheonggukjang	7	6.7–236.4[3]	ND–40.8	4.7–121.3	2.1–20.2	1.3–54.3	0.7–483.1	39.6–59.2	7.1–14.7	[28]
	102	NT[4]	NT	NT	NT	ND[5]–755.40	ND–1913.51	NT	NT	[30]
	13	NT	NT	NT	NT	NT	117.5–2539.0	NT	NT	[33]
Chunjang	4	13.3–19.9	2.2–11.8	9.2–11.7	1.7–6.6	11.6–22.4	29.7–54.6	1.4–12.8	ND–2.9	[28]
	4	19.57–31.35	ND–6.79	3.26–28.59	ND–2.04	1.85–272.55	19.78–131.27	0.24–11.63	ND–1.49	[25]
Doenjang	14	6.1–234.1	ND–529.2	9.9–1453.7	0.3–65.4	1.5–952.0	3.4–1430.7	4.2–23.4	ND–10.2	[28]
	10	ND–449.8	ND–544.0	28.8–1076.6	2.7–144.1	1.4–329.2	12.5–967.6	ND–30.3	ND–9.8	[32]
	23	ND–2808.1	ND–8704.6	ND–4292.3	ND–3235.5	ND–2794.8	ND–6616.1	ND–8804.0	ND–9729.5	[34]
	7	13.5–45.9	3.3–65.0	46.7–168.2	ND–12.9	71.1–382.4	46.4–1190.7	ND–24.7	NT	[29]
Doubanjiang	7	ND–62.43	1.43–185.61	1.15–129.17	ND–0.17	ND	ND–25.75	ND–0.18	ND–1.69	[26]
Douchi	26	ND–440	ND–239	ND–596	ND–191	ND–808	ND–529	ND–719	ND–242	[36]
Gochujang	5	17.9–36.6	0.7–9.1	2.5–3.2	ND–1.1	0.6–1.3	2.1–4.9	1.6–3.4	1.4–1.8	[28]
	7	ND–8.1	1.5–24.8	10.4–36.4	ND–18.1	2.2–59.0	2.9–126.8	ND–14.5	NT	[29]
Miso	5	21.6–23.7	0.7–8.1	16.4–23.2	2.8–3.2	0.8–1.1	2.0–95.3	9.5–21.9	1.3–3.1	[28]
	40	ND–762	ND	ND–12	ND–201	ND–221	ND–49	ND	ND–216	[31]
	22	ND–9.71	2.38–11.76	2.69–14.09	ND–1.31	ND–24.42	ND–66.66	ND–28.31	ND–2.85	[27]
Natto	39	ND–301.0	ND	ND–27.0	ND–42.0	ND–457.0	ND–45.0	ND–124.0	ND–71.0	[35]
	21	ND–45.80	ND–51.50	ND–43.10	ND–36.80	ND–34.40	ND–300.20	246.50–478.10	18.80–80.10	[37]
Soy sauce	11	ND–45.8	1.5–121.6	2.5–1007.5	0.7–32.3	3.9–398.8	26.8–794.3	1.5–53.1	ND–16.1	[28]

[1] Quantity of samples examined; [2] TRP: tryptamine, PHE: β-phenylethylamine, PUT: putrescine, CAD: cadaverine, HIS: histamine, TYR: tyramine, SPD: spermidine, SPM: spermine; [3] the range from minimum to maximum (the same number of digits is used after the decimal point in the values, as was presented in the corresponding references); [4] NT: not tested; [5] ND: not detected.

It is also worth pointing out that some specimens of fermented soybean food products have been found to contain relatively high levels of putrescine and cadaverine (Table 2). The putrefactive biogenic amines have been known to enhance the toxicity of vasoactive biogenic amines in foods [9]. Therefore, comprehensive monitoring and reduction strategies are required to reduce the risk of ingesting putrefactive biogenic amines as well as vasoactive biogenic amines in fermented soybean foods, which may come from the understanding of why there are differences in the amounts and diversity of biogenic amines between the types or batches of the food products. It is probably that the differences may be attributed to (i) the ratio of ingredients used in raw material, (ii) physicochemical and/or microbial contribution, and (iii) conditions and periods of the entire food supply chain [5]. Since fermented soybean foods have their own unique raw materials, physicochemical properties, and production processes, the present review focuses on bacterial contribution to biogenic amine formation conserved across most fermented soybean foods.

5. Bacterial Activity to Produce Biogenic Amines in Fermented Soybean Foods

It has been known that most fermented soybean foods, except for several types of soybean foods prepared by mold fermentation, are mainly fermented (or contaminated) by *Bacillus* species (particularly *B. subtilis*) [5,39,40], which, in turn, leads to biogenic amine formation in the fermented foods, although the abilities of *Bacillus* strains to produce biogenic amines are diverse depending on the types and/or batches of the food products from which the strains are isolated (refer to Table 3) [25,26,31,35–37]. In the studies, the reported ranges (mean ± standard deviation; minimum–maximum) of biogenic amines produced by *Bacillus* spp. in assay media, when cultured for 24 h with proper precursor amino acids, are as follows: histamine 0.22 ± 0.65–29.9 ± 13.4 µg/mL, tyramine 0.3 ± 0.5–30.6 ± 21.7 µg/mL, β-phenylethylamine not detected (ND)—11.2 ± 9.17 µg/mL, tryptamine 0.20 ± 0.45–6.17 ± 3.98 µg/mL, putrescine ND—7.59 ± 3.06 µg/mL, cadaverine ND—1.8 ± 1.1 µg/mL, spermidine 0.40 ± 0.20–9.26 ± 5.73 µg/mL, spermine 1.29 ± 0.86–27.2 ± 12.7 µg/mL. Among the *Bacillus* strains reported, *B. subtilis* strains isolated from Natto exhibited the strongest abilities to produce respective biogenic amines. Table 3 reveals the abilities to produce biogenic amines of different bacterial species isolated from representative types of fermented soybean food products.

Table 3. Production of biogenic amines by bacteria isolated from fermented soybean food products.

Fermented Soybean Products	Isolates	N [1]	Biogenic Amines (µg/mL) [2]								Ref.
			TRP	PHE	PUT	CAD	HIS	TYR	SPD	SPM	
Chunjang	*Bacillus* spp. [3]	89	0.45 ± 0.32 [4]	0.85 ± 0.23	0.95 ± 0.55	ND [5]	1.34 ± 1.19	1.41 ± 0.32	9.26 ± 5.73	2.17 ± 1.09	[25]
Doubanjiang	*Bacillus subtilis*	18	0.20 ± 0.45	0.67 ± 1.42	3.45 ± 1.29	1.03 ± 0.46	0.22 ± 0.65	0.59 ± 0.65	0.40 ± 0.20	1.29 ± 0.86	[26]
Douchi	*Bacillus subtilis*	4	NT [6]	2.3 ± 4.5	NT	0.5 ± 0.6	18.7 ± 9.3	0.3 ± 0.5	NT	4.5 ± 5.2	[36]
	Staphylococcus pasteuri	1	NT	ND	NT	1.2	20.0	ND	NT	ND	
	Staphylococcus capitis	3	NT	5.4 ± 9.3	NT	1.1 ± 0.9	375.3 ± 197.0	1.1 ± 1.9	NT	2.7 ± 3.4	
Miso	*Staphylococcus pasteuri*	1	NT	6.4	ND	ND	28.1	NT	ND	NT	[31]
	Bacillus sp.	1	NT	2.7	1.6	2.1	15.3	NT	8.6	NT	
	Bacillus amyloliquefaciens	2	NT	1.6 ± 2.2	ND	1.8 ± 1.1	16.5 ± 8.6	NT	4.2 ± 5.9	NT	
	Bacillus subtilis	2	NT	ND	0.5 ± 0.7	0.6 ± 0.8	29.9 ± 13.4	NT	5.0 ± 7.1	NT	
	Bacillus megaterium	2	NT	7.7 ± 0.4	ND	ND	14.6 ± 2.8	NT	4.7 ± 6.6	NT	
Natto	*Bacillus subtilis*	80	6.17 ± 3.98	11.2 ± 9.17	7.59 ± 3.06	0.94 ± 1.67	9.91 ± 1.61	30.6 ± 21.7	3.34 ± 1.82	27.2 ± 12.7	[37]
	Bacillus subtilis	2	NT	2.4 ± 3.3	NT	1.5 ± 0.1	15.5 ± 2.9	NT	NT	NT	[35]
	Staphylococcus pasteuri	2	NT	ND	NT	1.1 ± 0.1	15.0 ± 1.3	NT	NT	NT	

[1] Quantity of bacterial samples examined; [2] TRP: tryptamine, PHE: β-phenylethylamine, PUT: putrescine, CAD: cadaverine, HIS: histamine, TYR: tyramine, SPD: spermidine, SPM: spermine; [3] *Bacillus* spp. were identified to be *B. subtilis* (91.0%), *B. coagulans* (4.5%), *B. licheniformis* (1.1%) and *B. firmus* (1.1%); [4] mean ± standard deviation (the same number of digits is used after the decimal point in the values, as was presented in the corresponding references); [5] ND: not detected; [6] NT: not tested.

In addition to the aforementioned *Bacillus* spp., *Lactobacillus* sp. and *Enterococcus faecium*, which had been isolated from raw materials of Miso, were proposed to produce histamine and tyramine in Miso, respectively, through a qualitative detection using BCP (Bromo-cresol purple) agar plates and subsequently a quantitative test using liquid media [41,42]. In the quantitative test with incubation for 90 days, the strains of *Lactobacillus* sp. and *E. faecium* produced histamine and tyramine up to approximately 100 μg/mL and 150 μg/mL, respectively. Although *Lactobacillus* species are not commonly involved in the preparation of fermented soybean foods, diverse species of *Lactobacillus* have also been reported to be responsible for the formation of biogenic amines, including histamine, in lactic fermented foods [12]. *E. faecium* and *E. faecalis* have been found to possess *tdc* gene and produce tyramine in fermented foods, including dairy products, fermented sausages, wine and fermented soybean foods [12]. Thus, *E. faecium* strains have been used as target organisms for studies on the reduction of tyramine in fermented soybean foods [33,43], even though *Enterococcus* spp. are present as contaminants at relatively low levels (maximum up to 10^6 CFU/g) in the foods [44–46]. In the meantime, the absence of *hdc* gene encoding histidine decarboxylase was reported in both *E. faecium* and *E. faecalis* in one study [47], while histidine decarboxylase-positive *E. faecium* and *E. faecalis* strains were detected by a PCR (polymerase chain reaction) method in another study [48]. It is interesting to note that the PCR screening method used in the latter study employed the primers developed in the former study, which makes it difficult to conclude whether the species possess *hdc* gene or not.

As shown in Table 4, at present the Gene Bank database of the National Centre for Biotechnology Information (NCBI, National Center for Biotechnology Information, U.S. National Library of Medicine, Bethesda, MD, USA) provides the sequences of *tdc*, *odc*, and *ldc* genes in *E. faecium* and *tdc* and *ldc* genes in *E. faecalis*, while *hdc* gene sequence of both species is not available in the database. In contrast, the sequences of *hdc* gene in *B. licheniformis* and *B. coagulans* (this sequence is completely conserved between the two species) and *ldc* in *B. subtilis* have been deposited in the database, while *tdc* gene sequence is unavailable for the three species of *Bacillus*. Nevertheless, it has lately been suggested that *Bacillus* spp. are as significant as *Enterococcus* spp. for tyramine formation in fermented soybean foods [49]. The deposited genes encoding amino acid decarboxylases of *Bacillus* spp. and *Enterococcus* spp., the most important species related to biogenic amine formation in fermented soybean products, are listed in Table 4 (exceptionally, *odc*-Az encodes an antizyme inhibitor devoid of ornithine decarboxylase activity). All the bacteria and genes mentioned above should be targeted for preventive interventions to reduce biogenic amine formation in fermented soybean foods. Meanwhile, yeasts have been considered to produce only negligible amounts of biogenic amines [50,51]. Fungal distribution to biogenic amine accumulation is remained to be further studied because there appears to be but little literature available dealing with fungal formation of biogenic amines [52].

It is well known that various vasoactive and putrefactive biogenic amines are commonly formed by microbial decarboxylation of amino acids in fermented foods [6,7]. As such, it has been found that soybean fermentation results in an increase in the amount of spermine (and other biogenic amines), but a decrease in that of spermidine [26]. Since spermidine is essential for the growth and development of plants [53,54], this polyamine is abundantly present in soybean and non-fermented soybean foods such as Tofu (a curd product made from soy milk) [26,55,56] and degraded by bacterial enzymes during fermentation [57]. Consequently, fermented soybean foods contain a lower level of spermidine than their raw material, soybean [26,58]. This indicates that development and application of biogenic amine-degrading starter cultures are possible (and necessary) to reduce the contents of biogenic amines in fermented soybean foods. Identifying and understanding the dominant contributors to the formation of biogenic amines may facilitate the development of starter cultures for delaying or avoiding biogenic amine formation in the fermented foods. Taken together, it is clear that distinct and diverse bacterial community and/or capability of producing (and degrading) biogenic amines decisively determine the amounts and diversity of biogenic amines in fermented soybean foods.

Table 4. Genes encoding amino acid decarboxylases in *Bacillus* spp. and *Enterococcus* spp. registered in the NCBI database.

Species	Strain [1]	Source	Gene for Amino Acid Decarboxylase [2]	No. of Amino Acids	Locus Name	Accession (Version)	Size (bp)
B. subtilis	*B. subtilis* subsp. *subtilis* strain 168	Isolated strain	*odc*-Az	331	BACYACA	L77246.1	996
			ldc	490	AF012285	AF012285.1	1473
B. licheniformis/ *B. coagulans*	*B. licheniformis* A5/ *B. coagulans* SL5	Isolated strain	*hdc*	146	AB553282 /AB553281	AB553282.1 /AB553281.1	441
E. faecium	*E. faecium* strain 993	Isolated strains	*odc*	235	PDLZ01000281	PDLZ01000281.1	707
	E. faecium ATCC 700221	ATCC	*ldc*	191	CP014449	CP014449.1	576
	E. faecium ATCC 700221	ATCC	*tdc*	611	CP014449	CP014449.1	1836
E. faecalis	*E. faecalis* ATCC 51299	ATCC	*ldc*	194	JSES01000022	JSES01000022.1	585
	E. faecalis ATCC 19433	Type strain	*tdc*	620	KB944589	KB944589.1	1863

[1] Genes found in a single strain of each *Bacillus* species have been registered, while those of *Enterococcus* spp. found in multiple strains have been separately assigned to different loci, of which a representative locus is presented in the table; [2] *odc*-Az: 37.0% identity over 119 amino acids to the *E. coli* ornithine decarboxylase antizyme, *odc*: gene for ornithine decarboxylase, *ldc*: gene for lysine decarboxylase, *hdc*: gene for histidine decarboxylase, *tdc*: gene for tyrosine decarboxylase; ATCC: the American Type Culture Collection.

6. Control Strategies for Reducing Biogenic Amines in Fermented Soybean Foods

Regarding intervention measures that reduce biogenic amine formation in fermented soybean foods, to date, only a few reports are available in literature as follows: the use of irradiation [59], addition of nicotinic acid as a tyrosine decarboxylase inhibitor [43], and use of *Bacillus* starter cultures [60–63]. However, when extended to other fermented foods, a review of the relevant literature reveals that several types of intervention methods have been developed and used to reduce biogenic amine contents in the foods (mainly fermented sausage and cheese), which involve chemical intervention, such as the use of food additives and natural antimicrobial compounds [43,64–67], physical intervention, such as the use of irradiation [59,68], high hydrostatic pressure [69,70] and modified atmosphere packaging [71,72], and biological intervention, particularly such as the use of starter cultures [60–63,73–76]. The biological intervention methods also involve the control or adjustment of intrinsic and extrinsic factors, such as alterations of temperature, pH, a_w, and Eh, which have been well reviewed in literature [77–79].

Up to this day, thousands of additives have been used to extend shelf life of foods because of their antimicrobials, antioxidants, and antibrowning properties. Natural additives have lately been of great interest in food industry due to consumers' health concerns [80]. Apart from being used as food preservatives, numerous food additives and natural antimicrobial compounds, including glycine [64], nicotinic acid [43], potassium sorbate, sodium benzoate [67], sodium chloride [64,66], clove [65–67], garlic [65], etc., have been found to be effective in suppressing bacterial ability to produce biogenic amines in foods. Among the compounds, nicotinic acid is only one compound proven to practically inhibit the formation of biogenic amines (particularly tyramine) in a fermented soybean food, viz., Cheonggukjang [43]. In the report, the addition of nicotinic acid at concentrations of 0.15% and 0.20% resulted in significant reductions, by approximately 70% and 83%, respectively, compared to the control, of tyramine content in the treated Cheonggukjang samples after 24 h of fermentation. In addition, it is worth noting that even though a successful reduction of biogenic amines in a food product can be achieved by the addition of any of a variety of compounds, some of the additives may cause organoleptic alterations, such as an atypical taste and flavor, in the final food product, especially in the case of fermented soybean foods [5,49]. Therefore, sensory evaluation should be incorporated as an integral part of a program investigating effective inhibitors of biogenic amine formation in fermented soybean foods.

Besides the chemical intervention measures described above, a variety of physical intervention processes have been developed and applied for food preservation, which involve not only well-known classical processes, for instance, heating, refrigeration, and freezing, but also emerging novel processes such as microwave heating, ohmic heating and pulsed electric fields developed during the past 25 to 35 years [81]. Among the physical intervention methods, irradiation, high hydrostatic pressure and modified atmosphere packaging have been relatively recently reported to successfully inhibit biogenic amine formation in fermented foods, which have been achieved mostly by reducing microbial population, for instance, lactic acid bacteria, closely related to the fermentation of foods [59,69–72]. Despite the technological progress that has been made, as for fermented soybean foods, there has been only a single report describing biogenic amine reduction in the food treated by one of the physical intervention processes. In the report, γ-irradiation of raw materials with doses of 5, 10, and 15 kGy significantly reduced the contents of histamine, putrescine, tryptamine and spermidine by approximately 20–50% (but not tyramine, β-phenylethylamine, cadaverine, spermine and agmatine) in the final product of a fermented soybean food, viz., probably Doenjang [59]. However, it needs here to be noted that the irradiation even with the lowest dose resulted in an immediate and significant decrease in the numbers of *Bacillus* spp. and lactic acid bacteria, known as dominant bacteria in the food, by up to about 3 log CFU/g and 2 log CFU/g, respectively. As is well known, many of the physical intervention processes prevent the growth of fermenting microorganisms, as well as of biogenic amine-producing microorganisms, which may in turn not only delay fermentation, but also lead to abnormal fermentation caused by undesirable microorganisms resistant to the treatments [82]. Thus, introducing the processes would be somewhat challenging in the case of fermented soybean foods, considering the presence of fermenting and/or beneficial bioactive microorganisms in the foods.

The use of starter cultures has been suggested to be a successful way to enhance not only the quality and safety, but also the healthy functions, of fermented foods, causing less adverse organoleptic and unhealthy alterations [83–85]. Thus, with that a variety of microorganisms have been compared and screened for the ability to degrade biogenic amines and/or inability to produce biogenic amines in fermented foods, not only at the level of genus, species, or both, but also at the level of individual strain [73–76]. As for the fermentation of soybean, *Bacillus* strains have been steadily proposed as starter cultures to improve the sensory quality, but not the safety of fermented soybean foods [86,87]. On the contrary, less attention has been given to starter cultures for preventing or reducing biogenic amine formation in fermented soybean foods. As mentioned above, *Bacillus* spp. have been known as fermenting (or contaminating) microorganisms responsible for biogenic amine formation in different types of fermented soybean foods. Therefore, it is imperative to screen proper starter cultures (particularly *Bacillus* starter culture) with no or less ability to produce biogenic amines for the production of fermented soybean foods [62]. With respect to this, there have been a few reports in literature in which Doenjang and Cheonggukjang samples prepared with *B. subtilis* and *B. licheniformis* starter cultures, respectively, with low abilities to produce biogenic amines (the data on individual strains were not presented in the reports) contained lower levels of biogenic amines than those of previous studies [61,62]. Alternatively, the use of starter cultures that can degrade biogenic amines may facilitate the reduction of biogenic amines in fermented soybean foods [77]. At present, only two reports of biogenic amine-degrading starter cultures for the production of fermented soybean foods are available in literature as described below. In one study, *B. subtilis* and *B. amyloliquefaciens* strains which had been isolated from traditionally fermented soybean products degraded significant amounts of histamine (up to 71% of its initial concentration by *B. amyloliquefaciens*), tyramine (up to 70% by *B. amyloliquefaciens*), putrescine (up to 92% by *B. subtilis*) and cadaverine (up to 93% by *B. subtilis*) in cooked soybean after 10 days of fermentation [60]. In another study, *B. subtilis* and *B. amyloliquefaciens* strains which had been isolated from commercial fermented soybean products degraded 30–40% of tyramine in a phosphate buffer and probably thereby reduced tyramine content by 40–65% in the final product of Cheonggukjang, as compared to the control [63]. In addition, *B. subtilis* and *B. idriensis* strains isolated from a traditional fermented soybean food have been reported to be not only capable of degrading of, but also incapable of producing histamine and tyramine in vitro (but not applied to practical fermentation of soybean in the study) [88]. Consequently, it is feasible to screen *Bacillus* strains capable of degrading and/or incapable (or less capable) of producing biogenic amines, which would, in turn, make it possible to use them as starter cultures for reducing biogenic amine contents in fermented soybean foods. In the meantime, it is also necessary to fully identify and characterize *Bacillus* genes involved in the formation and degradation of biogenic amines, which would be helpful not only in selecting starter culture candidates but also in providing strategies to efficiently regulate the expression of these genes encoding relevant enzymes. Such molecular genetic studies would be further needed for better understanding of mechanisms by which intervention methods influencing intrinsic and extrinsic factors and/or microbial growth inhibit biogenic amine formation, at the level of gene. It is noteworthy that in addition to the aforementioned *Bacillus* starter cultures, strains of *E. faecium* and *L. plantarum* have also been proposed as starter cultures for fermented soybean foods because of their abilities to produce bacteriocin or to degrade biogenic amines, respectively [46,89]. Considering that the species are present as contaminants at relatively low levels in fermented soybean foods, as described above, further research is required prior to practical application to the fermentation of soybean in food industry.

Aside from the use of starter cultures, the production of biogenic amines has been known to be dependent on intrinsic and extrinsic factors of foods [77–79]. Furthermore, the factors may provide combined effects, especially in connection with technology applied, viz., the chemical and physical intervention measures described above [90]. As of now, however, the alterations of temperature, pH, a_w, and Eh (as another important, but classical, biological intervention strategy), seem to be less preferable for studies on the reduction of biogenic amines in fermented soybean foods than other alternatives, considering that there are no relevant reports available, which might be because of the need to consider strict demands of consumers and governments on unique sensory properties and manufacturing processes of fermented soybean foods. Nonetheless, it is expected that the changes of the intrinsic and extrinsic factors within narrow ranges would be applicable, depending on the types of fermented soybean foods, if organoleptic evaluation is preceded. The intrinsic and extrinsic factors influencing biogenic amine formation in foods have been well reviewed in a recent article [8]. Biogenic amine reduction strategies, including chemical, physical and biological intervention methods, are summarized in Table 5.

Table 5. Biogenic amine reduction strategies for food products.

Parameter Categories		Highly Effective Strategies
Chemical intervention		Nicotinic acid [43], glycine [64], garlic [65], clove [65], clove and sodium chloride [66], clove with potassium sorbate and sodium benzoate [67]
Physical intervention		Irradiation [59,68], high hydrostatic pressure [69,70], modified atmospheric packaging and temperature [71,72]
Biological intervention	Starter cultures	Lactic acid bacteria [84], *Lactobacillus sake* + *Pediococcus pentosaceus* + *Staphylococcus carnosus* + *S. xylosus* [73], *S. carnosus* [74], *S. xylosus* [74–76], *L. plantarum* [89], *Bacillus subtilis* [60,63,88], *B. amyloliquefaciens* [60,63], *B. licheniformis* [62], *B. idriensis* [88], *B. subtilis* + *Aspergillus oryzae* + *Mucor racemosus* [61]
	Intrinsic and extrinsic factors	Temperature, pH, a_w, Eh [77–79]

7. Conclusions

The presence of histamine in fish is of concern in many countries due to its toxic potential and implications. Accordingly, there are specific legislations regarding the histamine content in fish and

Foods **2019**, *8*, 85

fish products in US, EU, and other countries. In contrast, the significance of biogenic amines in fermented soybean foods has been overlooked despite the presence not only of abundant precursor amino acids of biogenic amines in soybean, but also of microorganisms capable of producing biogenic amines during the fermentation of soybean. Fortunately, the studies published to date indicate that the amounts of biogenic amines in most fermented soybean food products are within the safe levels for human consumption. However, it should be pointed out that the contents of vasoactive biogenic amines in certain types and/or batches of fermented soybean food products are greater than toxic levels. Nonetheless, lack of both legislation and guidelines on the contents of biogenic amines in fermented soybean food products may lead to serious (or unnecessary) concerns about the safety of the fermented foods. Therefore, it is required to establish guidance levels of biogenic amines in fermented soybean food products based on information about the national daily intake of the fermented foods per person and the amounts of biogenic amines in different types of fermented soybean foods commonly consumed in each country.

Meanwhile, many efforts have been made to reduce biogenic amines in various fermented foods, particularly fermented sausage and cheese, whereas less attention has given to biogenic amines in fermented soybean foods. Consequently, there is at present a little information available regarding intervention methods to reduce biogenic amines in fermented soybean foods. Although empirical data on controlling biogenic amines in fermented soybean foods are not much in literature, several reports have suggested that the use of starter cultures capable of degrading and/or incapable of producing biogenic amines is a preferable way to biocontrol biogenic amines in fermented soybean foods because it probably causes less adverse organoleptic and unhealthy alterations as well as little changes in bacterial communities in the foods. Alterations of intrinsic and extrinsic factors, such as temperature, pH, a_w, and Eh, in fermentation and manufacturing processes are also needed to be taken into consideration when biocontrol strategy is employed. With a successful reduction of biogenic amines in addition to significant health benefits, consumers may place a much higher value on fermented soybean foods.

Author Contributions: Conceptualization, J.-H.M., and H.-J.H.; Literature data collection, J.-H.M., Y.K.P., Y.H.J., and J.-H.L.; Writing-original draft preparation, J.-H.M., Y.K.P.; Writing-review and editing, J.-H.M.; Supervision, J.-H.M.

Funding: This work was supported by the National Research Foundation of Korea (NRF) grant funded by the Korea government (MSIT) (No. 2016R1A2B4012161).

Acknowledgments: The authors thank Jae Hoan Lee and Junsu Lee of Department of Food and Biotechnology at Korea University for English editing and technical assistance, respectively.

Conflicts of Interest: The authors declare no conflict of interest.

References

1. Joosten, H.M.L.J.; Nuñez, M. Prevention of histamine formation in cheese by bacteriocin-producing lactic acid bacteria. *Appl. Environ. Microbiol.* **1996**, *62*, 1178–1181. [PubMed]
2. Taylor, S.L.; Guthertz, L.S.; Leatherwood, H.; Tillman, F.; Lieber, E.R. Histamine production by foodborne bacterial species. *J. Food Saf.* **1978**, *1*, 173–187. [CrossRef]
3. Shin, D.; Jeong, D. Korean traditional fermented soybean products: *Jang*. *J. Ethn. Foods* **2015**, *2*, 2–7. [CrossRef]
4. Toro-Funes, N.; Bosch-Fuste, J.; Latorre-Moratalla, M.L.; Veciana-Nogués, M.T.; Vidal-Carou, M.C. Biologically active amines in fermented and non-fermented commercial soybean products from the Spanish market. *Food Chem.* **2015**, *173*, 1119–1124. [CrossRef] [PubMed]
5. Mah, J.-H. Fermented soybean foods: Significance of biogenic amines. *Austin J. Nutr. Food Sci.* **2015**, *3*, 1058.
6. Brink, B.; Damirik, C.; Joosten, H.M.L.J.; Huis In't Veld, J.H.J. Occurrence and formation of biologically active amines in foods. *Int. J. Food Microbiol.* **1990**, *11*, 73–84. [CrossRef]
7. Halász, A.; Baráth, Á.; Simon-Sarkadi, L.; Holzapfel, W. Biogenic amines and their production by microorganisms in food. *Trends Food Sci. Technol.* **1994**, *5*, 42–49. [CrossRef]

8. Gardini, F.; Özogul, Y.; Suzzi, G.; Tabanelli, G.; Özogul, F. Technological factors affecting biogenic amine content in foods: A review. *Front. Microbiol.* **2016**, *7*, 1218. [CrossRef] [PubMed]

9. Stratton, J.E.; Hutkins, R.W.; Taylor, S.L. Biogenic amines in cheese and other fermented foods: A review. *J. Food Prot.* **1991**, *54*, 460–470. [CrossRef]

10. Rice, S.L.; Eitenmiller, R.R.; Koehler, P.E. Biologically active amines in foods: A review. *J. Milk Food Technol.* **1976**, *39*, 353–358. [CrossRef]

11. Fernández-Reina, A.; Urdiales, J.; Sánchez-Jiménez, F. What we know and what we need to know about aromatic and cationic biogenic amines in the gastrointestinal tract. *Foods* **2018**, *7*, 145. [CrossRef] [PubMed]

12. Barbieri, F.; Montanari, C.; Gardini, F.; Tabanelli, G. Biogenic amine production by lactic acid bacteria: A Review. *Foods* **2019**, *8*, 17. [CrossRef] [PubMed]

13. Byun, B.Y.; Bai, X.; Mah, J.-H. Bacterial contribution to histamine and other biogenic amine content in *Juk* (Korean traditional congee) cooked with seafood. *Food Sci. Biotechnol.* **2013**, *22*, 1675–1681. [CrossRef]

14. U.S. Food and Drug Administration. Defect action levels for histamine in tuna: Availability of guide. *Fed. Reg.* **1982**, *47*, 40487–40488.

15. European Economic Commission (EEC). Council Directive 91/493/EEC of 22 July 1991 laying down the health conditions for the production and the placing on the market of fishery products. *Off. J. Eur. Comm.* **1991**, *268*, 15–34.

16. Shalaby, A.R. Significance of biogenic amines to food safety and human health. *Food Res. Int.* **1996**, *29*, 675–690. [CrossRef]

17. U.S. Food and Drug Administration (FDA). *Fish and Fishery Products Hazards and Controls Guidance*, 4th ed.; Center for Food Safety and Applied Nutrition: Rockville, MD, USA, 2011.

18. European Commision (EC). Commission Regulation No. 2073/2005 of 15th November 2005 on microbiological criteria for Foodstuffs. *Off. J. Eur. Union* **2005**, *L338*, 1–25.

19. European Food Safety Authority (EFSA). Scientific opinion on risk based control of biogenic amine formation in fermented foods. *EFSA J.* **2011**, *9*, 2393–2486. [CrossRef]

20. Food and Agriculture Organization of the United Nations/World Health Organization (FAO/WHO). *Joint FAO/WHO Expert Meeting on the Public Health Risk of Histamine and Other Biogenic Amines from Fish and Fishery Products*; FAO/WHO: Rome, Italy, 2013.

21. Codex. Discussion paper histamine. In *Joint FAO/WHO Food Standards Programme, Codex Committee on Fish and Fishery Products, Thirty-Second Session (CX/FFP 12/32/14), Bali, Indonesia, 1–5 October 2012*; FAO/WHO: Rome, Italy, 2012.

22. Food Standards Australia New Zealand (FSANZ). *Imported Food Risk Statement Fish and Fish Products from the Families Specifies and Histamine*; FSANZ: Canberra, Australia, 2016.

23. The Ministry of Food and Drug Safety (MFDS). *Food Code, Notification No. 2017-57*; MFDS: Osong, Korea, 2017.

24. China National Standards. *GB 2733-2015 National Food Safety Standards for Fresh and Frozen Animal Aquatic Products*; Standards Press of China: Beijing, China, 2016.

25. Bai, X.; Byun, B.Y.; Mah, J.-H. Formation and destruction of biogenic amines in *Chunjang* (a black soybean paste) and *Jajang* (a black soybean sauce). *Food Chem.* **2013**, *141*, 1026–1031. [CrossRef] [PubMed]

26. Byun, B.Y.; Bai, X.; Mah, J.-H. Occurrence of biogenic amines in *Doubanjiang* and *Tofu*. *Food Sci. Biotechnol.* **2013**, *22*, 55–62. [CrossRef]

27. Byun, B.Y.; Mah, J.-H. Occurrence of biogenic amines in *Miso*, Japanese traditional fermented soybean paste. *J. Food Sci.* **2012**, *77*, T216–T223. [CrossRef] [PubMed]

28. Cho, T.-Y.; Han, G.-H.; Bahn, K.-N.; Son, Y.-W.; Jang, M.-R.; Lee, C.-H.; Kim, S.-H.; Kim, D.-B.; Kim, S.-B. Evaluation of biogenic amines in Korean commercial fermented foods. *Korean J. Food Sci. Technol.* **2006**, *38*, 730–737.

29. Kim, T.-K.; Lee, J.-I.; Kim, J.-H.; Mah, J.-H.; Hwang, H.-J.; Kim, Y.-W. Comparison of ELISA and HPLC methods for the determination of biogenic amines in commercial *Doenjang* and *Gochujang*. *Food Sci. Biotechnol.* **2011**, *20*, 1747–1750. [CrossRef]

30. Ko, Y.-J.; Son, Y.-H.; Kim, E.-J.; Seol, H.-G.; Lee, G.-R.; Kim, D.-H.; Ryu, C.-H. Quality properties of commercial *Chungkukjang* in Korea. *J. Agric. Life Sci.* **2012**, *46*, 1–11.

31. Kung, H.-F.; Tsai, Y.-H.; Wei, C.-I. Histamine and other biogenic amines and histamine-forming bacteria in miso products. *Food Chem.* **2007**, *101*, 351–356. [CrossRef]

32. Lee, H.T.; Kim, J.H.; Lee, S.S. Analysis of microbiological contamination and biogenic amines content in traditional and commercial Doenjang. *J. Food Hyg. Saf.* **2009**, *24*, 102–109.

33. Oh, S.-J.; Mah, J.-H.; Kim, J.-H.; Kim, Y.-W.; Hwang, H.-J. Reduction of tyramine by addition of *Schizandra chinensis* Baillon in Cheonggukjang. *J. Med. Food* **2012**, *15*, 1109–1115. [CrossRef] [PubMed]

34. Shukla, S.; Park, H.-K.; Kim, J.-K.; Kim, M. Determination of biogenic amines in Korean traditional fermented soybean paste (Doenjang). *Food Chem. Toxicol.* **2010**, *48*, 1191–1195. [CrossRef] [PubMed]

35. Tsai, Y.-H.; Chang, S.-C.; Kung, H.-F. Histamine contents and histamine-forming bacteria in natto products in Taiwan. *Food Control* **2007**, *28*, 1026–1030. [CrossRef]

36. Tsai, Y.-H.; Kung, H.-F.; Chang, S.-C.; Lee, T.-M.; Wei, C.-I. Histamine formation by histamine-forming bacteria in douchi, a Chinese traditional fermented soybean product. *Food Chem.* **2007**, *103*, 1305–1311. [CrossRef]

37. Kim, B.; Byun, B.Y.; Mah, J.-H. Biogenic amine formation and bacterial contribution in *Natto* products. *Food Chem.* **2012**, *135*, 2005–2011. [CrossRef] [PubMed]

38. Shukla, S.; Kim, J.-K.; Kim, M. Occurrence of biogenic amines in soybean food products. In *Soybean and Health*; El-Shemy, H., Ed.; InTech: Rijeka, Croatia, 2011; pp. 182–184.

39. Onda, T.; Yanagida, F.; Shinohara, T.; Yokotsuka, K. Time series analysis of aerobic bacterial flora during Miso fermentation. *Lett. Appl. Microbiol.* **2003**, *37*, 162–168. [CrossRef] [PubMed]

40. Tamang, J.P. Diversity of fermented foods. In *Fermented Foods and Beverages of the World*; Tamang, J.P., Kailasapathy, K., Eds.; CRC Press: Boca Raton, FL, USA, 2010; pp. 41–84.

41. Ibe, A.; Nishima, T.; Kasai, N. Formation of tyramine and histamine during soybean paste (Miso) fermentation. *Jpn. J. Toxicol. Environ. Health* **1992**, *38*, 181–187. [CrossRef]

42. Ibe, A.; Nishima, T.; Kasai, N. Bacteriological properties of and amine-production conditions for tyramine-and histamine-producing bacterial strains isolated from soybean paste (Miso) starting materials. *Jpn. J. Toxicol. Environ. Health* **1992**, *38*, 403–409. [CrossRef]

43. Kang, H.-R.; Kim, H.-S.; Mah, J.-H.; Kim, Y.-W.; Hwang, H.-J. Tyramine reduction by tyrosine decarboxylase inhibitor in *Enterococcus faecium* for tyramine controlled *cheonggukjang*. *Food Sci. Biotechnol.* **2018**, *27*, 87–93. [CrossRef] [PubMed]

44. Oh, E.J.; Oh, M.-H.; Lee, J.M.; Cho, M.S.; Oh, S.S. Characterization of microorganisms in *Eoyukjang*. *Korean J. Food Sci. Technol.* **2008**, *40*, 656–660.

45. Sarkar, P.; Tamang, J.P.; Cook, P.E.; Owens, J. Kinema-a traditional soybean fermented food: Proximate composition and microflora. *Food Microbiol.* **1994**, *11*, 47–55. [CrossRef]

46. Yoon, M.Y.; Kim, Y.J.; Hwang, H.-J. Properties and safety aspects of *Enterococcus faecium* strains isolated from *Chungkukjang*, a fermented soy product. *LWT-Food Sci. Technol.* **2008**, *41*, 925–933. [CrossRef]

47. Coton, E.; Coton, M. Multiplex PCR for colony direct detection of Gram-positive histamine-and tyramine-producing bacteria. *J. Microbiol. Methods* **2005**, *63*, 296–304. [CrossRef] [PubMed]

48. Komprda, T.; Sládková, P.; Petirová, E.; Dohnal, V.; Burdychová, R. Tyrosine-and histidine-decarboxylase positive lactic acid bacteria and enterococci in dry fermented sausages. *Meat sci.* **2010**, *86*, 870–877. [CrossRef] [PubMed]

49. Jeon, A.R.; Lee, J.H.; Mah, J.-H. Biogenic amine formation and bacterial contribution in *Cheonggukjang*, a Korean traditional fermented soybean food. *LWT-Food Sci. Technol.* **2018**, *92*, 282–289. [CrossRef]

50. Caruso, M.; Fiore, C.; Contursi, M.; Salzano, G.; Paparella, A.; Romano, P. Formation of biogenic amines as criteria for the selection of wine yeasts. *World J. Microbiol. Biotechnol.* **2002**, *18*, 159–163. [CrossRef]

51. Izquierdo-Pulido, M.; Font-Fábregas, J.; Vidal-Carou, C. Influence of *Saccharomyces cerevisiae* var. *uvarum* on histamine and tyramine formation during beer fermentation. *Food Chem.* **1995**, *54*, 51–54.

52. Nout, M.K.R.; Ruiker, M.M.W.; Bouwmeester, H.M.; Beljaars, P.R. Effect of processing conditions on the formation of biogenic amines and ethyl carbamate in soybean tempe. *J. Food Saf.* **1993**, *33*, 293–303. [CrossRef]

53. Fuell, C.; Elliott, K.A.; Hanfrey, C.C.; Franceschetti, M.; Michael, A.J. Polyamine biosynthetic diversity in plants and algae. *Plant Physiol. Biochem.* **2010**, *48*, 513–520. [CrossRef] [PubMed]

54. Kumar, A.; Taylor, M.A.; Arif, S.A.M.; Davies, H.V. Potato plants expressing antisense and sense S-adenosylmethionine decarboxylase (SAMDC) transgenes show altered levels of polyamines and ethylene: Antisense plants display abnormal phenotypes. *Plant J.* **1996**, *9*, 147–158. [CrossRef]

55. Kalaè, P.; Krausová, P. A review of dietary polyamines: Formation, implications for growth, and health and occurrence in foods. *Food Chem.* **2005**, *90*, 219–230.

56. Liu, Z.-F.; Wei, Y.-X.; Zhang, J.-J.; Liu, D.-H.; Hu, Y.-Q.; Ye, X.-Q. Changes in biogenic amines during the conventional production of stinky tofu. *Int. J. Food Sci. Technol.* **2011**, *46*, 687–694. [CrossRef]

57. Woolridge, D.P.; Vazquez-Laslop, N.; Markham, P.N.; Chevalier, M.S.; Gerner, E.W.; Neyfakh, A.A. Efflux of the natural polyamine spermidine facilitated by the *Bacillus subtilis* multidrug transporter Blt. *J. Biol. Chem.* **1997**, *272*, 8864–8866. [CrossRef] [PubMed]

58. Righetti, L.; Tassoni, A.; Bagni, N. Polyamines content in plant derived food: A comparison between soybean and Jerusalem artichoke. *Food Chem.* **2008**, *111*, 852–856. [CrossRef]

59. Kim, J.-H.; Ahn, H.-J.; Kim, D.-H.; Jo, C.; Yook, H.-S.; Park, H.-J.; Byun, M.-W. Irradiation effects on biogenic amines in Korean fermented soybean paste during fermentation. *J. Food Sci.* **2003**, *68*, 80–84. [CrossRef]

60. Kim, Y.S.; Cho, S.H.; Jeong, D.Y.; Uhm, T.-B. Isolation of biogenic amines-degrading strains of *Bacillus subtilis* and *Bacillus amyloliquefaciens* from traditionally fermented soybean products. *Korean J. Microbiol.* **2012**, *48*, 220–224. [CrossRef]

61. Shukla, S.; Lee, J.S.; Park, H.-K.; Yoo, J.-A.; Hong, S.-Y.; Kim, J.-K.; Kim, M. Effect of novel starter culture on reduction of biogenic amines, quality improvement, and sensory properties of *Doenjang*, a traditional Korean soybean fermented sauce variety. *J. Food Sci.* **2015**, *80*, M1794–M1803. [CrossRef] [PubMed]

62. Kim, S.-Y.; Kim, H.-E.; Kim, Y.-S. The potentials of *Bacillus licheniformis* strains for inhibition of *B. cereus* growth and reduction of biogenic amines in *cheonggukjang* (Korean fermented unsalted soybean paste). *Food Control* **2017**, *79*, 87–93.

63. Kang, H.-R.; Lee, Y.-L.; Hwang, H.-J. Potential for application as a starter culture of tyramine-reducing strain. *J. Korean Soc. Food Sci. Nutr.* **2017**, *46*, 1561–1567. [CrossRef]

64. Mah, J.-H.; Hwang, H.-J. Effects of food additives on biogenic amine formation in *Myeolchi-jeot*, a salted and fermented anchovy (*Engraulis japonicus*). *Food Chem.* **2009**, *114*, 168–173. [CrossRef]

65. Mah, J.-H.; Kim, Y.J.; Hwang, H.-J. Inhibitory effects of garlic and other spices on biogenic amine production in *Myeolchi-jeot*, Korean salted and fermented anchovy product. *Food Control* **2009**, *20*, 449–454. [CrossRef]

66. Wendakoon, C.N.; Sakaguchi, M. Combined effect of sodium chloride and clove on growth and biogenic amine formation of *Enterobacter aerogenes* in mackerel muscle extract. *J. Food Prot.* **1993**, *56*, 410–413. [CrossRef]

67. Wendakoon, C.N.; Sakaguchi, M. Combined effects of cloves with potassium sorbate and sodium benzoate on the growth and biogenic amine production of *Enterobacter aerogenes*. *Biosci. Biotechnol. Biochem.* **1993**, *57*, 678–679. [CrossRef]

68. Rabie, M.A.; Siliha, H.; el-Saidy, S.; el-Badawy, A.A.; Malcata, F.X. Effects of γ-irradiation upon biogenic amine formation in Egyptian ripened sausages during storage. *Innov. Food Sci. Emerg. Technol.* **2010**, *11*, 661–665. [CrossRef]

69. Novella-Rodriguez, S.; Veciana-Nogués, M.T.; Saldo, J.; Vidal-Carou, M.C. Effects of high hydrostatic pressure treatments on biogenic amine contents in goat cheeses during ripening. *J. Agric. Food Chem.* **2002**, *50*, 7288–7292. [CrossRef] [PubMed]

70. Ruiz-Capillas, C.; Colmenero, F.J.; Carrascosa, A.V.; Muñoz, R. Biogenic amine production in Spanish dry-cured "chorizo" sausage treated with high-pressure and kept in chilled storage. *Meat Sci.* **2007**, *77*, 365–371. [CrossRef] [PubMed]

71. Patsias, A.; Chouliara, I.; Paleologos, E.K.; Savvaidis, I.; Kontominas, M.G. Relation of biogenic amines to microbial and sensory changes of precooked chicken meat stored aerobically and under modified atmosphere packaging at 4 °C. *Eur. Food Res. Technol.* **2006**, *223*, 683–689. [CrossRef]

72. Ruiz-Capillas, C.; Pintado, T.; Jiménez-Colmenero, F. Biogenic amine formation in refrigerated fresh sausage "chorizo" keeps in modified atmosphere. *J. Food Biochem.* **2012**, *36*, 449–457. [CrossRef]

73. Ayhan, K.; Kolsarici, N.; Özkan, G.A. The effects of a starter culture on the formation of biogenic amines in Turkish soudjoucks. *Meat Sci.* **1999**, *53*, 183–188. [CrossRef]

74. Bover-Cid, S.; Izquierdo-Pulido, M.; Vidal-Carou, M.C. Effect of proteolytic starter cultures of *Staphylococcus* spp. on biogenic amine formation during the ripening of dry fermented sausages. *Int. J. Food Microbiol.* **1999**, *46*, 95–104. [CrossRef]

75. Gardini, F.; Martuscelli, M.; Crudele, M.A.; Pararella, A.; Suzzi, G. Use of *Staphylococcus xylosus* as a starter culture in dried sausages: Effect on the biogenic amine content. *Meat Sci.* **2002**, *61*, 275–283. [CrossRef]

76. Mah, J.-H.; Hwang, H.-J. Inhibition of biogenic amine formation in a salted and fermented anchovy by *Staphylococcus xylosus* as a protective culture. *Food Control* **2009**, *20*, 796–801. [CrossRef]

77. Alvarez, M.A.; Moreno-Arribas, M.V. The problem of biogenic amines in fermented foods and the use of potential biogenic amine-degrading microorganisms as a solution. *Trends Food Sci. Technol.* **2014**, *39*, 146–155. [CrossRef]

78. Naila, A.; Flint, S.; Fletcher, G.; Bremer, P.; Meerdink, G. Control of biogenic amines in food-existing and emerging approaches. *J. Food Sci.* **2010**, *75*, R139–R150. [CrossRef] [PubMed]

79. Stadnik, J. Significance of biogenic amines in fermented foods and methods of their control. In *Beneficial Microbes in Fermented and Functional Foods*; Rai, V.R., Bai, J.A., Eds.; CRC Press: Boca Raton, FL, USA, 2014; pp. 149–163.

80. Carocho, M.; Barreiro, M.F.; Morales, P.; Ferreira, I.C. Adding molecules to food, pros and cons: A review on synthetic and natural food additives. *Compr. Rev. Food Sci. Food Saf.* **2014**, *13*, 377–399. [CrossRef]

81. Floros, J.D.; Newsome, R.; Fisher, W.; Barbosa-Cánovas, G.V.; Chen, H.; Dunne, C.P.; German, J.B.; Hall, R.L.; Heldman, D.R.; Karwe, M.V.; et al. Feeding the world today and tomorrow: The importance of food science and technology. *Compr. Rev. Food Sci. Food Saf.* **2010**, *9*, 572–599. [CrossRef]

82. Fang, S.-H.; Lai, Y.-J.; Chou, C.-C. The susceptibility of *Streptococcus thermophilus* 14085 to organic acid, simulated gastric juice, bile salt and disinfectant as influenced by cold shock treatment. *Food Microbiol.* **2013**, *33*, 55–60. [CrossRef] [PubMed]

83. Hansen, E.B. Commercial bacterial starter cultures for fermented foods of the future. *Int. J. Food Microbiol.* **2002**, *78*, 119–131. [CrossRef]

84. Leroy, F.; De Vuyst, L. Lactic acid bacteria as functional starter cultures for the food fermentation industry. *Trends Food Sci. Technol.* **2004**, *15*, 67–78. [CrossRef]

85. Leroy, F.; Verluyten, J.; De Vuyst, L. Functional meat starter cultures for improved sausage fermentation. *Int. J. Food Microbiol.* **2006**, *106*, 270–285. [CrossRef] [PubMed]

86. Omafuvbe, B.O.; Abiose, S.H.; Shonukan, O.O. Fermentation of soybean (*Glycine max*) for soy-*daddawa* production by starter cultures of *Bacillus*. *Food Microbiol.* **2002**, *19*, 561–566. [CrossRef]

87. Tamang, J.P.; Nikkuni, S. Selection of starter cultures for the production of kinema, a fermented soybean food of the Himalaya. *World J. Microbiol. Biotechnol.* **1996**, *12*, 629–635. [CrossRef] [PubMed]

88. Eom, J.S.; Seo, B.Y.; Choi, H.S. Biogenic amine degradation by *Bacillus* species isolated from traditional fermented soybean food and detection of decarboxylase-related genes. *J. Microbiol. Biotechnol.* **2015**, *25*, 1519–1527. [CrossRef] [PubMed]

89. Yi-Chen, L.; Hsien-Feng, K.; Ya-Ling, H.; Chien-Hui, W.; Yu-Ru, H.; Yung-Hsiang, T. Reduction of biogenic amines during miso fermentation by *Lactobacillus plantarum* as a starter culture. *J. Food Prot.* **2016**, *79*, 1556–1561.

90. Ruiz-Capillas, C.; Herrero, A.M. Impact of biogenic amines on food quality and safety. *Foods* **2019**, *8*, 62. [CrossRef] [PubMed]

foods

MDPI

Article

Quality Assessment of Fresh Meat from Several Species Based on Free Amino Acid and Biogenic Amine Contents during Chilled Storage

Mehdi Triki [1,*], Ana M. Herrero [2], Francisco Jiménez-Colmenero [2] and Claudia Ruiz-Capillas [2,*]

[1]   Ministry of Public Health, P.O. Box 42, Doha, Qatar
[2]   Department of Products, Institute of Food Science, Technology and Nutrition, ICTAN-CSIC,
     Ciudad Universitaria, 28040 Madrid, Spain; ana.herrero@ictan.csic.es (A.M.H.);
     fjimenez@ictan.csic.es (F.J.-C.)
*   Correspondence: mrmehditriki@gmail.com (M.T.); claudia@ictan.csic.es (C.R.-C.); Tel.: +974-44-07-0226 (M.T.);
    +34-91-549-23-00 (C.R.-C.); Fax: +974-44-07-0824 (M.T.); +34-91-549-36-27 (C.R.-C.)

Received: 19 July 2018; Accepted: 23 August 2018; Published: 25 August 2018

Abstract: This paper studies the changes that occur in free amino acid and biogenic amine contents of raw meats (beef, pork, lamb, chicken and turkey) during storage (2 °C, 10 days). The meat cuts samples were harvested from a retail outlet (without getting information on the animals involved) as the following: Beef leg (four muscles), pork leg (five muscles), lamb leg (seven muscles), turkey leg (four muscles), and chicken breast (one muscle). Meat composition varied according to meat types. In general, pH, microbiology counts, biogenic amine (BA), and free amino acid (FAA) contents were also affected by meat types and storage time ($p < 0.05$). Chicken and turkey presented the highest levels ($p < 0.05$) of FAAs. Total free amino acids (TFAA) were higher ($p < 0.05$) in white meats than in red ones. The behavior pattern, of the total free amino acids precursors (TFAAP) of Bas, was saw-toothed, mainly in chicken and turkey meat during storage, which limits their use as quality indexes. Spermidine and spermine contents were initially different among the meats. Putrescine was the most prevalent BA ($p < 0.05$) irrespective of species. In general, chicken and turkey contained the highest ($p < 0.05$) levels of BAs, and TFAAP of BAs. In terms of the biogenic amine index (BAI), the quality of chicken was the worst while beef meat was the only sample whose quality remained acceptable through the study. This BAI seems to be more suitable as a quality index for white meat freshness than for red meat, especially for beef.

Keywords: meat species; free amino acid; biogenic amines; quality index

1. Introduction

Meat and meat products constitute an important protein group of foods, that can be consumed directly or as products after undergoing different processes. Consumers nowadays are asking for safe and high quality meat products. This quality is influenced by various factors, and complex interactions between the biological traits of the live animal, including mainly the biological processes that occur postmortem as muscles conversion to meat, processing and storage phases, etc. [1,2]. These meat and meat products, especially when they are fresh, undergo spoilage even during refrigerated storage [3]. This deterioration is associated with major proteolysis and microbial growth. Due to proteolysis, peptides, dipeptides, and free amino acids (FAA) are formed and used by microorganisms for their growth. Following these processes, different compounds, including biogenic amines (BA), are formed by amino acid decarboxylase action from microbial origin (Figure 1). The biogenic amine content depends on a number of interrelated factors such as the raw material (meat composition, pH, handling and hygienic conditions, etc.), additives (salt, sugar, nitrites), etc. These factors affect free amino acid availability, microbiological aspects (bacterial species and strain, bacterial growth, etc.), technical processing of the meat or meat products (e.g., steaks, roasts and hams, and ground, restructured, comminuted, fresh, cooked, smoked, and fermented meats, etc.), and storage conditions (time/temperature, packaging, temperature abuse, etc.). The combined action of all these factors will determine the final biogenic amine profile and concentrations by directly or indirectly determining substrate and enzyme presence and activity.

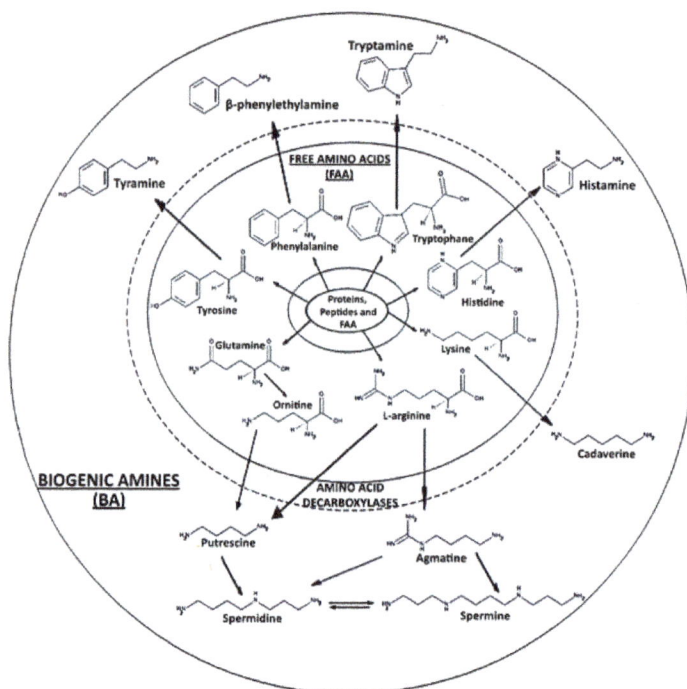

Figure 1. Biogenic amine formation from free amino acids.

Therefore, biogenic amines are of particular concern in food hygiene. Indeed, they have been used as quality indices, mainly in fish and meat, under different processing and storage conditions, whether considered individually or in combined forms [4–9]. In this regard, Hernández-Jover et al. [10] suggested a biogenic amine index (BAI) as a sum of tyramine, histamine, putrescine and cadaverine, with four-scale classification intended for cooked meat products which were based on the sum of total BA concentration in the BAI. Tyramine is also widely used individually as a quality indicator for vacuum-packed beef and cooked ham; this is also the case for spermidine and spermine [11]. Other authors as Vinci & Antonelli [12] proposed the use of cadaverine and tyramine concentrations to assess beef and chicken deterioration during storage. Moreover, some BAs were studied for their potential toxicity for consumers, especially tyramine, histamine, cadaverine and putrescine [5,13,14]. Based on these considerations, regulations have been introduced to limit BA intake levels in various kinds of food [15,16].

Consequently, the determination of biogenic amines and free amino acid fractions can provide useful information for the industry regarding freshness or spoilage and sanitary quality of fresh muscle that could be consumed directly or used as raw material for meat products preparation. High concentrations of certain amines in food may be interpreted as a consequence of poor quality of the raw materials used, contamination, or inappropriate conditions during food processing and storage.

In this regard, numerous studies were carried out in order to understand how FAA and biogenic amine formation is associated with microorganisms development in various meat products, mainly in fermented ones [5,7,8,17,18], and rarely in fresh meat [12]. On the other hand, "already-formed BAs" in the raw meat materials cannot be destroyed by thermal action during meat product processing or cooking, and this can lead to higher amine levels at the end of the pool [5,9]. However, some authors reported that BA formation not only depends on the conservation method used (refrigeration, protective atmosphere, etc.) or the type of processing (fermented, cooked, fresh, etc.), but it also depends on the type of raw material or animal species studied. Delgado-Pando et al. [19] and Triki et al. [20,21] observed differences in BA levels in reformulated (pork) frankfurters and fresh (beef) "merguez" sausages. Thus, FAA and BA production and microbial growth in fresh meat is of great interest for understanding and controlling the influence of raw meat as a factor in the final quality of fresh meat products (hamburger, fresh sausages, etc.). This approach might improve the safety as well as hygiene aspects of raw meat whether when they are used in the preparation for other meat products or employed for direct consumption or during the chilled storage of such foodstuffs. The aim of this study is, then, to assess the changes that take place in free amino acid and biogenic amine contents during chilled storage of fresh meats from some of the most frequently consumed species (beef, pork, lamb, chicken and turkey), which are used in the preparation of meat products as raw material.

2. Material and Methods

2.1. Fresh Meat Samples

Approximately 4 kg of commercial cuts of each type of fresh lean muscle meat from five species were purchased from a local supermarket. Leg cuts were taken from the four species: Beef (*Rectus femoris* M., *Semitendinosus* M., *Flexor digitorum longus* M., *Gastrocnemius* M.), pork (*Biceps femoris* M., *Semimembranosus* M., *Semitendinosus* M., *Gracilis* M., *Adductor* M.), turkey (*Flexor perforans* M., *Gastrocnemius pars external* M., *Gastrocnemius pars internal* M., *Fiburalis longus* M.), and lamb (*Quadriceps femoris* M., *Biceps femoris* M., *Semimembranosus* M., *Gluteus medius* M., *Gastrocnemius* M., *Adductor* M., *Semitendinosus* M.). Breast cuts were taken from chicken (*Pectoralis Major*). Two hundred to two hundred and fifty grams of each type of meat cut were representative of the pieces. Then meat cuts were placed on expanded polystyrene (EPS) trays (Type 89 white SPT-Linpac Packaging Pravia, S.A., Pravia, Spain) and covered with oxygen-permeable cling film (LINPAC Plastics, Pontivy, France) in aerobic conditions. From the 15 trays of each meat type that were kept in chilled storage (0 to 4 °C), three were taken periodically for further analysis. First was taken the sample for the microbiological analysis and then the sample was homogenized for the other analysis (protein content, pH, FAA and BA). Samples were assessed at 0, 3, 6, and 10 days of chilled storage.

2.2. Protein Content and pH Determination

Protein content was measured in quadruplicate with a LECO FP-2000 Nitrogen Analyzer (Leco Corporation, St. Joseph, MI, USA). For pH determination, 10 g homogenate samples in 100 mL of distilled water were prepared using a pH meter (827 pH Lab Methrom, Herisau, Switzerland). Both analyses followed the methodology used by Triki et al. [20]. Three measurements were performed per sample.

2.3. Microbiological Analysis

Ten grams of each representative sample were taken and placed in a sterile plastic bag with 90 mL of peptone water (0.1%) with 0.85% NaCl. After 2 min in a stomacher blender (Stomacher Colworth 400, Seward, UK), appropriate decimal dilutions were pour-plated (1 mL) on the following media: Plate Count Agar (PCA) (Merck, Darmstadt, Germany) for the total viable count (TVC) (30 °C for 72 h); De Man, Rogosa, Sharpe Agar (MRS) (Merck, Darmstadt, Germany) for lactic acid bacteria (LAB) (30 °C for 3–5 days); and Violet Red Bile Glucose Agar (VRBG) (Merck, Darmstadt, Germany) for

Enterobacteriaceae (37 °C for 24 h). All microbial counts were converted to logarithms of colony-forming units per gram (Log cfu/g), following the methodology used by Triki et al. [20].

2.4. Determination of Free Amino Acids (FAA)

Free amino acids extracts were prepared with 5 g of fresh meat samples from each species. They were homogenized with 10 mL of perchloric acid 6% (w/v) (to extract the FAA and precipitate the proteins and peptides) in an Ultraturrax homogenizer (IKA-Werke, Janke, & Kunkel, Staufen, Germany) then centrifuged at 27,000× g (Sorvall RTB6000B, DuPont, Wilmington, DE, USA) for 10 min at 4 °C. 2 mL of KOH 1M were added to the centrifugation tube and the whole was centrifuged again. Afterwards, the supernatant was filtered through a Millipore filter (45 μm) (Millipore, Ireland) and put into vials until use [6].

Free amino acids (FAA) were determined by cation-exchange chromatography, using a Biochron 20 automatic amino acid analyser (Amersham Pharmacia LKB, Biotech Biocom, Uppsala, Sweden) with an Ultropac high-resolution cation-exchange resin column (9 ± 0.5 μm particle size, Pharmacia, Biotech) 200 × 4.6 mm. Amino acids were determined and measured using a ninhydrin derivative reagent at 570 nm, while proline was measured at 440 nm. It should be noted that the derivatization used did not allow the determination of tryptophan. Results are means of at least three determinations.

Total FAA precursors (TFAAP) of BA was calculated by summing the levels of the following FAA: Tyrosine + Phenylalanine + Histidine + Lysine + Arginine

Total FAA (TFAA) was calculated by summing the levels of the following FAAs: Aspartic acid + Threonine+ Serine + Glutamic acid + Glycine + β alanine + Cysteine + Valine + Methionine + Isoleucine + Leucine + TFAAP

2.5. Determination of Biogenic Amines (BA)

Tyramine, phenylethylamine, histamine, putrescine, cadaverine, tryptamine, agmatine, spermidine, and spermine were determined using an acid based extraction prepared with trichloroacetic acid (7.5%) in fresh meat samples from each specie. They were analyzed in a HPLC model 1022 with a Pickering PCX 3100 post-column system (Pickering Laboratories, Mountain View, Ca, USA) following the methodology of Triki et al. [22] by ion-exchange chromatography. Briefly, 15 g of/from each sample were mixed with 30 mL of 7.5% trichloroacetic acid in an omnimixer (Omni Internacional, Waterbury, CT, USA) (20,000 rpm, 3 min) and centrifuged at 5000 g for 15 min at 4 °C in a desktop centrifuge (Sorvall RTB6000B, DuPont, Wilmington, DE, USA) (for proteins and peptides precipitation and BA extraction in the supernatant). The supernatants were filtered through a Whatman No. 1 filter, passed back through a 0.22 μm Nylon filter (Millipore, Ireland), and then placed in opaque vials in the auto-sampler of the HPLC. The results are averages of at least 3 determinations.

Biogenic amine index (BAI) was calculated by summing tyramine, histamine, putrescine and cadaverine levels in the different meat types according to Hernández-Jover et al. [10]. When one of the BA involved was not detected (ND), its value was considered as being 0. BAI < 5 mg/kg means good meat quality; between 5–20 mg/kg means acceptable meat quality; between 20–50 mg/kg means poor meat quality; and BAI > 50 mg/kg means spoiled meat.

2.6. Statistical Analysis

A One-way ANOVA analysis of variance was performed in order to evaluate the statistical significance ($p < 0.05$) of the meat type effect. Analysis of the main effect of each independent variable and any interaction between them were carried out with two-way ANOVA, which was performed as a function of meat type and storage days, using SPSS Statistics general linear model (GLM) procedure (v.14, SPSS Inc.; Chicago, IL, USA). The types of meat and storage time were assigned as fixed effects and the replication (samples were taken from meat types of different animals) was considered as a random effect. Least squares differences were used for comparison between the mean values among meat types and Tukey's HSD (honestly significant difference) test was used to identify significant differences ($p < 0.05$) between sample type and storage time. The error terms used throughout this study are standard deviations (SD). For the presented tables throughout the study: Different superscript letters in the same row indicate significant difference ($p < 0.05$) between storage days for the same meat type. Different superscript numbers in the same column indicate significant difference ($p < 0.05$) between meat types for the same storage day.

3. Results and Discussion

3.1. Protein Content and pH

Protein contents of meat (pork, 21.30 ± 0.16; lamb 19.32 ± 0.39; beef, 21.78 ± 0.25; turkey 20.02 ± 0.21; chicken 25.55 ± 0.36) were within the normal ranges for each species and type of meat cut [23]. Differences between protein amounts were due to the nature of species and cuts. As reported previously, protein is an important precursor of various compounds involved in BA formation.

The initial pH levels were significantly higher in pork (6.70) followed by turkey, chicken, lamb and beef (Table 1). This variance could be due to post mortem metabolism difference between species. As a matter of fact, both intrinsic (species, animal age, type of muscle and position of the muscle, concentration of glycogen etc.) and extrinsic (pre-slaughter stress, slaughter conditions, post-slaughter handling and temperature) factors can affect the extent of post-mortem glycolysis, and consequently the ultimate pH [1,24].

Table 1. pH values of the fresh different meat types during chilled storage at 2 °C.

Meat Type	Chilling Storage at 2 °C			
	Day 0	Day 3	Day 6	Day 10
Pork leg (5 muscles)	6.70 ± 0.02 [c5]	5.89 ± 0.02 [a2]	6.26 ± 0.03 [b3]	6.32 ± 0.01 [b3]
Lamb leg (7 muscles)	5.90 ± 0.01 [b2]	6.05 ± 0.01 [c3]	5.77 ± 0.01 [a2]	6.01 ± 0.03 [c1]
Turkey leg (4 muscles)	6.54 ± 0.01 [a4]	6.89 ± 0.01 [b5]	6.63 ± 0.02 [a4]	7.34 ± 0.01 [c5]
Chicken breast (1 muscle)	6.39 ± 0.26 [b3]	6.30 ± 0.02 [a, b4]	6.26 ± 0.01 [a3]	6.67 ± 0.00 [c4]
Beef leg (4 muscles)	5.71 ± 0.01 [a1]	5.71 ± 0.10 [a1]	5.52 ± 0.02 [a1]	6.20 ± 0.01 [b2]

Each value is the mean of three replicates per meat sample and storage day ± standard deviation (SD). Different superscript letters in the same row indicate significant difference ($p < 0.05$) between storage days for the same meat type. Different superscript numbers in the same column indicate significant difference ($p < 0.05$) between meat types for the same storage day.

Initial pH levels for fresh beef and lamb meats were considered normal in comparison with reported levels in the literature [25,26]. However, they were high in pork, turkey and chicken compared with other studies [27–29]. These initial values were related to the kind of the cut for each species and its quality according to the rigor's resolution. The pH increased in all meat types during storage, except for pork, and at the end of the storage they were between 6.01 and 7.34 (Table 1). These results agree with previous reports in different types of meats, also during refrigerated storage [30]. The increases were associated with the production of nitrogenized basic compounds, mainly aminic, which are the main results of microbial spoilage and are conditioned by the type of packaging. On the other hand, some studies reported lower pH values particularly for pork, turkey and chicken [31].

3.2. Microbiology

Microbial growth (TVC, LAB and *Enterobaceteriaceae*) was affected by meat type and storage time ($p < 0.05$) (Table 2). At the beginning of the experiment, the highest level of TVC was observed in lamb (5.64 Log cfu/g) and the lowest in chicken (4.13 Log cfu/g). Similar behavior was observed in LAB counts. Initial levels of *Enterobacteriaceae* were the lowest in pork (2.95 Log cfu/g). Generally, similar amounts were also reported by other authors for meat and meat products based on the different meat types [17,20,21,32,33].

Table 2. Microbial counts Log (cfu/g) of the fresh different meat types during chilled storage at 2 °C.

Microorganisms	Meat Type	Chilling Storage at 2 °C			
		Day 0	Day 3	Day 6	Day 10
Total Viable Count (TVC)	Pork leg (5 muscles)	5.53 ± 0.12 [a4]	7.79 ± 0.05 [b3]	9.74 ± 0.01 [c4]	10.04 ± 0.01 [d1, 2]
	Lamb leg (7 muscles)	5.64 ± 0.25 [a4]	7.54 ± 0.40 [b2]	9.99 ± 0.17 [c5]	10.04 ± 0.02 [c2]
	Turkey leg (4 muscles)	4.98 ± 0.01 [a3]	6.45 ± 0.13 [b1]	8.96 ± 0.05 [c1]	9.91 ± 0.09 [d1]
	Chicken breast (1 muscle)	4.13 ± 0.17 [a1]	7.79 ± 0.03 [b3]	9.55 ± 0.02 [c3]	10.04 ± 0.05 [d1, 2]
	Beef leg (4 muscles)	4.74 ± 0.10 [a2]	6.55 ± 0.09 [b1]	9.25 ± 0.04 [c2]	10.22 ± 0.03 [d2]
Lactic acid bacteria (LAB)	Pork leg (5 muscles)	3.76 ± 0.07 [a3]	5.64 ± 0.01 [b3]	7.57 ± 0.03 [c5]	8.34 ± 0.08 [d2]
	Lamb leg (7 muscles)	4.24 ± 0.01 [a4]	5.70 ± 0.05 [b3]	6.05 ± 0.11 [c3]	8.26 ± 0.09 [d2]
	Turkey leg (4 muscles)	4.65 ± 0.03 [a5]	5.61 ± 0.10 [b3]	7.40 ± 0.02 [c4]	8.08 ± 0.13 [d1]
	Chicken breast (1 muscle)	2.96 ± 0.02 [a1]	5.38 ± 0.06 [b2]	6.26 ± 0.02 [b1]	8.99 ± 0.09 [c1]
	Beef leg (4 muscles)	3.20 ± 0.14 [a2]	4.34 ± 0.06 [b1]	5.89 ± 0.01 [c2]	8.04 ± 0.06 [d1]
Enterobacteriaceae	Pork leg (5 muscles)	2.95 ± 0.00 [a1]	4.95 ± 0.04 [b1]	6.11 ± 0.05 [c1]	7.53 ± 0.10 [d1]
	Lamb leg (7 muscles)	4.71 ± 0.00 [a4]	6.55 ± 0.05 [b3, 4]	6.37 ± 0.02 [b3]	7.99 ± 0.04 [c2]
	Turkey leg (4 muscles)	4.24 ± 0.02 [a3]	5.88 ± 0.06 [b2]	6.79 ± 0.02 [c4]	7.68 ± 0.03 [d4]
	Chicken breast (1 muscle)	4.07 ± 0.76 [a2, 3]	6.73 ± 0.03 [b4]	6.72 ± 0.04 [b2]	7.27 ± 0.09 [c1]
	Beef leg (4 muscles)	3.92 ± 0.03 [a2]	6.42 ± 0.12 [b3]	6.62 ± 0.01 [c2]	7.66 ± 0.05 [d3]

Each value is the mean of three replicates per meat sample and storage day ± standard deviation (SD). For every type of microorganism: Different superscript letters in the same row indicate significant difference ($p < 0.05$) between storage days for the same meat type, and different superscript numbers in the same column indicate significant difference ($p < 0.05$) between meat types for the same storage day.

Microbial growth (TVC, LAB, *Enterobacteriaceae*) increased by 1 and 2 logarithmic units during refrigerated storage (Table 2). After three days, all meat samples registered 6 Log cfu/g of TVC. These levels reached more than 8.9 Log cfu/g on day six and up to one more Log cfu/g unit until day 10. Other authors reported similar microbial behavior during chilled storage of fresh meat [17,20,21]. TVC values are commonly associated with meat spoilage when it reaches levels higher than 6 Log cfu/g [34]. The high levels reported during the experiment were in relation of the high pH levels of the samples (Table 1).

Microorganisms levels in raw meat are influenced by many factors which directly affect meat quality such as animal stress susceptibility, pre- and post-slaughter handling, processing, transport, packaging, storage, composition, etc. The experiment could have been finalized at microorganism levels that limit meat consumption, but samples were analyzed for a longer period in order to better understand BA and FAA formation and the relationship between microbial counts, FAA, and BA levels.

3.3. Free Amino Acids (FAA)

During storage, free amino acid levels and their behavior varied considerably depending on the type of meat (Table 3 and Figure 2).

Initially, the most abundant FAA in all meat samples was β-alanine, with levels of 18.70–37.64 mg/100 g, followed by the glutamic acid (4.42–28.95 mg/100 g) (Table 3). High levels of glycine, threonine and aspartic acid were also detected at the beginning of the storage. The lowest amounts were observed in cysteine, while histidine was not detected in any sample. Serine was highly present in turkey and chicken (37.63 and 27.67 mg/100 g, respectively) but it is not present in pork as much as in the aforementioned meat types (5.18 mg/100 g). However, it was detected neither in lamb nor beef until the sixth and tenth days of storage respectively (Table 3).

In general, chicken and turkey presented the highest levels of total free amino acids (TFAA) and TFAAP of Bas, while the lowest levels were observed in lamb which are followed by beef and pork (Figure 2a,b). These differences seem to be related to the type of meat (poultry or mammals). Indeed, white meats (chicken and turkey) registered approximately twice the amounts of TFAAs (234.30 and 201.33 mg/100 g respectively) as lamb (86.56 mg/100 g) and three times more than pork (77.50 mg/100 g) or beef (67.5 mg/100 g) at the beginning of storage (Figure 2a). These results are consistent with reports by the USDA [35] and other authors such as Leggio et al. [18], who reported levels between 47.5 and 93.07 mg/kg of TFAA of industrial "sopressata" pork sausage. Even though higher levels have been reported in chicken [36], low concentrations were found by Cowieson et al. [37] and Rabie et al. [7] in chicken, and in other meats (horse, beef, and turkey).

These differences could be due to many variables possibly influencing the formation and destruction of FAAs in the various meat types. These include factors such as the intrinsic properties of the product (biological factors relating to species, breed, sex, etc.; physiological aspects—genetic background, stress responses, etc.; and production practice—feeding, finishing weight, age at slaughter, etc.), handling, and processing conditions, etc. [5,38].

Table 3. Free amino acids (FAA) concentration (mg/100 g) of the fresh different meat types during chilled storage at 2 °C.

	FAA	Meat Type	Chilling Storage at 2 °C			
			Day 0	Day 3	Day 6	Day 10
FAA no precursors of BA	Aspartic acid	Pork leg (5 muscles)	2.96 ± 0.04 [a2]	6.82 ± 0.86 [c3]	9.46 ± 2.59 [d2]	4.87 ± 0.02 [b1]
		Lamb leg (7 muscles)	1.72 ± 0.00 [a1]	1.80 ± 0.31 [a1]	4.00 ± 0.12 [b1]	5.98 ± 0.04 [c2]
		Turkey leg (4 muscles)	7.95 ± 0.41 [a3]	13.61 ± 0.32 [b4]	20.68 ± 0.02 [d3]	15.92 ± 0.39 [c3]
		Chicken breast (1 muscle)	15.38 ± 0.83 [b3]	23.87 ± 0.16 [c5]	29.18 ± 0.29 [d4]	4.57 ± 0.11 [a1]
		Beef leg (4 muscles)	2.37 ± 1.00 [a1,2]	5.15 ± 0.18 [b2]	3.99 ± 0.00 [b1]	4.96 ± 0.11 [b1]
	Threonine	Pork leg (5 muscles)	2.21 ± 0.04 [a2]	4.23 ± 0.23 [b3]	11.23 ± 2.71 [c5]	12.30 ± 0.04 [c2]
		Lamb leg (7 muscles)	2.67 ± 0.04 [a b2]	2.10 ± 0.02 [a1]	4.08 ± 0.05 [b c2]	4.97 ± 0.19 [c1]
		Turkey leg (4 muscles)	8.30 ± 0.80 [a3]	11.06 ± 0.48 [b4]	12.40 ± 0.18 [b3]	24.41 ± 0.14 [c3]
		Chicken breast (1 muscle)	13.71 ± 0.59 [a4]	21.17 ± 0.34 [b5]	30.82 ± 0.23 [c4]	12.51 ± 0.26 [a2]
		Beef leg (4 muscles)	1.40 ± 0.04 [a1]	2.58 ± 0.07 [b2]	2.75 ± 0.07 [b1]	5.10 ± 0.10 [c1]
	Serine	Pork leg (5 muscles)	5.18 ± 0.45 [a1]	9.12 ± 0.74 [b1]	10.51 ± 2.41 [c2]	8.45 ± 0.06 [b3]
		Lamb leg (7 muscles)	ND	ND	5.84 ± 0.12 [b1]	4.36 ± 0.00 [a1]
		Turkey leg (4 muscles)	37.63 ± 1.04 [c5]	41.59 ± 0.32 [d2]	31.69 ± 0.17 [b3]	15.26 ± 0.19 [a4]
		Chicken breast (1 muscle)	27.67 ± 1.42 [b2]	41.41 ± 0.52 [d2]	36.99 ± 0.73 [c4]	8.85 ± 0.28 [a3]
		Beef leg (4 muscles)	ND	ND	ND	7.40 ± 0.04 [a2]
	Glutamic acid	Pork leg (5 muscles)	8.09 ± 0.96 [a2]	16.78 ± 0.94 [b3]	48.67 ± 1.93 [d3]	28.97 ± 0.19 [c1]
		Lamb leg (7 muscles)	5.94 ± 0.14 [a1]	11.05 ± 0.26 [b2]	19.09 ± 0.39 [c2]	27.66 ± 1.73 [d1]
		Turkey leg (4 muscles)	28.95 ± 1.66 [a4]	50.35 ± 0.09 [b5]	55.64 ± 0.04 [b4]	87.58 ± 0.19 [c5]
		Chicken breast (1 muscle)	21.07 ± 0.88 [a3]	36.66 ± 0.54 [c4]	51.58 ± 0.73 [d3]	30.52 ± 0.66 [b1]
		Beef leg (4 muscles)	4.42 ± 0.38 [a1]	8.81 ± 0.08 [b1]	15.06 ± 0.30 [c1]	65.41 ± 0.14 [d2]
	Glycine	Pork leg (5 muscles)	6.17 ± 0.37 [a2]	12.65 ± 0.77 [b c2]	13.71 ± 3.14 [c2]	11.23 ± 0.06 [b2]
		Lamb leg (7 muscles)	18.10 ± 0.02 [a4]	19.89 ± 0.06 [b3]	17.88 ± 0.33 [a3]	17.69 ± 0.42 [a3]
		Turkey leg (4 muscles)	18.24 ± 0.99 [a4]	20.88 ± 0.22 [b3,4]	18.62 ± 0.18 [a3]	28.28 ± 0.01 [c4]
		Chicken breast (1 muscle)	16.76 ± 0.75 [b3]	21.93 ± 0.44 [c4]	27.41 ± 0.45 [d4]	11.74 ± 0.21 [a2]
		Beef leg (4 muscles)	3.14 ± 0.07 [a1]	8.23 ± 0.08 [c1]	6.73 ± 0.18 [b1]	7.04 ± 0.03 [b1]
	β-alanine	Pork leg (5 muscles)	18.70 ± 1.06 [a1]	29.68 ± 1.51 [b1]	38.51 ± 8.70 [c2]	27.69 ± 0.03 [b1]
		Lamb leg (7 muscles)	30.83 ± 0.00 [a3]	36.65 ± 0.09 [c2]	32.65 ± 0.68 [b1]	41.50 ± 0.87 [d2]
		Turkey leg (4 muscles)	37.64 ± 2.14 [a4]	42.34 ± 0.10 [b3]	41.90 ± 0.45 [b2]	48.33 ± 0.11 [c4]
		Chicken breast (1 muscle)	22.68 ± 0.86 [a2]	38.59 ± 0.67 [c2]	52.54 ± 0.87 [d3]	28.75 ± 0.72 [b1]
		Beef leg (4 muscles)	29.09 ± 0.20 [a3]	47.69 ± 0.34 [c4]	41.59 ± 0.77 [b2]	42.99 ± 0.30 [b2]
	Cysteine	Pork leg (5 muscles)	0.53 ± 0.12 [a2]	0.92 ± 0.04 [b2]	1.61 ± 0.27 [c3]	1.71 ± 0.03 [c3]
		Lamb leg (7 muscles)	0.20 ± 0.04 [a1]	1.05 ± 0.09 [b2]	1.17 ± 0.00 [b2]	1.20 ± 0.04 [b1]
		Turkey leg (4 muscles)	0.33 ± 0.00 [a1]	0.39 ± 0.00 [a1]	0.42 ± 0.04 [a1]	2.01 ± 0.09 [b4]
		Chicken breast (1 muscle)	0.22 ± 0.00 [a1]	0.33 ± 0.00 [a1]	3.95 ± 0.04 [c4]	1.74 ± 0.09 [b3]
		Beef leg (4 muscles)	0.98 ± 0.04 [a3]	0.98 ± 0.12 [a2]	1.06 ± 0.00 [a2]	1.42 ± 0.03 [b2]
	Valine	Pork leg (5 muscles)	5.77 ± 0.04 [a2]	8.17 ± 0.45 [b3]	13.16 ± 2.74 [c3]	16.25 ± 0.06 [d3]
		Lamb leg (7 muscles)	2.59 ± 0.08 [a1]	4.64 ± 0.08 [b1]	7.58 ± 0.14 [c1]	7.09 ± 0.11 [c1]
		Turkey leg (4 muscles)	8.29 ± 0.37 [a3]	11.28 ± 0.43 [b4]	11.49 ± 0.18 [b2]	23.19 ± 0.14 [c4]
		Chicken breast (1 muscle)	12.97 ± 0.47 [a4]	20.28 ± 0.34 [c5]	23.45 ± 0.18 [d4]	16.67 ± 0.61 [b3]
		Beef leg (4 muscles)	5.30 ± 0.08 [a2]	6.23 ± 0.56 [b2]	8.26 ± 0.16 [c1]	9.37 ± 0.04 [d2]
	Methionine	Pork leg (5 muscles)	2.67 ± 0.29 [a2]	3.17 ± 0.10 [a2]	9.31 ± 1.94 [b3]	12.30 ± 0.04 [c3]
		Lamb leg (7 muscles)	1.10 ± 0.00 [a1]	1.44 ± 0.30 [a1]	3.97 ± 0.01 [b1]	4.12 ± 0.01 [b1]
		Turkey leg (4 muscles)	3.19 ± 0.07 [a2]	4.84 ± 0.34 [b3]	5.00 ± 0.04 [b2]	16.79 ± 0.01 [c4]
		Chicken breast (1 muscle)	7.10 ± 0.30 [a3]	11.58 ± 0.27 [b4]	14.63 ± 0.21 [c4]	12.34 ± 0.29 [b3]
		Beef leg (4 muscles)	1.31 ± 0.00 [a1]	2.20 ± 0.68 [a1,2]	4.09 ± 0.05 [b1]	6.62 ± 0.08 [c2]
	Isoleucine	Pork leg (5 muscles)	2.98 ± 0.12 [a3]	4.36 ± 0.25 [b2]	8.28 ± 1.74 [c2]	9.88 ± 0.08 [d3]
		Lamb leg (7 muscles)	1.73 ± 0.13 [a1]	2.40 ± 0.10 [b1]	3.92 ± 0.08 [c1]	4.05 ± 0.00 [c1]
		Turkey leg (4 muscles)	5.25 ± 0.21 [a4]	7.24 ± 0.18 [b3]	6.97 ± 0.08 [b2]	16.33 ± 0.00 [c4]
		Chicken breast (1 muscle)	8.16 ± 0.35 [a5]	13.98 ± 0.23 [c4]	15.75 ± 0.40 [d3]	10.08 ± 0.20 [b3]
		Beef leg (4 muscles)	2.21 ± 0.04 [a2]	3.02 ± 0.51 [a b1]	3.90 ± 0.05 [b1]	5.88 ± 0.07 [c2]
	Leucine	Pork leg (5 muscles)	5.35 ± 0.08 [a2]	6.88 ± 0.38 [b3]	14.18 ± 2.92 [c3]	16.65 ± 0.06 [d3]
		Lamb leg (7 muscles)	3.09 ± 0.34 [a1]	4.20 ± 0.14 [b1]	7.05 ± 0.19 [c1]	7.22 ± 0.04 [c1]
		Turkey leg (4 muscles)	7.66 ± 0.40 [a3]	11.28 ± 0.10 [b4]	11.48 ± 0.07 [b2]	26.19 ± 0.06 [c4]
		Chicken breast (1 muscle)	14.29 ± 0.44 [a4]	24.10 ± 0.46 [c5]	26.76 ± 0.46 [d4]	16.86 ± 0.60 [b3]
		Beef leg (4 muscles)	3.79 ± 0.04 [a1]	5.65 ± 0.54 [b2]	6.99 ± 0.05 [c1]	9.99 ± 0.09 [d2]

Table 3. *Cont.*

	FAA	Meat Type	Chilling Storage at 2 °C			
			Day 0	Day 3	Day 6	Day 10
FAA precursors of BA	Tyrosine	Pork leg (5 muscles)	3.82 ± 0.24 [a3]	5.02 ± 0.35 [b3]	7.90 ± 1.79 [c3]	8.94 ± 0.27 [c3]
		Lamb leg (7 muscles)	2.13 ± 0.06 [a2]	2.36 ± 0.04 [b1]	2.62 ± 0.17 [c1]	3.09 ± 0.47 [c1]
		Turkey leg (4 muscles)	6.21 ± 0.34 [a4]	8.75 ± 0.13 [b4]	8.20 ± 0.41 [b3]	9.24 ± 0.07 [a3]
		Chicken breast (1 muscle)	11.23 ± 0.54 [b5]	18.14 ± 0.43 [d5]	15.80 ± 0.20 [c4]	17.17 ± 0.27 [c4]
		Beef leg (4 muscles)	1.63 ± 0.18 [a1]	2.96 ± 0.17 [b2]	5.22 ± 0.18 [c2]	6.04 ± 0.56 [d2]
	Phenylalanine	Pork leg (5 muscles)	3.94 ± 0.33 [a3]	4.39 ± 0.37 [a3]	14.82 ± 3.31 [b3]	17.41 ± 0.22 [b3]
		Lamb leg (7 muscles)	2.36 ± 0.22 [a2]	2.50 ± 0.12 [a1]	4.86 ± 0.63 [b1]	5.82 ± 0.59 [b1]
		Turkey leg (4 muscles)	4.10 ± 0.46 [a3]	6.87 ± 0.28 [b4]	6.45 ± 0.15 [b2]	19.92 ± 0.05 [c4]
		Chicken breast (1 muscle)	8.24 ± 0.12 [a4]	13.70 ± 0.35 [b5]	15.74 ± 0.34 [c3]	18.41 ± 0.78 [d3]
		Beef leg (4 muscles)	1.72 ± 0.27 [a1]	3.54 ± 0.06 [b2]	4.72 ± 0.22 [c1]	13.18 ± 0.00 [d2]
	Histidine	Pork leg (5 muscles)	ND	3.26 ± 0.16 [a1]	3.85 ± 0.91 [a1]	5.00 ± 0.03 [b1]
		Lamb leg (7 muscles)	ND	ND	ND	ND
		Turkey leg (4 muscles)	ND	ND	ND	ND
		Chicken breast (1 muscle)	ND	ND	ND	5.62 ± 0.78 [a1]
		Beef leg (4 muscles)	ND	ND	ND	ND
	Lysine	Pork leg (5 muscles)	4.96 ± 0.29 [a1]	9.92 ± 056 [b1]	24.46 ± 5.63 [c3]	25.84 ± 0.04 [c3]
		Lamb leg (7 muscles)	7.69 ± 0.00 [a3]	9.64 ± 0.06 [b1]	12.69 ± 0.16 [c1]	12.76 ± 0.24 [c1]
		Turkey leg (4 muscles)	15.09 ± 0.92 [a4]	20.66 ± 0.17 [b3]	26.19 ± 0.26 [c3]	60.33 ± 0.04 [d5]
		Chicken breast (1 muscle)	32.38 ± 1.74 [b5]	37.50 ± 0.73 [c4]	69.68 ± 0.89 [d4]	27.13 ± 0.52 [a4]
		Beef leg (4 muscles)	5.91 ± 0.05 [a2]	12.09 ± 0.17 [b2]	13.58 ± 0.16 [c2]	18.56 ± 1.63 [d2]
	Arginine	Pork leg (5 muscles)	4.19 ± 0.40 [b1]	7.56 ± 0.58 [c1]	0.12 ± 0.01 [a2]	ND
		Lamb leg (7 muscles)	6.39 ± 0.28 [b2]	7.06 ± 0.12 [c1]	0.08 ± 0.01 [a1]	ND
		Turkey leg (4 muscles)	12.49 ± 0.89 [b3]	16.17 ± 0.13 [c2]	1.45 ± 0.23 [a3,4]	0.72 ± 0.04 [d5]
		Chicken breast (1 muscle)	22.44 ± 1.21 [b4]	26.32 ± 0.62 [c3]	1.41 ± 0.06 [a3]	ND
		Beef leg (4 muscles)	4.27 ± 0.11 [b1]	6.74 ± 0.33 [c1]	1.57±0.06 [a4]	ND
TFAAP of BA		Pork leg (5 muscles)	16.91 ± 1.25 [a2]	30.15 ± 0.55 [b3]	51.15 ± 11.76 [c3]	57.18 ± 0.04 [c3]
		Lamb leg (7 muscles)	18.58 ± 0.56 [a2]	21.56 ± 0.02 [c1]	20.25 ± 0.85 [b1]	20.68 ± 0.12 [b1]
		Turkey leg (4 muscles)	37.89 ± 2.61 [a3]	52.46 ± 0.70 [c4]	42.29 ± 0.52 [b3]	90.21 ± 0.27 [d5]
		Chicken breast (1 muscle)	74.29 ± 3.60 [b4]	95.66 ± 2.12 [c5]	102.63 ± 1.48 [d4]	62.71 ± 1.37 [a4]
		Beef leg (4 muscles)	13.53 ± 0.61 [a1]	25.33 ± 0.27 [b2]	25.09 ± 0.62 [b2]	37.79 ± 1.07 [c2]
TFAA		Pork leg (5 muscles)	77.50 ± 4.35 [a2]	132.94 ± 6.63 [b3]	229.78 ± 42.86 [c3]	207.48 ± 0.23 [c3]
		Lamb leg (7 muscles)	86.56 ± 0.10 [a3]	106.78 ± 0.98 [b1]	127.48 ± 2.46 [c2]	147.51 ± 3.50 [d1]
		Turkey leg (4 muscles)	201.33 ± 10.70 [a4]	267.32 ± 3.10 [c4]	257.94 ± 1.62 [b3]	394.50 ± 0.36 [d5]
		Chicken breast (1 muscle)	234.30 ± 10.49 [b5]	349.56 ± 6.08 [c5]	415.68 ± 6.08 [d4]	217.34 ± 5.40 [a4]
		Beef leg (4 muscles)	67.54 ± 0.94 [a1]	117.96 ± 2.37 [b2]	119.51 ± 2.12 [b1]	203.98 ± 0.26 [c2]

ND: Not Detected. Each value is the mean of three replicates per meat sample and storage day ± standard deviation (SD). For every type of free amino acid: Different superscript letters in the same row indicate significant difference ($p < 0.05$) between storage days for the same meat type, and different superscript numbers in the same column indicate significant difference ($p < 0.05$) between meat types for the same storage day.

In general, the relative initial differences in TFAAs and TFAA precursors of BAs in chicken and turkey were maintained until the end of storage. Contents were at their highest ($p < 0.05$) in these species, although they were smaller in chicken than in turkey at the end of the experiment following a decrease after day six (Figure 2a,b). Glutamic acid was the most prevalent FAA at the end of the storage (reaching 87.58 mg/100 g in turkey and 65.41 mg/100 g in beef), followed by β-alanine (48.33 mg/100 g in turkey and 41.50 mg/100 g in lamb), and lysine (27.13 mg/100 g in chicken and 60.33 mg/100 g in turkey), which is a precursor of cadaverine. Serine registered a considerable decrease in all samples while valine, methionine and isoleucine increased from the initial levels (Table 3). The concentration of histidine, a precursor of histamine, was very low or beneath the threshold of detection throughout the study since it is not typically present in meat, but it is rather one of the characteristics of fish products [6]. On the other hand, Arginine was the most prevalent FAA precursor at the beginning of the storage and went undetected at its end. This decrease in arginine was due to agmatine and putrescine formation, which can also lead to spermidine and spermine production since the formation of these three amines is interrelated [13,39].

Some meat types (mainly chicken and turkey) presented a saw-toothed pattern for TFAA and TFAAP of BA over storage (Figure 2a,b). The saw-toothed pattern of the FAAs observed during the experiment is typical of the one reported in myosystems such as meat [35] and in various research studies on fish and seafood [40]. This pattern is related to both formation and destruction of FAAs [6,41], which are associated with meat proteolysis (breakdown of proteins into small peptides and free amino acids) during storage. This hydrolysis of the peptide bonds may be of endogenous (endogenous proteolytic enzymes as exopeptidases) or exogenous origin. The latter origins are associated with microbial activity and the transformation of FAAs into other compounds through chemical and metabolic reactions. Other authors [36] also reported significant increases of amino acid levels in chicken during refrigerated storage that were associated with proteolysis. The saw-toothed pattern could limit the use of FAAs as reliable quality indexes for fresh refrigerated meat, as reported elsewhere [6].

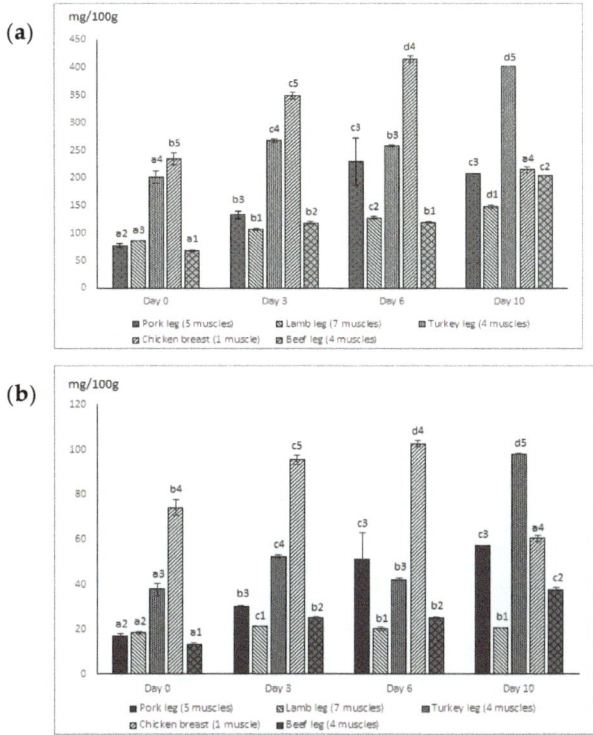

Figure 2. Total free amino acid (TFAA) (a) and Total free amino acid precursors (TFAAP) (b) of biogenic amines in mg/100 g of the different meat types during chilled storage at 2 °C. (Each value is the mean of three replicates per meat sample and storage day ± standard deviation (SD). Different letters indicate significant difference ($p < 0.05$) between storage days for the same meat type and different numbers indicate significant difference ($p < 0.05$) between meat types for the same storage day).

Nevertheless, the presence of certain amino acids such as glutamic acid, β-alanine, and phenylalanine were associated with typical flavors in meat and myosystems [42]. They are also very important as potential flavor and odor precursors through their interactions during heating, which contributes to the flavor and/or odor of cooked meat [43]. Indeed, glutamic acid and glycine are used as flavor enhancing additives in the food industry [44,45].

3.4. Biogenic Amines

Biogenic amine contents of the different meat types were affected ($p < 0.05$) by storage (Table 4). Except for the physiological amines spermidine and spermine (Spd and Spm), the initial levels of BAs were very low, and in some cases they were not detected. Spermidine and spermine presented the highest ($p < 0.05$) concentrations of BAs at the beginning of the experiment in all meat types, with spermine as the most abundant one (27.60–45.03 mg/kg). Levels of Spd and Spm were significantly lower in pork and beef as opposed to the rest of the meats, with the highest levels detected in chicken.

The reported amounts of these amines in the literature are wide-ranging. Similar values of Spd and Spm were reported in meat products formulated with pork and beef [5,7,12,20–22,39], in raw turkey [46], lamb, and sheep liver [47]. On the other hand, Rokka et al. [48] reported higher levels of Spm and Spd in chicken in comparison with our results, while other authors observed smaller amounts for Spm [12,32]. In addition, significantly lower levels of the physiological amines were observed in lamb and mutton as in our study [47].

Table 4. Biogenic amines (BA) concentration (mg/kg) of fresh different meat types during chilled storage at 2 °C.

BA	Meat Type	Chilling Storage at 2 °C			
		Day 0	Day 3	Day 6	Day 10
Tyramine	Pork leg (5 muscles)	0.67 ± 0.03 [a3]	1.10 ± 0.02 [b3]	11.20 ± 0.04 [c4]	16.58 ± 0.04 [d4]
	Lamb leg (7 muscles)	0.10 ± 0.00 [a1]	0.19 ± 0.01 [a1]	7.05 ± 0.25 [b3]	10.71 ± 0.01 [c3]
	Turkey leg (4 muscles)	ND	0.40 ± 0.04 [a2]	1.72 ± 0.02 [b2]	6.88 ± 0.14 [c2]
	Chicken breast (1 muscle)	ND	0.47 ± 0.07 [a2]	27.54 ± 0.86 [b5]	35.16 ± 0.36 [c5]
	Beef leg (4 muscles)	0.34 ± 0.04 [a2]	0.42 ± 0.02 [a2]	0.53 ± 0.03 [b1]	1.57 ± 0.07 [c1]
Histamine	Pork leg (5 muscles)	ND	ND	ND	ND
	Lamb leg (7 muscles)	ND	ND	ND	ND
	Turkey leg (4 muscles)	ND	ND	ND	ND
	Chicken breast (1 muscle)	0.53 ± 0.03 [a1]	1.23 ± 0.03 [b2]	1.73 ± 0.03 [c2]	2.11 ± 0.01 [d2]
	Beef leg (4 muscles)	ND	0.10 ± 0.00 [a1]	0.21 ± 0.01 [b1]	0.50 ± 0.02 [c1]
Phenylethylamine	Pork leg (5 muscles)	ND	0.77 ± 0.03 [a1]	1.28 ± 0.02 [b1]	1.66 ± 0.06 [c1]
	Lamb leg (7 muscles)	0.76 ± 0.00 [a3]	4.80 ± 0.08 [b3]	7.57 ± 0.27 [c3]	9.05 ± 0.15 [d3]
	Turkey leg (4 muscles)	0.21 ± 0.05 [a1]	15.07 ± 0.01 [d4]	12.85 ± 0.47 [c4]	11.33 ± 0.29 [b4]
	Chicken breast (1 muscle)	ND	16.87 ± 0.13 [c5]	17.99 ± 0.15 [b5]	12.81 ± 0.05 [a5]
	Beef leg (4 muscles)	0.47 ± 0.01 [a2]	2.33 ± 0.05 [b2]	2.47 ± 0.05 [b2]	2.62 ± 0.02 [c2]
Putrescine	Pork leg (5 muscles)	0.57 ± 0.09 [a1]	0.72 ± 0.02 [a1]	5.10 ± 0.48 [b2]	14.55 ± 0.09 [c3]
	Lamb leg (7 muscles)	1.19 ± 0.07 [a2]	3.22 ± 0.00 [b4]	6.40 ± 0.24 [c3]	10.11 ± 0.01 [d2]
	Turkey leg (4 muscles)	1.23 ± 0.03 [a2]	4.70 ± 0.08 [b5]	8.44 ± 0.44 [c5]	68.72 ± 0.02 [d5]
	Chicken breast (1 muscle)	1.23 ± 0.03 [a2]	1.83 ± 0.01 [b2]	7.47 ± 0.07 [c4]	51.99 ± 0.29 [d4]
	Beef leg (4 muscles)	1.34 ± 0.02 [a3]	2.07 ± 0.03 [b3]	3.99 ± 0.03 [c1]	7.40 ± 0.04 [d1]
Cadaverine	Pork leg (5 muscles)	ND	ND	1.11 ± 0.07 [a1]	16.16 ± 0.28 [b4]
	Lamb leg (7 muscles)	ND	ND	3.42 ± 0.02 [a2]	5.08 ± 0.12 [b1]
	Turkey leg (4 muscles)	ND	ND	1.27 ± 0.03 [a1]	13.25 ± 0.27 [b2]
	Chicken breast (1 muscle)	ND	ND	3.98 ± 0.10 [a3]	14.31 ± 0.11 [b3]
	Beef leg (4 muscles)	ND	ND	ND	ND
Tryptamine	Pork leg (5 muscles)	ND	ND	6.37 ± 0.07 [a1]	6.56 ± 0.10 [a2]
	Lamb leg (7 muscles)	ND	ND	ND	ND
	Turkey leg (4 muscles)	ND	ND	ND	ND
	Chicken breast (1 muscle)	3.82 ± 0.12 [b1]	15.78 ± 0.16 [d1]	7.47 ± 0.07 [c2]	0.37 ± 0.01 [a1]
	Beef leg (4 muscles)	ND	ND	ND	ND
Agmatine	Pork leg (5 muscles)	ND	ND	ND	ND
	Lamb leg (7 muscles)	ND	0.15 ± 0.01 [a1]	1.14 ± 0.02 [b1]	2.30 ± 0.08 [c1]
	Turkey leg (4 muscles)	ND	ND	ND	ND
	Chicken breast (1 muscle)	ND	ND	ND	ND
	Beef leg (4 muscles)	ND	ND	ND	ND
Spermidine	Pork leg (5 muscles)	2.63 ± 0.13 [a2]	2.70 ± 0.00 [a1]	3.18 ± 0.14 [b1]	3.88 ± 0.18 [c1]
	Lamb leg (7 muscles)	8.09 ± 0.13 [a4]	11.99 ± 0.13 [c4]	8.69 ± 0.21 [a4]	10.16 ± 0.12 [b5]
	Turkey leg (4 muscles)	7.33 ± 0.05 [a3]	18.27 ± 0.07 [d5]	12.14 ± 0.10 [c5]	9.67 ± 0.09 [b4]
	Chicken breast (1 muscle)	9.78 ± 0.06 [d5]	6.69 ± 0.13 [b3]	6.24 ± 0.20 [a3]	7.02 ± 1.10 [c3]
	Beef leg (4 muscles)	2.29 ± 0.05 [a1]	3.39 ± 0.05 [b2]	4.35 ± 0.01 [c2]	5.39 ± 0.05 [d2]
Spermine	Pork leg (5 muscles)	27.60 ± 0.96 [b1]	27.10 ± 0.12 [b1]	25.23 ± 1.17 [a1]	26.87 ± 0.31 [b2]
	Lamb leg (7 muscles)	31.36 ± 0.58 [a2]	40.85 ± 0.37 [c3]	31.60 ± 0.98 [a3]	36.42 ± 0.00 [b4]
	Turkey leg (4 muscles)	35.44 ± 1.28 [b3]	49.20 ± 0.40 [d4]	36.95 ± 0.53 [c4]	32.55 ± 0.09 [a3]
	Chicken breast (1 muscle)	45.03 ± 0.81 [b4]	53.60 ± 0.24 [d5]	47.26 ± 0.54 [c5]	41.92 ± 0.16 [a5]
	Beef leg (4 muscles)	30.86 ± 0.36 [b2]	33.02 ± 0.24 [c2]	29.54 ± 0.40 [b2]	25.08 ± 0.04 [a1]
BAI	Pork leg (5 muscles)	1.24	1.82	17.41	47.29
	Lamb leg (7 muscles)	1.29	3.41	16.87	25.90
	Turkey leg (4 muscles)	1.23	5.10	11.43	88.85
	Chicken breast (1 muscle)	1.76	3.53	40.72	103.57
	Beef leg (4 muscles)	1.68	2.59	4.73	9.47

Each value is the mean of three replicates per meat sample and storage day ± standard deviation (SD). ND: Not detected (average limit of detection being 0.065 mg/L). For every biogenic amine: Different superscript letters in the same row indicate significant difference ($p < 0.05$) between storage days for the same meat type, and different superscript numbers in the same column indicate significant difference ($p < 0.05$) between meat types for the same storage day.

During refrigerated storage, the levels of physiological amines fluctuated in turkey for Spd and in turkey and chicken for Spm following a saw-toothed pattern (Table 4). Irrespective of the meat type. Both amines increased, peaking after the third day (18.27 mg/kg of Spd in turkey and 53.60 mg/kg of Spm in chicken) which was followed by a decrease until the end of the storage. This pattern, which is reported in other meat products in chilled storage [7], could reflect the relationship between these FAAs and the evolution pattern of their precursor, arginine (Figure 1). Initial arginine levels were considerable (Table 3). In fact, these levels increased up to day six and then decreased. Finally, they disappeared in all the samples except in turkey, which registered very low levels at the end of the experiment.

Putrescine (Put) was also detected throughout the study in all samples. Initial levels of these physiological amines were low in pork (0.57 mg/kg) while the other meat types contained the double of that amount (Table 4). Put levels increased significantly in all meats during the experiment but not at the same rate for all samples. For instance, in lamb and turkey, levels increased considerably by day three but the highest Put levels were recorded in chicken and turkey at the end of the study (51.99 mg/kg and 68.72 mg/kg, respectively). These were the highest levels among all BAs, even higher than tyramine, which is the most prevalent BA in meat products. At the end of the storage, only in the case of pork and lamb, similar levels of putrescine and tyramine were observed (Table 4). These putrescine levels were mainly related to *Enterobacteriaceae* growth (Table 2), which registered the highest levels. In addition, put production is associated with a reduction in arginine, which was also observed in all samples (Table 3) as noted earlier. Put is identified as one of the toxic biogenic amines, together with cadaverine (Cad), since they favor intestinal absorption of HIS and Tyr and contribute to catabolism reduction, thus enhancing their toxicity [5,14].

Cadaverine (Cad) is another important amine that was not detected in the meat samples until day six of chilled storage, except for beef, where it was not detected throughout the study. Other authors [20] reported that cadaverine was undetectable in beef whereas the meats in which it was detected, concentrations rose quickly, except for lamb. However, considerable levels were observed in pork, chicken, and turkey (16.16, 14.31 and 13.25 mg/kg, respectively). In most cases, no clear relationship was observed regarding the levels of its FAA lysine precursor, except for chicken and turkey, in which there was some correlation (coefficient > 0.92). It is worth noticing that at the end of storage, lysine was the precursor with the highest levels, peaking in turkey and chicken at day six of the experiment (Table 3). High levels of the FAA precursors that did not correlate with high levels of their corresponding BAs could be due to the lower BA decarboxylation capacity of the microorganisms that grow in this type of meat. These amines are also associated with *Enterobacteriaceae* growth; but, in this study, there was no clear relationship. However, there are other factors that can also affect BA presence in meat or meat products such as processing, meat matrix nature, etc. [9,20–22,49].

Agmatine (Agm) was present only in lamb meat with very low levels (0.15–2.30 mg/kg) (Table 4) starting from the third day of storage. However, its precursor arginine was widely represented in all meats, especially chicken, throughout the study (Table 3). This inverse behavior in lamb could be due to the weakness or lack of aminogenetic capacity of the microbiota [8] and the fact that arginine can be transformed into Putrescine (Figure 1) depending on the microbial flora type present in the matrix. As explained before, put levels were indeed consistent with the arginine levels observed in this study.

In general, the other BAs' levels increased, from day three until the end of the storage. Levels of toxic amines such as tyramine (Tyr) and β-phenylethylamine (Phe) were very low (<0.8 mg/kg) at the beginning of the study and were not even detected in some meats such as in chicken and turkey (Tyr), and in pork and chicken (Phe). However, they were detected afterwards in all the studied kinds of meats during storage. These amounts increased during the experiment until day 10, but levels remained below 36 mg/kg in the case of Tyr, which reached its highest levels in chicken and lowest in beef (1.57 mg/kg), followed by turkey (6.88 mg/kg), lamb (10.71 mg/kg), and pork (16.58 mg/kg). Phe levels registered lower amounts in all species at the end of the storage. Chicken also contained the highest levels of Phe (12.81 mg/kg), followed by turkey, lamb, beef, and pork (Table 4) at the end of the experiment. These levels are very low in terms of toxicological limits, especially for Tyr (800 mg/kg) [50] while the limit for Phe is 30 mg/kg, which is twice the level found in this study [51]. These BAs are formed from the decarboxylation reactions of tyrosine and phenylalanine, respectively. The highest levels of tyrosine were observed in chicken, which can be associated with the high levels of tyramine in this species. However, no clear relationship was observed for the other species regarding tyramine and tyrosine even though the evolution of this FAA was clear throughout the storage and was species-dependent (Table 3). As a matter of fact, in some species, such as pork, lamb, beef or turkey, tyramine increased over storage, and these changes correlated with tyrosine production. On the other hand, the evolution in chicken followed a saw-toothed pattern, thus the possibility for establishing a correlation became more unlikely with Tyr production. Several authors reported similar evolutions of these FAAs and their relationship with BAs [6].

Phenylalanine levels (Table 3) were higher than tyrosine's although its corresponding biogenic amine (Phe) levels were significantly lower than tyramine in some species. Phenylalanine levels were significantly higher at the end of the storage in all meat types, particularly turkey, chicken, and pork (Table 3). Other authors also found little correlation between Phe and its FAA precursor, which seems to be closely related to the nature of the flora and its aminogenic capacity [8,52]. Lamb, for example, contained higher levels of Phe (9.05 mg/kg) at the end of the experiment than of phenylalanine (5.82 mg/100 g) suggesting that its flora has high Phenylalanine decarboxylation capacity. The rest of the meats presented the opposite pattern suggesting that their flora presented less phenylalanine decarboxylation capacity.

On the other hand, histamine (HIS), another toxicological amine, was not detected in the majority of meat types (pork, lamb and turkey), except in chicken and beef where levels were very low (2.11 and 0.50 mg/kg, respectively at the end of storage) (Table 4). This was consistent with histidine levels, the FAA precursor of HIS (Table 3), which were not initially detected in any type of meat, except for pork and chicken, with very low levels (<6 mg/100 g). Final HIS levels in the meat types samples were significantly lower than the legal limit of 50 mg/kg [15], meaning that there is no potential health risk after consumption of these meats in their fresh status. Moreover, HIS levels are related to those of its FAA precursor, which is poorly represented in the studied meats (Table 3). Some authors also reported very low to undetectable HIS levels in fresh beef sausages and dry fermented pork sausages [20,21,39]. These low amounts are consistent with the type of samples analyzed since HIS is not typically found in fresh meat and meat products [5].

Tryptamine (Trp) was detected only in chicken and pork. While its presence in chicken was observed from the beginning of the storage, in the case of pork it was only detected after six days.

In general, higher BA levels (Table 4) were observed in chicken, turkey and pork, and these results were associated with the total free amino acid BA precursors (TFAAP) (Table 3). Given the importance of BAs as quality indexes, the application of the biogenic amines index (BAI) in this study showed a clear increase over storage, which was related to the meat type (Table 4). At the outset, all the BAIs registered less than 2 mg/kg, with the highest levels in chicken. According to the index's classification, all meat types presented good quality, which was maintained until day six, afterwards there was a considerable increase in BAI levels, with the highest registered once more in chicken. The latter reached levels of 40.72 mg/kg, followed by pork (17.41 mg/kg), lamb (16.87 mg/kg), turkey (11.43 mg/kg), and beef (4.73 mg/kg). According to this index, chicken was considered of poor quality at day six while the other meats were still classified as acceptable. However, this classification is not related to microorganism levels on the same day of analysis (Table 2). As a matter of fact, a delay was observed in the formation of biogenic amines with respect to microbial growth. Such a delay was also reported by other authors [21,53] and constituted one of the factors that some authors used to set the limit of 6 Log/cfu for TVC as indicating unfitness for consumption.

At the end of storage, the highest BAI levels were observed in chicken (103.87 mg/kg) and turkey (88.85 mg/kg), which were spoiled, and the lowest BAI level was found in beef (9.47 mg/kg), which was still the only meat in the range of BAI acceptability (Table 4), while pork and lamb were considered as exhibiting poor quality.

Several authors demonstrated that beef is less spoilable than the other types of meats and that chicken is the first to undergo deterioration reactions [12,54]. BAI levels in turkey, beef, and pork presented a high correlation with TFAAP (0.97, 0.92, and 0.86, respectively) as well as with TFAA levels in lamb (0.97), beef (0.96, and turkey (0.95) throughout the storage (Table 4).

These results support the theory of rapid white meat spoilage compared with red meat [12,54], as shown in the FAA section. This behavior was also observed in tuna, where amine levels were generally higher in the white than in the red muscle [40]. In this study; the BAI levels of white meats (turkey and chicken) close to and above 50, at 10 days of chilling storage.

These BAI levels reflect the rate of deterioration of each type of meat and thus provide useful information when planning meat product processing and manufacture strategies. These BAI levels showed better results than some BA contents such as Tyramine or Spermidine and Spermine individually. However, BAI results did not show any clear relationship with microbial levels, which exceeded the permitted limits of microorganisms in all types of meats from day three of storage (Table 2). Therefore, in this case, the BAI index seems to be a more suitable indicator for white meat freshness than for red ones, especially beef, which was also reported by other authors [10].

4. Conclusions

The evolutions of free amino acids (FAA) and biogenic amines (BA) were clearly influenced by the meat type. The largest amounts were observed in chicken and turkey followed by the other meat types. Even though a clear relationship was observed for certain meat species between the biogenic amine index (BAI) and total free amino acids (TFAA), this index did not correlate with microbial growth. The relationship between overall TFAAs and BAIs was closer between an FAA precursor and its corresponding biogenic amine when considered individually. BAIs showed that only beef maintained acceptable quality throughout the study (<10 at day 10 of chilling storage), while chicken presented a poor quality (103.57) followed by turkey, lamb and pork. Overall, BA levels were higher in white meats than in red ones. During storage, some TFAAP of BA followed a saw-toothed pattern mainly in chicken and turkey meat. This limits its use as a quality index for fresh meat during chilled storage. Finally, the BAI index seems to be more suitable as a quality index for white meat freshness than for red meat, especially for beef.

Author Contributions: Conceived and designed the experiments; C.R.-C. and M.T.; performed the experiments: M.T., C.R-C and A.M.H. analyzed the data; M.T., C.R-C, A.M.H., F.J.-C.; wrote the paper: M.T., C.R-C, A.M.H., F.J.-C.

Funding: This research was supported by projects AGL 2011-29644-C02-01 of the Plan Nacional de Investigación Científica, Desarrollo e Innovación Tecnológica (I+D+I) Ministerio de Ciencia y Tecnología, the Intramural Project 2014470E073 and the project MEDGAN-CM S2013/ABI2913 of CAM and ESI financial.

Acknowledgments: The authors wish to thank the AECID-MAE for Mehdi Triki's outstanding scholarship.

Conflicts of Interest: The authors declare no conflict of interest.

References

1. Jiménez-Colmenero, F.; Herrero, A.; Cofrades, S.; Ruiz-Capillas, C. Meat: Eating Quality and Preservation. In *The Encyclopedia of Food and Health*; Caballero, B., Finglas, P., Toldrá, F., Eds.; Oxford Academic Press: Kidlington, UK, 2016; Volume 3, pp. 685–692.

2. Shabbir, M.A.; Raza, A.; Anjum, F.M.; Khan, M.R.; Suleria, H.A.R. Effect of Thermal Treatment on Meat Proteins with Special Reference to Heterocyclic Aromatic Amines (HAAs). *Crit. Rev. Food Sci. Nutr.* **2015**, *55*, 82–93. [CrossRef] [PubMed]

3. Jiménez Colmenero, F.; Herrero, A.M.; Ruiz-Capillas, C.; Cofrades, S. Meat and functional foods. In *Handbook of Meat and Meat Processing*; Hui, Y.H., Ed.; CRC Press. Taylor & Francis Group: Boca Raton, FL, USA, 2012; pp. 225–248.

4.   Mietz, J.L.; Karmas, E. Polyamine and histamine content of rockfish, salmon, lobster, and shrimp as an indicator of decomposition. *J. Assoc. Off. Anal. Chem.* **1978**, *61*, 139–145.

5.   Ruiz-Capillas, C.; Jiménez-Colmenero, F. Biogenic amines in meat and meat products. *Crit. Rev. Food Sci. Nutr.* **2004**, *44*, 489–499. [CrossRef] [PubMed]

6.   Ruiz-Capillas, C.; Moral, A. Changes in free amino acids during chilled storage of hake (*Merluccius merluccius* L.) in controlled atmospheres and their use as a quality control index. *Eur. Food Res. Technol.* **2001**, *212*, 302–307. [CrossRef]

7.   Rabie, M.A.; Peres, C.; Malcata, F.X. Evolution of amino acids and biogenic amines throughout storage in sausages made of horse, beef and turkey meats. *Meat Sci.* **2014**, *96*, 82–87. [CrossRef] [PubMed]

8.   Latorre-Moratalla, M.L.; Bover-Cid, S.; Bosch-Fusté, J.; Veciana-Nogués, M.T.; Vidal-Carou, M.C. Amino acid availability as an influential factor on the biogenic amine formation in dry fermented sausages. *Food Control* **2014**, *36*, 76–81. [CrossRef]

9.   Jairath, G.; Singh, P.K.; Dabur, R.S.; Rani, M.; Chaudhari, M. Biogenic amines in meat and meat products and its public health significance: A review. *J. Food Sci. Technol.* **2015**, *52*, 6835–6846. [CrossRef]

10. Hernández-Jover, T.; Izquierdo-Pulido, M.; Veciana-Nogués, M.T.; Vidal-Carou, M.C. Biogenic Amine Sources in Cooked Cured Shoulder Pork. *J. Agric. Food Chem.* **1996**, *44*, 3097–3101. [CrossRef]

11. Silva, C.M.G.; Glória, M.B.A. Bioactive amines in chicken breast and thigh after slaughter and during storage at 4 ± 1 °C and in chicken-based meat products. *Food Chem.* **2002**, *78*, 241–248. [CrossRef]

12. Vinci, G.; Antonelli, M.L. Biogenic amines: Quality index of freshness in red and white meat. *Food Control* **2002**, *13*, 519–524. [CrossRef]

13. Halász, A.; Baráth, Á.; Simon-Sarkadi, L.; Holzapfel, W. Biogenic amines and their production by microorganisms in food. *Trends Food Sci. Technol.* **1994**, *5*, 42–49. [CrossRef]

14. Önal, A. A review: Current analytical methods for the determination of biogenic amines in foods. *Food Chem.* **2007**, *103*, 1475–1486. [CrossRef]

15. European Food Safety Authority (EFSA). Scientific Opinion on risk based control of biogenic amine formation in fermented foods. *EFSA J.* **2011**, *9*, 2393. [CrossRef]

16. World Health Organization. Joint FAO/WHO Expert Meeting on the Public Health Risks of Histamine and other Biogenic Amines from Fish and Fishery Products: Meeting Report. Available online: http://www.fao. org/fileadmin/user_upload/agns/news_events/Histamine_Final_Report.pdf (accessed on 10 July 2018).

17. Cofrades, S.; López-López, I.; Ruiz-Capillas, C.; Triki, M.; Jiménez-Colmenero, F. Quality characteristics of low-salt restructured poultry with microbial transglutaminase and seaweed. *Meat Sci.* **2011**, *87*, 373–380. [CrossRef] [PubMed]

18. Leggio, A.; Belsito, E.L.; De Marco, R.; Di Gioia, M.L.; Liguori, A.; Siciliano, C.; Spinella, M. Dry fermented sausages of Southern Italy: A comparison of free amino acids and biogenic amines between industrial and homemade products. *J. Food Sci.* **2012**, *77*, S170–S175. [CrossRef] [PubMed]

19. Delgado-Pando, G.; Cofrades, S.; Ruiz-Capillas, C.; Solas, M.T.; Triki, M.; Jiménez-Colmenero, F. Low-fat frankfurters formulated with a healthier lipid combination as functional ingredient: Microstructure, lipid oxidation, nitrite content, microbiological changes and biogenic amine formation. *Meat Sci.* **2011**, *89*, 65–71. [CrossRef] [PubMed]

20. Triki, M.; Herrero, A.M.; Jiménez-Colmenero, F.; Ruiz-Capillas, C. Storage stability of low-fat sodium reduced fresh merguez sausage prepared with olive oil in konjac gel matrix. *Meat Sci.* **2013**, *94*, 438–446. [CrossRef] [PubMed]

21. Triki, M.; Herrero, A.M.; Jiménez-Colmenero, F.; Ruiz-Capillas, C. Effect of preformed konjac gels, with and without olive oil, on the technological attributes and storage stability of merguez sausage. *Meat Sci.* **2013**, *93*, 351–360. [CrossRef] [PubMed]

22. Triki, M.; Jiménez-Colmenero, F.; Herrero, A.M.; Ruiz-Capillas, C. Optimisation of a chromatographic procedure for determining biogenic amine concentrations in meat and meat products employing a cation-exchange column with a post-column system. *Food Chem.* **2012**, *130*, 1066–1073. [CrossRef]

23. ANSES Ciqual Table de Composition Nutritionnelle Des Aliments. Available online: https://ciqual.anses.fr/ (accessed on 10 July 2018).

24. Obanor, F.O. Biochemical Basis of the Effect of Pre-Slaughter Stress and Post-Slaughter Processing Conditions on Meat Tenderness. Master's Thesis, Lincoln University, Christchurch, New Zealand, 2002.

25. Page, J.K.; Wulf, D.M.; Schwotzer, T.R. A survey of beef muscle color and pH. *J. Anim. Sci.* **2001**, *79*, 678–687. [CrossRef] [PubMed]

26. Fleck, C.; Kozačinski, Ž.L.; Njari, B.; Marenčić, D.; Mršić, G.; Špiranec, K.; Špoljarić, D.; Čop, M.J.; Živković, M.; Popović, M. Technological properties and chemical composition of the meat of sheep fed with *Agaricus bisporus* supplement. *Vet. Arh.* **2015**, *85*, 591–600.

27. Van Laack, R.L.J.M.; Kauffman, R.G.; Sybesma, W.; Smulders, F.J.M.; Eikelenboom, G.; Pinheiro, J.C. Is color brightness (*L*-value) a reliable indicator of water holding capacity in porcine muscle? *Meat Sci.* **1994**, *38*, 193–201. [CrossRef]

28. Patterson, B.A.; Matarneh, S.K.; Stufft, K.M.; England, E.M.; Scheffler, T.L.; Preisser, R.H.; Shi, H.; Stewart, E.C.; Eilert, S.; Gerrard, D.E. Pectoralis major muscle of turkey displays divergent function as correlated with meat quality. *Poult. Sci.* **2017**, *96*, 1492–1503. [PubMed]

29. Cortez-Vega, W.R.; Pizato, S.; Prentice, C. Quality of raw chicken breast stores at 5 °C and packaged under different modified atmospheres. *J. Food Saf.* **2012**, *32*, 360–368. [CrossRef]

30. Yang, C.C.; Chen, T.C. Effects of Refrigerated Storage, pH Adjustment, and Marinade on Color of Raw and Microwave Cooked Chicken Meat. *Poult. Sci.* **1993**, *72*, 355–362. [CrossRef]

31. Debut, M.; Berri, C.; Baeza, E.; Sellier, N.; Arnould, C.; Guemene, D.; Jehl, N.; Boutten, B.; Jego, Y.; Beaumont, C. Variation of chicken technological meat quality in relation to genotype and preslaughter stress conditions. *Poult. Sci.* **2003**, *82*, 1829–1838. [CrossRef] [PubMed]

32. Lázaro, C.A.; Conte-Júnior, C.A.; Canto, A.C.; Monteiro, M.L.G.; Costa-Lima, B.; da Cruz, A.G.; Mársico, E.T.; Franco, R.M. Biogenic amines as bacterial quality indicators in different poultry meat species. *LWT Food Sci. Technol.* **2015**, *60*, 15–21. [CrossRef]

33. Laranjo, M.; Gomes, A.; Agulheiro-Santos, A.C.; Potes, M.E.; Cabrita, M.J.; Garcia, R.; Rocha, J.M.; Roseiro, L.C.; Fernandes, M.J.; Fraqueza, M.J.; et al. Impact of salt reduction on biogenic amines, fatty acids, microbiota, texture and sensory profile in traditional blood dry-cured sausages. *Food Chem.* **2017**, *218*, 129–136. [CrossRef] [PubMed]

34. Commission Regulation (EC) No 2073/2005 of 15 November 2005 on Microbiological Criteria for Foodstuffs. Available online: https://eur-lex.europa.eu/LexUriServ/LexUriServ.do?uri=CONSLEG: 2005R2073:20060101:EN:PDF (accessed on 14 August 2018).

35. USDA. USDA National Nutrient Database for Standard Reference. Available online: https://ndb.nal.usda.gov/ndb/search/list (accessed on 14 August 2018).

36. Niewiarowicz, A. Meat Anomalies in Broilers. *Poult. Int.* **1978**, *17*, 50–51.

37. Cowieson, A.J.; Acamovic, T.; Bedford, M.R. The effects of phytase and phytic acid on the loss of endogenous amino acids and minerals from broiler chickens. *Br. Poult. Sci.* **2004**, *45*, 101–108. [CrossRef] [PubMed]

38. Iida, F.; Miyazaki, Y.; Tsuyuki, R.; Kato, K.; Egusa, A.; Ogoshi, H.; Nishimura, T. Changes in taste compounds, breaking properties, and sensory attributes during dry aging of beef from Japanese black cattle. *Meat Sci.* **2016**, *112*, 46–51. [CrossRef] [PubMed]

39. Ruiz-Capillas, C.; Triki, M.; Herrero, A.M.; Jiménez-Colmenero, F. Biogenic amines in low- and reduced-fat dry fermented sausages formulated with konjac gel. *J. Agric. Food Chem.* **2012**, *60*, 9242–9248. [CrossRef] [PubMed]

40. Ruiz-Capillas, C.; Moral, A. Free amino acids and biogenic amines in red and white muscle of tuna stored in controlled atmospheres. *Amino Acids* **2004**, *26*, 125–132. [CrossRef] [PubMed]

41. Freiding, S.; Gutsche, K.A.; Ehrmann, M.A.; Vogel, R.F. Genetic screening of *Lactobacillus sakei* and *Lactobacillus curvatus* strains for their peptidolytic system and amino acid metabolism, and comparison of their volatilomes in a model system. *Syst. Appl. Microbiol.* **2011**, *34*, 311–320. [CrossRef] [PubMed]

42. Subramaniyan, S.A.; Kang, D.R.; Belal, S.A.; Cho, E.-S.-R.; Jung, J.-H.; Jung, Y.-C.; Choi, Y.-I.; Shim, K.-S. Meat Quality and Physicochemical Trait Assessments of Berkshire and Commercial 3-way Crossbred Pigs. *Korean J. Food Sci. Anim. Resour.* **2016**, *36*, 641–649. [CrossRef] [PubMed]

43. Dermiki, M.; Phanphensophon, N.; Mottram, D.S.; Methven, L. Contributions of non-volatile and volatile compounds to the umami taste and overall flavour of shiitake mushroom extracts and their application as flavour enhancers in cooked minced meat. *Food Chem.* **2013**, *141*, 77–83. [CrossRef] [PubMed]

44. Dashdorj, D.; Amna, T.; Hwang, I. Influence of specific taste-active components on meat flavor as affected by intrinsic and extrinsic factors: An overview. *Eur. Food Res. Technol.* **2015**, *241*, 157–171. [CrossRef]

45. Ruiz-Capillas, C.; Nollet, L.M.L. *Flow Injection Analysis of Food Additives*; CRC Press. Taylor & Francis Group: Boca Raton, FL, USA, 2016.
46. Fraqueza, M.J.; Alfaia, C.M.; Barreto, A.S. Biogenic amine formation in turkey meat under modified atmosphere packaging with extended shelf life: Index of freshness. *Poult. Sci.* **2012**, *91*, 1465–1472. [CrossRef] [PubMed]
47. Dadáková, E.; Pelikánová, T.; Kalač, P. Concentration of biologically active polyamines in meat and liver of sheep and lambs after slaughter and their changes in mutton during storage and cooking. *Meat Sci.* **2011**, *87*, 119–124. [CrossRef] [PubMed]
48. Rokka, M.; Eerola, S.; Smolander, M.; Alakomi, H.-L.; Ahvenainen, R. Monitoring of the quality of modified atmosphere packaged broiler chicken cuts stored in different temperature conditions: B. Biogenic amines as quality-indicating metabolites. *Food Control* **2004**, *15*, 601–607. [CrossRef]
49. Roig-Roig-Sagués, A.X.; Ruiz-Capillas, C.; Espinosa, D.; Hernández, M. The decarboxylating bacteria present in foodstuffs and the effect of emerging technologies on their formation. In *Biological Aspects of Biogenic Amines, Polyamines and Conjugates*; Dandrifosse, G., Ed.; Transworld Research Network: Kerala, India, 2009.
50. Ten Brink, B.; Damink, C.; Joosten, H.M.; Huis in't Veld, J.H. Occurrence and formation of biologically active amines in foods. *Int. J. Food Microbiol.* **1990**, *11*, 73–84. [CrossRef]
51. Gardini, F.; Martuscelli, M.; Caruso, M.C.; Galgano, F.; Crudele, M.A.; Favati, F.; Guerzoni, M.E.; Suzzi, G. Effects of pH, temperature and NaCl concentration on the growth kinetics, proteolytic activity and biogenic amine production of Enterococcus faecalis. *Int. J. Food Microbiol.* **2001**, *64*, 105–117. [CrossRef]
52. Curiel, J.A.; Ruiz-Capillas, C.; de Las Rivas, B.; Carrascosa, A.V.; Jiménez-Colmenero, F.; Muñoz, R. Production of biogenic amines by lactic acid bacteria and enterobacteria isolated from fresh pork sausages packaged in different atmospheres and kept under refrigeration. *Meat Sci.* **2011**, *88*, 368–373. [CrossRef] [PubMed]
53. Ruiz-Capillas, C.; Herrero, A.; Triki, M.; Jiménez-Colmenero, F. Biogenic Amine Formation in Reformulated Cooked Sausage Without Added Nitrite. *J. Nutr. Med. Diet Care* **2017**, *3*, 2–6. [CrossRef]
54. James, S.; James, C. Raw material selection: meat and poultry. In *Chilled Foods: A Comprehensive Guide*, 3rd ed.; Brown, M., Ed.; Woodhead Publishing: Cambridge, England, 2008; pp. 61–82.

![foods logo] *foods*

MDPI

Article

The Determination of Some Microbiological and Chemical Features in Herby Cheese

Kamil Ekici [1],*, Hayrettin Okut [2], Ozgur Isleyici [1], Yakup Can Sancak [1] and Rabia Mehtap Tuncay [1]

[1] Department of Food Hygiene and Technology, Veterinary College, Yuzuncu Yıl University, Van 65080, Turkey; oisleyici@hotmail.com (O.I.); ycsancak@yyu.edu.tr (Y.C.S.); r.m.gunes@hotmail.com (R.M.T.)

[2] Departement of Preventive Medicine, School of Medical, University of Kansas, Kansas City, KS 66160, USA; hokut@kumc.edu

* Correspondence: kekici@yyu.edu.tr; Tel.: +90-432-225-1128 (ext. 21560)

Received: 7 December 2018; Accepted: 7 January 2019; Published: 11 January 2019

Abstract: The objective of this study is to measure the amounts of biogenic amines, microbial counts, values of pH, titratable acidity, dry matter, and salt (%) in herby cheese, a very popular staple in the Turkish diet, and to evaluate the concentration of biogenic amines in terms of public health risks. A high-performance liquid chromatography (HPLC) method was used for the determination of eight biogenic amines in 100 herby cheeses sold in the local markets of Van. The bacterial load of the herby cheeses ranged between 4.0 and 8.90 log CFU/g for viable total aerobic mesophilic bacteria (TAMB), <1 and 7.0 log CFU/g for lactic bacteria (LAB), <1 and 6.08 log CFU/g for coliform bacteria, <1 and 5.81 log CFU/g for Enterobacteriaceae, <1 and 2.60 log CFU/g for *Staphylococcus aureus*, and 3.70 and 8.05 log CFU/g for yeasts and molds. The results obtained suggested significant changes in the pH, titratable acidity, dry matter, and salt contents of the examined herby cheese samples. The detection levels of biogenic amines in the samples ranged from <0.025 to 33.36 mg/kg for tryptamine, from <0.038 to 404.57 mg/kg for β-phenylethylamine, from 0.03 to 426.35 mg/kg for putrescine, from <0.039 to 1438.22 mg/kg for cadaverine, from <0.033 to 469 mg/kg for histamine, from <0.309 to 725.21 mg/kg for tyramine, from <0.114 to 1.70 mg/kg for spermidine, and from <0.109 to 1.88 mg/kg for spermine. As a result, these cheeses are fit for consumption in terms of the amounts of biogenic amines they contain.

Keywords: biogenic amines; histamine; herby cheese; HPLC; decarboxylase enzymes

---

1. Introduction

Biogenic amines (BAs) are organic bases with an aliphatic, aromatic, or heterocyclic structure, which have been found in many foods, such as fish products, cheese, wine, beer, and other fermented foods [1–3]. Biogenic amine accumulation in foods usually results from the decarboxylation of amino acids by enzymes of bacterial origin, which is associated with food hygiene and technology [3–5]. The term "biogenic amines" defines decarboxylation products such as histamine, serotonin, tyramine, phenylethylamine, tryptamine, and also aliphatic polyamines [6].

Numerous bacteria, both intentional and adventitious, have been reported as being capable of producing biogenic amines. These are *Escherichia*, *Enterobacter*, *Salmonella*, *Shigella*, *Clostridium perfringens*, *Streptococcus*, *Lactobacillus*, and *Leuconostoc* [7–9]. The main biogenic amine producers in cheese are Gram-positive bacteria, with LAB being the main histamine and tyramine producers [10]. *Leuconostoc mesenteroides* has a high potential to form tyramine or histamine in wine [11,12]. The presence of biogenic amines in food constitutes a potential public health concern due to their psychological and toxicological effects [13]. Biogenic amines may also be considered as carcinogens because they are able to react with nitrites to form potentially carcinogenic nitrosamines [14]. Consuming contaminated fish of the Scombridae family is the most common type of fish poisoning in Europe and worldwide, as it is consumed in large quantities (tuna, bonito, and mackerel), thereby causing a pseudoallergic poisoning known as the scombroid syndrome. This syndrome may be triggered by various species from the family, such as sea urchins, bluefish, herring, anchovies, sardines, and dolphin fish. Poisoning resulting from the ingestion of this fish accounts for 40% and up to 5% of all food poisoning cases reported in the United States [15]. Histamine poisoning is the most common and toxic form of poisoning. Histamine intoxication, also termed scombroid poisoning, is an important foodborne disease common all over the world [8]. The intake of foods with high concentrations of biogenic amines can cause migraines, headaches, gastric and intestinal problems, and pseudoallergic responses [9,16].

Cheese represents an ideal environment for biogenic amine production. Several factors may contribute to biogenic amine formation in cheese. The utilization of raw or pasteurized milk in cheesemaking, higher ripening temperature, excessive proteolysis, high pH, and low salt concentration may contribute to the ability of an organism to produce biogenic amines [17–19].

Herby cheese, which has a semihard texture and a salty taste, is produced in small family businesses for their needs and for commercial purposes in Van city. Van city is located by the shores of Lake Van, in the eastern part of Turkey, bordering Iran. In addition, it is produced in well-equipped factories. It is made from raw sheep milk in the eastern and southeastern parts of Turkey. If sheep milk is not available, a mixture of sheep and cow or sheep and goat milk can be used for cheesemaking. Herby cheese, named "Otlu peynir" in Turkish, is homemade in villages and by some small local producers for many years. Most people consume it as a part of almost every meal [20–22]. There has been a continuous demand for this cheese in recent years, and this may further increase in the future, since the market of herby cheese has been spreading to the big cities in the country [23].

This study was undertaken to determine the amounts of biogenic amines in herby cheese, since biogenic amines are important with regard to toxicological effects. Besides the content of biogenic amines, microbial counts, values of pH, titratable acidity, dry matter, and salt parameters were also measured to provide complementary information on the microbiological and biochemical features of herby cheese, focusing on hygiene and the consumer health aspects of herby cheese.

2. Materials and Methods

2.1. Sample Origin

In the present study, 100 samples of herby cheese, collected from various sales points in Van city center, were used in this study. The samples were placed in sterile jars and brought to the laboratory in a cooler (3–6 °C) and analyzed immediately. Assays were done on duplicate samples with the results being averaged.

2.2. Microbiological Analysis

In brief, 10 g of herby cheese samples were weighed in stomacher bags, and then 90 mL of sterile physiologic water with peptone (0.85% NaCl + 0.1% peptone) was added. Then, samples were homogenized in stomacher bags for 2 min. After homogenization, serial decimal dilutions were prepared until dilutions of 10^9 CFU/mL were reached, and from these dilutions, Petri dishes were inoculated. To assess the viable total aerobic mesophilic bacteria (TAMB), plates of plate count agar (Oxoid CM325) were incubated for 48 h at 35 °C; for lactic acid bacteria (LAB), plates of de Man, Rogosa and Sharpe (MRS) agar (Oxoid CM 361) were incubated for two days at 35 °C; for yeasts and molds (YM), plates of potato dextrose agar (Oxoid CM139) were incubated for five days at 21 °C; for coliform bacteria, plates of violet red bile lactose agar (Oxoid CM 107) were incubated for 24 h at 35 °C; for Enterobacteriaceae, plates of violet red bile glucose agar (Oxoid CM 485) were incubated for

48 h at 32 °C, and for *S. aureus*, plates of Baird–Parker agar (Oxoid CM 275) were incubated for 48 h at 35 °C. After inoculation on Petri dishes, colonies formed were counted [24].

2.3. Chemical Analysis

Herby cheese samples were analyzed for titratable acidity, also known as lactic acid (LA%), and dry matter, salt (%), and pH according to the method described by Tekinsen et al. [25].

2.4. Biogenic Amine Analysis

Sample preparation and biogenic amine analysis were done according to the high-performance liquid chromatography (HPLC) method described by Eerola et al. [26].

2.4.1. Sample Preparation and Homogenization Procedure

One hundred herby cheese samples were collected, where each one weighed approximately 750 g. Then, the herby cheese samples were sliced with a clean stainless steel knife and were grated and homogenized thoroughly, and from each cheese sample, a 2 g sample was weighed (to the nearest 0.001 g) and transferred into a plastic Falcon tubes, then homogenized with a metallic staff homogenizer tool (T-25 digital Ultra-Turrax from IKA®-Works, Inc., Wilmington, NC, USA) for about 2 min. The homogenization was done by adding 125 µL of an internal standard (1.7-diaminoheptane) Sigma-Aldrich Company, St. Louis, MO, USA) with 10 mL of 0.4 M perchloric acid (Merck, Darmstadt, Germany) to the sample. In the next step, the homogenized samples were centrifuged ($1210\times g$ for 10 min under 4 °C) by high-speed refrigerated centrifuge (Hitachi Koki Co., Ltd., Tokyo, Japan), and then the extraction solvents were transferred and filtered with filter paper (Schleicher and Schuell, 589 Black ribbon Ø 70 mm, Dassel, Germany) into a volumetric flask. The remaining (supernatant) part was again centrifuged with 10 mL of perchloric acid and filtered into the same volumetric flask, then supplemented to 25 mL with 0.4 M perchloric acid. An aliquot of 1 mL of the final extract was then used for analysis after derivatization, while the remaining volume was stored at 4 °C for no more than one week.

2.4.2. Derivatization of Extracts and Standards

Eight aqueous standard solutions containing cadaverine dihydrochloride (purchased from Aldrich Company, Buchs, Switzerland), putrescine dihydrochloride, tyramine hydrochloride, histamine dihydrochloride, tryptamine hydrochloride, 2-phenylethylamine hydrochloride, spermidine, spermine, or 1.7-diaminoheptane (as the internal standard) (all purchased from Sigma-Aldrich, St. Louis, MO, USA) were obtained. The standards used were prepared by our group with materials purchased from the respective suppliers. The dansylated derivatives of the amines were formed by adding 1 mL of sample extract or standard solution to 200 µL of 2 N NaOH (Merck, Germany) and 300 µL of saturated $NaHCO_3$ (Merck) solution and mixing by vortex (Heidolph D-91126, Reax top, Schwabach, Germany), and then 2 mL of dansyl chloride solution and 2 mg of dansyl chloride per mL in acetone (Sigma) were added and the solution was again vortexed. Fresh dansyl chloride solutions were prepared each time immediately before use. After shaking, samples were left in the incubator at 40 °C for 45 min. After the reaction time had passed, the residual dansyl chloride was removed by the addition of 100 µL of ammonia (Merck) 25% (*v/v*), followed by vortex mixing and holding for 30 min at room temperature. The derivatization was completed upon the addition of an ammonium acetate (Merck, Germany) and acetonitrile (Merck) mixture (1:1; *v/v*) and adjustment to 5 mL. Finally, the mixture was centrifuged (Hettich Zentrifugen, Werk Nr, Tuttlingen, Germany) at $1210\times g$ for 5 min under 4 °C and the supernatant was filtered through 0.45-µm pore-size filters (Millipore Co., Bedford, MA, USA).

2.4.3. Chromatographic Conditions

Two solvent reservoirs containing (A) ammonium acetate and (B) acetonitrile were used to separate all the amines with an HPLC elution program (Shimadzu, Kyoto, Japan). The gradient–elution system used 0.1 M ammonium acetate as solvent A and acetonitrile as solvent B. The gradient–elution program was started with 50% solvent B and ended at 90% solvent B after 25 min. The system was equilibrated for 10 min before the next analysis. The flow rate was 1.0 mL/min and the column temperature was 40 °C. A 20 μL sample was injected onto the column. The quantitative determinations were carried out by an internal standard (1.7-diaminoheptane) method using peak heights.

2.5. Statistical Analysis of Data

SAS version 9.4 (SAS Institute, Inc. Cary, NC, USA) [27] was used for all data analysis. PROC UNIVARIATE in SAS was used for the descriptive statistics for variables. The results were defined as the mean values ± standard error of the mean.

3. Results

Tables 1–3 show the results of the microbial counts (in log CFU/g), chemical results, and biogenic amine levels (mean ± SE (standard error of the mean), mg/kg wet weight) found for the herby cheese samples. Log_{10} transformations were applied on the microbiological data. Presumptive viable total aerobic mesophilic bacteria (TAMB), lactic bacteria (LAB), Enterobacteriaceae, coliform bacteria, *Staphylococcus aureus*, and yeast and mold counts were investigated as general microbiological quality parameters. As can be seen from Table 1, the bacterial load of herby cheese ranged between 4.0 and 8.90 log CFU/g for viable total aerobic mesophilic bacteria (TAMB), <10 and 7.0 log CFU/g for lactic bacteria (LAB), <10 and 6.08 log CFU/g for coliform bacteria, <10 and 5.81 log CFU/g for Enterobacteriaceae, <10 and 2.60 log CFU/g for *S. aureus*, and 3.7 and 8.05 log CFU/g for yeasts and molds.

Table 1. Results of microbial counts (in log CFU/g) in herby cheese samples.

Microorganisms	N	Mean ± SE	Min.	Max.
TAMB	100	5.52 ± 0.10	4.00	8.90
LAB	100	3.70 ± 0.17	<10	7.00
Coliform	100	3.57 ± 0.11	<10	6.08
Enterobacteriaceae	100	3.42 ± 0.13	<10	5.81
S. aureus	100	0.11 ± 0.04	<100	2.60
YM	100	5.31 ± 0.07	3.70	8.05

SE: standard error of the mean, TAMB: total mesophilic aerobic microorganisms, LAB: lactic acid bacteria, YM: yeasts and molds, N: number of analyzed samples.

Table 2. Results of chemical results found in herby cheese samples.

Chemical Features	N	Mean ± SE	Min.	Max.
Salt	100	8.64 ± 0.17 (%)	5.85	11.70
LA	100	2.01 ± 0.04 (%)	1.00	3.16
pH	100	5.47 ± 0.04	4.49	6.64
DM	100	60.37 ± 0.39 (%)	49.66	65.80

LA%: lactic acid, DM: dry matter, N: number of analyzed samples.

Table 3. Results of biogenic amine levels (mean ± SE, mg/kg wet weight) in herby cheese samples.

Biogenic Amines	N	Mean ± SE	Min.	Max.	LOD
TR	100	2.24 ± 0.47	ND	33.36	0.025
PEA	100	14.81 ± 4.53	ND	404.57	0.038
PUT	100	44.71 ± 8.06	0.03	426.35	0.028
CAD	100	98.42 ± 23.47	ND	1438.22	0.039
HI	100	54.20 ± 10.08	ND	469.00	0.033
TY	100	103.18 ± 17.10	ND	725.21	0.309
SPD	100	0.17 ± 0.03	ND	1.70	0.114
SPM	100	0.25 ± 0.03	ND	1.88	0.109

TR: tryptamine, PEA: β-phenylethylamine, PUT: putrescine, CAD: cadaverine, HI: histamine, TY: tyramine, SPD: spermidine, SPM: spermine, N: number of analyzed samples, ND: not detected.

4. Discussion

Various studies have been carried out on the microbiological quality of herby cheeses in Turkey; for example, Ozturk [28] determined the TMAB, coliform, *S. aureus*, and mould–yeast counts in herby cheese to be 7.14, 3.96, 3.29, and 3.48 log CFU/g, respectively. Isleyici and Akyuz [29] reported a herby cheese TMAB count of 7.82 log CFU/g, mould–yeast count of 5.81 log CFU/g, coliform bacteria count of 2.23 log CFU/g, staphylococci count of 3.93 log CFU/g, and LAB count of 8.08 log CFU/g. Sagun et al. [30] found the TMAB, coliform, LAB, and mould–yeast counts in herby cheese to have the mean values of 6.24 ± 0.66, 2.99 ± 2.27, 5.48 ± 0.61, and 4.60 ± 2.11 log CFU/g, respectively. Tekinsen [31] determined the TMAB, Enterobacteriaceae, coliform, *S. aureus*, and mould–yeast counts in herby cheese as being 8.53, 5.44, 4.61, 4.34, and 5.50 log CFU/g, respectively. Alemdar and Agaoglu [32] reported the TMAB and LAB counts in herby cheese as 8.45 and 8.61 log CFU/g, respectively. These diverse results from various researchers and the present study could be explained by the nonstandardized production of herby cheese and the sale of both ripened and unripened cheeses in different storage conditions in the market [33].

Today, society is increasingly aware of the importance of diet for health, and hence, any issue relating to food safety has a considerable impact on consumer behavior and official policy [16]. Among fermented food products, cheese is most commonly related with biogenic amines (mostly histamine, tyramine, cadaverine, and putrescine) intoxication [34,35]. Recently, the EFSA (European Food Safety Authority) Panel on Biological Hazards (BIOHAZ) conducted a qualitative risk assessment for biogenic amines (BAs) in fermented foods, and concluded that our present knowledge of their toxicity was limited and that further research was needed [36,37]. In normal circumstances, the human body is able to rapidly detoxify histamine and tyramine absorbed from foods through acetylation and oxidation mediated by the enzymes monoamine oxidase (MAO; EC 1.4.3.4), diamine oxidase (DAO; EC 1.4.3.6), and polyamine oxidase (PAO; EC 1.5.3.11) [16]. The ingestion of biogenic amine (BA)-rich food can cause adverse toxicological reactions and intoxications harmful to health [38]. In fact, the presence of these biogenic amines in food, especially in conjunction with other factors, such as the consumption of monoamine oxidase-inhibiting drugs, alcohol, and other food amines (e.g., spermine, spermidine, putrescine, and cadaverine), may cause food poisoning [39].

Regarding the content of biogenic amines in herby cheese, the highest level of cadaverine was observed to be 1438.22 mg/kg (Table 3). Standarová et al. [40] reported that Olomouc tvorogs contained, among other amines, the highest level of cadaverine, at up to 2413.00 mg/kg. According to Bonczar et al. [35], this amine is predominant in Harzer cheese and can occur at the level of 377.50 mg/kg, and Fiechter et al. [41] found this amine at the level of 1268.00 mg/kg in Harzer cheese. According to Andiç et al. [22], the levels of cadaverine ranged from not detected to 1844.50 mg/kg in herby cheese. Vale and Gloria [42] reported finding cadaverine at levels of up to 1110.00 mg/kg in Brazilian cheese. These results are higher than that found in this study and are less than that found by Andiç et al. [22]. According to the European Food Safety Authority [36], fresh cheeses can contain cadaverine at levels from 10.70 to 45.00 mg/kg and hard cheeses from 47.80 to 83.50 mg/kg.

The levels of cadaverine and putrescine are usually considered to be indicators of contamination and also markers of the hygiene standards of the production process. The representatives of the Enterobacteriaceae family and *Pseudomonas* genus are regarded as sources of cadaverine and putrescine [43].

The amount of histamine found in the herby cheese samples tested in this paper ranged from <0.033 to 469.00 mg/kg (Table 3). The histamine content of some cheeses varies widely; for example, Budak et al. [44] noted that the histamine concentration reached the level of 265.50 mg/kg after 90 days ripening at 10 °C, and it was detected in an amount higher than those of other biogenic amines in the Turkish white cheese samples investigated. Sancak et al. [45] reported that the histamine amount in 47 herby cheeses ranged between 25.62 and 957.62 mg/kg. Andiç et al. [22] reported that histamine levels ranged from 0 to 681.50 mg/kg. Antila et al. [46] found the amount of histamine in Emmental cheeses matured for 3 and 6 months to be 12.20 mg/100 g and 17.50 mg/100 g, respectively. According to Madejska et al. [47], the highest amount of histamine, of 730.47 ± 20.01 mg/kg, was found in in Gorgonzola Piccante cheese stored for 42 days at room temperature. The amount of histamine was less than what was reported previously by Madejska et al. [47] in Gorgonzola Piccante cheese and was in agreement with those reported previously by Andiç et al. [22] for herby cheese. The formation of histamine throughout the ripening period in herby cheese was studied by Sagun et al. [48], who reported that the concentration of histamine was 21.90 mg/kg on the first day of ripening, which then gradually increased and reached 46.20 mg/kg on the 90th day. Although the toxicity of histamine to man is a controversial subject, ingestion of 70–1000 mg histamine will usually cause the clinical symptoms of intoxication [49]. Food and Drug Administration (FDA) has established a hazard action concentration for histamine in tuna fish of 50 mg histamine/100 g [9,50]. The amounts of histamine found in the present research are lower than those that lead to clinical symptoms, which are 70–1000 mg [49]. The histamine content found in 10 of the 100 (10%) cheese samples was found to be higher than 200 mg/kg. The prevalence of this amine was high, but the levels detected were low and below its toxic threshold (50 mg histamine/100 g).

The amount of β-phenylethylamine found in the herby cheeses ranged from <0.038 to 404.57 mg/kg of sample (Table 3). The phenylethylamine level was previously reported to be 3.77 mg/kg in feta cheese by Valsamaki et al. [51]. Andiç et al. [22] reported that phenylethylamine levels in herby cheese ranged between 0 and 100 mg/kg, and Bonczar et al. [35] reported the mean phenylethylamine level of 8.76 ± 6.85 mg/kg in Emmental cheese.

The level of tryptamine determined in the herby cheeses ranged from <0.025 to 33.36 mg/kg of sample (Table 3). Andiç et al. [22] described tryptamine levels in the range of not detected to 172.60 mg/kg in herby cheese. According to Bonczar et al. [35], the mean amount of tryptamine was 48.91 ± 19.99 mg/kg in Harzer cheese. This variability within the same type of cheeses could be attributed to differences in the manufacturing process, such as the type of milk used (sheep or cow), heat treatment of the milk (such as pasteurization), ripening time, microflora, and cheese mass, as discussed by Andiç et al. [22].

The amount of tyramine found in the herby cheeses ranged from <0.309 to 725.21 mg/kg of sample (Table 3). According to Andiç et al. [22], tyramine levels in herby cheese ranged between 18.00 and 1125.50 mg/kg, and according to Fiechter et al. [41], were 51.60 mg/100 g in Harzer cheese with caraway seeds. Nout [52] pointed out that the maximum allowable level of tyramine in foods should be in the range of 100–800 mg/kg, and Shalaby [14] and Valsamaki et al. [51] stated that the safe summed level of histamine, tyramine, putrescine, and cadaverine should not significantly exceed the higher dose of 900 mg/kg. The concentrations of tyramine found in our study were lower than these levels.

Putrescine levels ranged from 0.03 to 426.35 mg/kg in the herby cheeses (Table 3). Bonczar et al. [35] described the mean putrescine level of 281.33 ± 114.90 in Harzer cheese. The biogenic amine (BA) content of cheese can be extremely variable and depends on the type of cheese, the ripening time, the manufacturing process, and the microorganisms present. The production of biogenic amines in cheese has often been associated with non-starter lactic acid bacteria and *Enterobacteriaceae* [53], so it may be a toxicological risk associated with the consumption of raw milk cheese, especially for sensitive individuals [34].

The mean level of spermine found in the herby cheeses was 0.25 ± 0.03 mg/kg of sample (Table 3). According to Vale and Gloria [42], spermine levels in Prato cheese ranged between 0.07 and 0.90 mg/kg, and according to Komprda et al. [1], the mean was 0.2 ± 0.1 mg/kg in Dutch-type hard cheese. According to Spizzirri et al. [54], the mean level of spermine in Parmigiano Reggiano cheese was 36.7 ± 2.3 mg/kg, and Bonczar et al. [35] reported the mean spermine level of 5.70 ± 1.48 mg/kg in Harzer cheese. The presence of these amines in herby cheese should be considered as a consequence of poor hygienic milk quality [22].

The mean content of spermidine found in herby cheese was 0.17 ± 0.03 mg/kg of sample (Table 3). According to El-Zahar [34], the mean level of spermidine in Mish cheese was 4 ± 0.63 mg/100 g, and Komprda et al. [1] reported the mean as being 0.3 ± 0.1 mg/kg in Dutch-type hard cheese. Spizzirri et al. [54] described that mean spermidine levels in Parmigiano Reggiano cheese were 73.1 ± 1.5 mg/kg, and Bonczar et al. [35] determined the mean spermidine content of Harzer cheese as being 7.74 ± 1.09 mg/kg.

The risk of biogenic amine poisoning could be controlled by applying basic good manufacturing and hygiene practices associated with an appropriate hazard analysis critical control point (HACCP) system [55]. In evaluating the risks of foodstuffs to the consumer's health, the safe daily biogenic amines (BAs) intake should be regarded as complex. Some substances, such as alcohol, can decrease the activity of enzymes that participate in the degradation of BAs in the human intestines [14]. Since cheeses are often served with alcoholic drinks such as beer or wine, even low concentrations of BAs in cheeses can cause adverse effects. Beer and wine often contain high amounts of BAs, which might intensify the negative impact of BAs in cheeses on human health [43,56].

The pH value is an important factor influencing amino acid decarboxylase activity, which is stronger in an acidic environment, with the optimum pH level being between 4.00 and 5.50 [9]. The pH values ranged from 4.49 to 6.64 in this paper (Table 2). According to Andic et al. [22], the pH of herby cheeses ranges from 4.03 to 6.09. Kavaz et al. [57] reported the mean pH value in herby cheeses as being 5.82 ± 0.07. According to Tarakci et al. [20], the pH values of herby cheeses range from 4.01 to 5.40. The ripened herby cheeses showed pH values close to being alkaline, which is caused by the change of lactic acid into carbon dioxide. Sagun et al. [30] reported the pH value of herby cheese of 4.59 ± 0.44 as mean. Tuncturk et al. [58] described the mean pH value of herby cheeses as being 5.32 ± 0.04. Higher levels of organic acids or lower pH in ripened herby cheeses is found to be the specific result for this kind of cheese.

The dry matter content ranged from 49.66 to 65.80% with the mean of $60.37 \pm 0.39\%$ in this study (Table 2). Andic et al. [22] found the mean dry matter content in herby cheese to be $54.3 \pm 1.03\%$. Kavaz et al. [57] reported the mean dry matter content as being 58.15 ± 0.28 in herby cheese, and Tarakci et al. [20] determined that the dry matter content of herby cheese ranges from 50.54 to 66.05%, with the mean being $55.41 \pm 4.454\%$. Isleyici [59] found the mean dry matter content in herby cheese to be $47.783 \pm 5.06\%$. Tekinsen [31] reported that the dry matter content in herby cheese ranges from 29.10 to 61.57%. According to Tuncturk et al. [58], the mean dry matter content in herby cheese was $46.01 \pm 0.09\%$.

Titratable acidity (TA), measured by in lactic acid (LA%), ranged from 1.00 to 3.16 with the mean of 2.01 ± 0.04 in this study (Table 2). Tarakci et al. [20] reported that the titratable acidity (LA%) of herby cheese ranges from 0.82 to 2.35 with the mean of 1.84 ± 0.374. Isleyici [59] found the mean TA in herby cheese to be $0.809 \pm 0.333\%$, and Sagun et al. [30] reported the TA in herby cheese of 1.18 ± 0.21. Tekinsen [31] reported that the TA of herby cheese ranges from 0.184 to 1.757. According to Tuncturk et al. [58], the mean TA of herby cheese is 1.47 ± 0.05.

The salt levels (%) ranged from 5.85 to 11.70% with the mean of 8.64 ± 0.17% in this study (Table 2). Sagun et al. [30] described that the mean salt level in herby cheese was 5.14 ± 0.61%. Isleyici [59] found the mean salt level in herby cheese to be 5.69 ± 1.11%. Kavaz et al. [57] reported the mean salt level to be 3.63 ± 0.14% in herby cheese. Tarakci et al. [20] found that the salt level of herby cheese ranged from 4.80 to 9.07% with the mean of 6.64 ± 1.190%. According to Andic et al. [22], the mean salt level in herby cheese was 9.01 ± 0.4%. Tuncturk et al. [58] described that the mean salt level in herby cheese was 4.45 ± 0.17%. There has been great variability in the results obtained for the chemical composition of the examined herby cheeses. This variability depends on many factors, such as the fresh milk quality, technological processes applied (pasteurization, starter culture addition, time and temperature of thermomechanical curd treatment, salting), and time and temperature of ripening, among many others [60].

5. Conclusions

The current study provides valuable information on the bacterial, chemical, and biogenic amine content in herby cheese. Taking into account all the microbiological results, it could be concluded that herby cheese can be highly prone to contamination, particularly with Enterobacteriaceae, coliform, and *S. aureus*, which is probably related to poor hygiene during cheesemaking manipulations. The use of raw milk and various herbs in the making of herby cheeses may result in pathogen contamination. Nevertheless, it can be said that the herby cheeses analyzed presented remarkably low bacterial densities. During the manufacture of herby cheese, it has been deemed necessary and highly appropriate to take several measures aimed at reducing the production of biogenic amines, to ensure sanitary conditions in making herby cheese, and to use starter cultures formed by lactic acid bacteria with acidifying capacity. The amines were not detected in every sample, and there was high variability in the amine levels among the samples analyzed. Histamine was only found in ten (10%) herby cheese samples at levels of up to 200 mg/kg, which can be considered toxicologically significant. The levels of spermine and spermidine were low and did not exceed the value of 35 mg/kg [52]. The levels of tyramine were also low and did not exceed the value of 800 mg/kg that is regarded as safe for the consumer's health. These cheeses are fit for consumption in terms of the amount of biogenic amines they contain. Moreover, susceptible individuals should be advised to consume cheeses with low biogenic amine contents. However, the handling of raw materials and production technology for herby cheeses are relatively primitive in Turkey. Particularly if the person is vulnerable (when the histamine detoxification mechanism is inhibited), biogenic amines in amounts much lower than those mentioned here may cause intoxications. For this reason, the sources and critical control points for biogenic amine formation during cheesemaking should be determined in order to limit amine formation and accumulation in herby cheese.

Author Contributions: Conceptualization, K.E., H.O.; Sample analysis, O.I., R.M.T.; Sample collection, Y.C.S.; Statistical analysis of data, H.O.

Funding: This research received no external funding.

Acknowledgments: The authors would like to thank Filiz Karatas and Tuncer Cakmak for helping with laboratory analysis, and give thanks for the English support provided by the Yüzüncü Yıl University, Distance Education Center Directorate (YÜSEM), and Marion Kelly from the Swan Training Institute and Trinity College Dublin, Ireland.

Conflicts of Interest: The authors declare no conflict of interest.

References

1. Komprda, T.; Smělá, D.; Novická, K.; Kalhotka, L.; Šustová, K.; Pechová, P. Content and distribution of biogenic amines in Dutch-type hard cheese. *Food Chem.* **2007**, *102*, 129–137. [CrossRef]
2. Pachlová, V.; Buňka, F.; Flasarová, R.; Válková, P.; Buňkova, L. The effect of elevated temperature on ripening of Dutch type cheese. *Food Chem.* **2012**, *132*, 1846–1854. [CrossRef]

3.   Renes, E.; Diezhandino, I.; Fernandez, D.; Ferrazza, V.R.E.; Tornadijo, M.E.; Fresno, J.M. Effect of autochthonous starter cultures on the biogenic amine content of ewe's milk cheese throughout ripening. *Food Microbiol.* **2014**, *44*, 271–277. [CrossRef] [PubMed]

4.   Bover-Cid, S.; Hernandez-Jover, T.; Miguelez-Arrizado, M.J.; Vidal-Carou, M.C. Contribution of contaminant enterobacteria and lactic acid bacteria to biogenic amine accumulation in spontaneous fermentation of pork sausages. *Eur. Food Res. Technol.* **2003**, *216*, 477–482. [CrossRef]

5.   Buňková, L.; Adamcova, G.; Hudcovac, K.; Velichova, H.; Pachlova, V.; Lorencova, E.; Bunka, F. Monitoring of biogenic amines in cheeses manufactured at small-scale farms and in fermented dairy products in the Czech Republic. *Food Chem.* **2013**, *141*, 548–551. [CrossRef] [PubMed]

6.   Righetti, L.; Tassoni, A.; Bagni, N. Polyamines content in plant derived food: A comparison between soybean and Jerusalem artichoke. *Food Chem.* **2008**, *111*, 852–856. [CrossRef]

7.   Edwards, S.T.; Sandine, W.E. Public health significance of amines in cheese. *J. Dairy Sci.* **1981**, *64*, 2431–2438. [CrossRef]

8.   Chong, C.Y.; Abu Bakar, F.; Russly, A.R.; Jamilah, B.; Mahyudin, N.A. The effects of food processing on biogenic amines formation. *Int. Food Res. J.* **2011**, *18*, 867–876.

9.   Stratton, J.E.; Hutkins, R.W.; Taylor, S.L. Biogenic amines in cheese and other fermented foods: A review. *J. Food Prot.* **1991**, *54*, 460–470. [CrossRef]

10. Linares, D.M.; Martín, M.C.; Ladero, V.; Álvarez, M.A.; Fernández, M. Biogenic amines in dairy products. *Crit. Rev. Food Sci. Nutr.* **2011**, *51*, 691–703. [CrossRef]

11. Moreno-Arribas, V.; Polo, M.C.; Jorganes, F.; Muñoz, R. Screening of biogenic amine production by lactic acid bacteria isolated from grape must and wine. *Int. J. Food Microbiol.* **2003**, *84*, 117–123. [CrossRef]

12. Landete, J.M.; Pardo, I.; Ferrer, S. Tyramine and phenylethylamine production among lactic acid bacteria isolated from wine. *Int. J. Food Microbiol.* **2007**, *115*, 364–368. [CrossRef] [PubMed]

13. Bover-Cid, S.; Schoppen, S.; Izquierdo-Pulido, M.; Vidal-Carou, M.C. Relationship between biogenic amine contents and the size of dry fermented sausages. *Meat Sci.* **1999**, *51*, 305–311. [CrossRef]

14. Shalaby, A.R. Significance of biogenic amines to food safety and human health. *Food Res. Int.* **1996**, *29*, 675–690. [CrossRef]

15. Tortorella, V.; Masciari, P.; Pezzi, M.; Mola, A.; Tiburzi, S.P.; Zinzi, M.C.; Scozzafava, A.; Verre, M. Histamine poisoning from ingestion of fish or scombroid syndrome. *Case Rep. Emerg. Med.* **2014**, *2014*, 1–4. [CrossRef] [PubMed]

16. Ruiz-Capillas, C.; Jimenez Colmenero, F. Biogenic amines in meat and meat products. *Crit. Rev. Food Sci. Nutr.* **2004**, *44*, 489–499. [CrossRef]

17. Joosten, H.M.L.J.; Van-Boekel, M.A.J.S. Condition allowing the formation of biogenic amines in cheese. 4. A study of the kinetics of histamine formation in an infected Gouda cheese. *Neth. Milk Dairy J.* **1988**, *42*, 3–24.

18. Schneller, R.; Good, P.; Jenny, M. Influence of pasteurized milk, raw milk and different ripening cultures on biogenic amine concentrations in semi soft cheeses during ripening. *Z. Lebensm. Unters. Forsch. A* **1997**, *204*, 265–272. [CrossRef]

19. Gardini, F.; Martuscelli, M.; Caruso, M.C.; Galgano, F.; Crudele, M.A.; Favati, F.; Guerzoni, M.E.; Suzzi, G. Effects of pH, temperature and NaCl concentration on the growth kinetics, proteolitic activity and biogenic amine production of *Enterococcus faecalis*. *Int. J Food Microbiol.* **2001**, *64*, 105–117. [CrossRef]

20. Tarakci, Z.; Coskun, H.; Tuncturk, Y. Some properties of fresh and ripened Herby cheese: A traditional variety produced in Turkey. *Food Technol. Biotechnol.* **2004**, *42*, 47–50.

21. Ekici, K.; Coskun, H.; Tarakci, Z.; Ondul, E.; Sekeroglu, R. The contribution of herbs to the accumulation of histamine in "otlu" cheese. *J. Food Biochem.* **2006**, *30*, 362–371. [CrossRef]

22. Andic, S.; Genccelep, H.; Köse, S. Determination of biogenic amines in Herby Cheese. *Int. J. Food. Prop.* **2010**, *13*, 1300–1314. [CrossRef]

23. Tuncturk, Y.; Coskun, H.; Ghosh, B.C. Nitrogen fractions in brine during ripening of Herby cheese (Otlu peynir). *Indian J. Dairy Sci.* **2003**, *56*, 208–212.

24. Pichhardt, K. *Lebensmittelmikrobiologie*; 3. Auflage; Springer: Berlin, Germany, 1993; pp. 117, 134, 193, 197. ISBN 978-3-642-97449-6.

25. Tekinşen, O.C.; Atasever, M.; Keleş, A. *Süt Ürünleri Üretimi ve Kontrolü*; S.Ü. Basımevi, Mimoza Basım Yayım ve Dağıtım AŞ: Konya, Turkey, 1997; ISBN 975-448-123-3.

26. Eerola, S.; Hinkkanen, R.; Lindfors, E.; Hirvi, T. Liquid chromatographic determination of biogenic amines in dry sausages. *J. AOAC Int.* **1993**, *76*, 575–577.

27. SAS. *SAS/STAT Software: Hangen and Enhanced*; Version 9.4; SAS, Inst. Inc.: Cary, NC, USA, 2014.

28. Ozturk, B. Van'da Faaliyet Gösteren süt İşletmelerinde Üretilen Peynirlerin Son Ürün Kalitesi Yönünden Değerlendirilmesi. Master's Thesis, Y.Y.Ü. Fen Bil. Enst., Van, Turkey, 2000.

29. Isleyici, O.; Akyuz, N. Van İlinde satışa sunulan otlu peynirlerde mikrofloranın ve laktik asit bakterilerinin belirlenmesi. *YYU Vet. Fak. Derg.* **2009**, *20*, 59–64.

30. Sagun, E.; Sancak, H.; Durmaz, H. Van'da kahvaltı salonlarında tüketime sunulan süt ürünlerinin mikrobiyolojik ve kimyasal kaliteleri üzerine bir araştırma. *Y.Y.Ü. Vet. Fak. Derg.* **2001**, *12*, 108–112.

31. Tekinşen, K.K. Hakkari ve çevresinde üretilen Otlu peynirlerin mikrobiyolojik ve kimyasal kalitesi. *Vet. Bil. Derg.* **2004**, *20*, 79–85.

32. Alemdar, S.; Ağaoğlu, S. Behavior of *E. coli* O157:H7 during the ripening of herby cheese manufactured from raw milk. *Food Health* **2016**, *2*, 49–56. [CrossRef]

33. Erkan, E.M.; Cıftcıoglu, G.; Vural, A.; Aksu, H. Some microbiological characteristics of Herbed cheeses. *J. Food Qual.* **2007**, *30*, 228–236. [CrossRef]

34. El-Zahar, K.M. Biogenic amines and microbiological profile of Egyptian cheeses. *Univers. J. Food Nutr. Sci.* **2014**, *2*, 18–26.

35. Bonczar, G.; Filipczak-Fiutak, M.; Pluta-Kubica, A.; Duda, I.; Walczycka, M.; Staruch, L. The range of protein hydrolysis and biogenic amines content in selected acid-and rennet-curd cheeses. *Chem. Pap.* **2018**, *72*, 2599–2606. [CrossRef] [PubMed]

36. European Food Safety Authority (EFSA). Scientific opinion on risk based control of biogenic amine formation in fermented foods. Panel on Biological Hazards (BIOHAZ). *EFSA J.* **2011**, *9*, 2393–2486. [CrossRef]

37. Linares, D.M.; del Rio, B.; Redruello, B.; Ladero, V.; Martin, M.C.; Fernandez, M.; Ruas-Madiedo, P.; Alvarez, M.A. Comparative analysis of the in vitro cytotoxicity of the dietary biogenic amines tyramine and histamine. *Food Chem.* **2016**, *197*, 658–663. [CrossRef]

38. Ladero, V.; Calles-Enriquez, M.; Fernandez, M.; Alvarez, M.A. Toxicological effects of dietary biogenic amines. *Curr. Nutr. Food Sci.* **2010**, *6*, 145–156. [CrossRef]

39. Parente, E.; Martuscelli, M.; Gadrini, F.; Grieco, S.; Crudele, M.A.; Suzzi, G. Evolution of microbial populations and biogenic amine production in dry sausages produced in Southern Italy. *J. Appl. Microbiol.* **2001**, *90*, 882–891. [CrossRef] [PubMed]

40. Standarová, E.; Vorlová, L.; Kordiovská, P.; Janštová, B.; Dračková, M.; Borkovcová, I. Biogenic amine production in olomouc curd cheese (olomoucké tvarůžky) at various storage conditions. *Acta Vet. Brno* **2010**, *79*, 147–156. [CrossRef]

41. Fiechter, G.; Sivec, G.; Mayer, H.K. Application of UHPLC for the simultaneous analysis of free amino acids and biogenic amines in ripened acid-curd cheeses. *J. Chromatogr. B* **2013**, *927*, 191–200. [CrossRef] [PubMed]

42. Vale, S.; Gloria, M.B.A. Determination of biogenic amines in cheese. *J. AOAC Int.* **1997**, *60*, 651–657.

43. Bunka, F.; Budinský, P.; Cechová, M.; Drienovský, V.; Pachlová, V.; Matoulková, D.; Kubán, V.; Bunková, L. Content of biogenic amines and polyamines in beers from the Czech Republic. *J. Inst. Brew.* **2012**, *118*, 213–216. [CrossRef]

44. Budak, G.N.F.; Karahan, A.G.; Çakmakçı, M.L. Factors affecting histamine and tyramine formation in Turkish White Cheese. *Hacettepe J. Biol. Chem.* **2008**, *36*, 197–206.

45. Sancak, Y.C.; Ekici, K.; Isleyici, Ö.; Sekeroglu, R.; Noyan, T. A study on the determination of histamine levels in Herby cheese. *Milchwissenschaft-Milk Sci. Int.* **2004**, *60*, 162–163.

46. Antila, P.; Antila, V.; Mattila, J.; Hakkarainen, H. Biogenic amine in cheese. I. Determination of biogenic amines in Finnish cheese using high performance liquid chromotography. *Milchwissenschaft* **1984**, *39*, 81–85.

47. Madejska, A.; Michalski, M.; Osek, J. Histamine content in rennet ripening cheeses during storage at different temperatures and times. *J. Vet. Res.* **2018**, *62*, 65–69. [CrossRef] [PubMed]

48. Sagun, E.; Ekici, K.; Durmaz, H. The formation of histamine in Herby cheese during ripening. *J Food Qual.* **2005**, *28*, 171–178. [CrossRef]

49. Henry, M. Dosage biologique de l'histamine dans les aliments. *Ann. Fals. Exp. Chim.* **1960**, *53*, 24–33.

50. Sumner, S.S.; Roche, F.; Taylor, S.L. Factors controlling histamine production in Swiss cheese inoculated with *Lactobacillus buchneri*. *J. Dairy Sci.* **1990**, *73*, 3050–3058. [CrossRef]

51. Valsamaki, K.; Michaelidou, A.; Polychroniadou, A. Biogenic amine production in Feta cheese. *Food Chem.* **2000**, *71*, 259–266. [CrossRef]

52. Nout, M.J.R. Fermented foods and food safety. *Food Res. Int.* **1994**, *27*, 291–298. [CrossRef]

53. Joosten, H.M.L.J.; Northolt, M.D. Conditions allowing the formation of biogenic amines in cheese. 2. Decarboxylative properties of some non-starter bacteria. *Neth. Milk Dairy J.* **1987**, *41*, 259–280.

54. Spizzirri, U.G.; Restuccia, D.; Curcio, M.; Parisi, O.I.; Iemma, F.; Nevio Picci, N. Determination of biogenic amines in different cheese samples by LC with evaporative light scattering detector. *J. Food Comp. Anal.* **2013**, *29*, 43–51. [CrossRef]

55. Visciano, P.; Schirone, M.; Tofalo, R.; Suzzi, G. Histamine poisoning and control measures in fish and fishery products. *Front. Microbiol.* **2014**, *5*, 1–3. [CrossRef] [PubMed]

56. Ancín-Azpilicueta, C.; González-Marco, A.; Jiménez-Moreno, N. Current knowledge about the presence of amines in wine. *Crit. Rev. Food Sci.* **2008**, *48*, 257–275. [CrossRef] [PubMed]

57. Kavaz, A.; Bakırcı, İ.; Kaban, G. Some physico-chemical properties and organic acid profiles of Herby cheeses. *Kafkas Univ. Vet. Fak. Derg.* **2013**, *19*, 89–95.

58. Tunçtürk, Y.; Ocak, E.; Köse, Ş. Farklı süt türlerinden üretilen Van otlu peynirlerinin fiziksel ve kimyasal özellikleri ile proteoliz profillerinde olgunlaşma sürecinde meydana gelen değişimler. *Gıda* **2014**, *39*, 163–170.

59. Isleyici, O. Otlu Peynir Mikroflorasındaki Laktik Asit Bakterilerininin Izolasyonu, Identifikasyonu ve bu Peynir Yapımında Kullanılabilecek Starter Kültürlerin Tespiti. Ph.D. Thesis, Y.Y.Ü. Fen Bilimleri Enstitüsü, Van, Turkey, 1999.

60. Fuentes, L.; Mateo, J.; Quinto, E.J.; Caro, I. Changes in quality of nonaged pasta filata Mexican cheese during refrigerated vacuum storage. *J. Dairy Sci.* **2015**, *98*, 2833–2842. [CrossRef] [PubMed]

foods

MDPI

Article

Screening Method to Evaluate Amino Acid-Decarboxylase Activity of Bacteria Present in Spanish Artisanal Ripened Cheeses

Diana Espinosa-Pesqueira, Artur X. Roig-Sagués * and M. Manuela Hernández-Herrero

CIRTTA—Departament de Ciència Animal i dels Aliments, Universitat Autònoma de Barcelona, Travessera dels Turons S/N, 08193 Barcelona, Spain; diespe@gmail.com (D.E.-P.); manuela.hernandez@gmail.com (M.M.H.-H.)
* Correspondence: arturxavier.roig@uab.cat; Tel.: +34-935-812-582

Received: 13 September 2018; Accepted: 31 October 2018; Published: 6 November 2018

Abstract: A qualitative microplate screening method, using both low nitrogen (LND) and low glucose (LGD) decarboxylase broths, was used to evaluate the biogenic amine (BA) forming capacity of bacteria present in two types of Spanish ripened cheeses, some of them treated by high hydrostatic pressure. BA formation in decarboxylase broths was later confirmed by High Performance Liquid Chromatography (HPLC). An optimal cut off between 10–25 mg/L with a sensitivity of 84% and a specificity of 92% was obtained when detecting putrescine (PU), tyramine (TY) and cadaverine (CA) formation capability, although these broths showed less capacity detecting histamine forming bacteria. TY forming bacteria were the most frequent among the isolated BA forming strains showing a strong production capability (exceeding 100 mg/L), followed by CA and PU formers. *Lactococcus*, *Lactobacillus*, *Enterococcus* and *Leuconostoc* groups were found as the main TY producers, and some strains were also able to produce diamines at a level above 100 mg/L, and probably ruled the BA formation during ripening. *Enterobacteriaceae* and *Staphylococcus* spp., as well as some *Bacillus* spp. were also identified among the BA forming bacteria isolated.

Keywords: biogenic amines; decarboxylase activity; screening method; artisanal cheese; high hydrostatic pressure

1. Introduction

Cheese is, after fish, the food product that most usually causes poisoning due to the presence of high amounts of biogenic amines (BA), compounds with psychoactive and vasoactive properties that can be formed in foodstuffs due to the microbial decarboxylation of amino acids [1–6]. Amino acid decarboxylase activity has been described for several groups of microorganisms, such as *Enterobacteriaceae*, *Pseudomonas* spp, *Enteroccoccus*, *Micrococcus* and Lactic Acid Bacteria (LAB). These BA-producing organisms may be part of the microbiota of the raw materials or may be introduced by contamination during or after processing of foodstuffs [4,7–14]. The specificity of the amino acid decarboxylases is strain dependent [11,15]. Lactic acid bacteria (LAB) have an important role in cheese elaboration and they are also the most important bacterial group that may build-up biogenic amine (BA), especially tyramine (TY) and putrescine (PU), but also cadaverine (CA) and histamine (HI) [13,16–18]. Sumner et al. [19] isolated a strain of *Lactobacillus buchneri* (strain St2A) from a Swiss cheese involved in an outbreak of HI poisoning occurred in the USA in 1980 that was able to form high amounts of HI. This LAB, later classified al *L. parabuchneri*, is able to grow and produce histamine at refrigeration temperatures [20]. *Enterobacteriaceae*, *Staphylococcus* spp. or *Bacillus* spp. have also been related to the accumulation of diamines in foods, including cheese, but also TY and/or HI [8,17,21,22].

Diverse qualitative and quantitative methods have been described in the literature to evaluate the amino acid decarboxylase activity of microorganisms isolated in food products. Different culture media have been proposed to be used as screening qualitative procedures, the most being formulated as a basal medium that include sources of carbon (glucose), nitrogen (peptone, yeast or meat extract), vitamins, salt, a relative high amount of one (or several) precursor amino acids and a pH indicator (e.g., bromocresol purple). Decarboxylase activity is then detected by the pH shift that changes the color of the medium when the carboxylic group is released from the amino acid(s) leaving in the medium the more alkaline BA(s) [23,24]. False-positive results have been described probably due to the formation of other alkaline compounds [9,22,25], but also false-negative responses are possible as a result of the fermentative activity of some bacteria, such as LAB, which produce acid that neutralize the alkalinity of BA [23,26].

In a previous work, the formation of BA in two artisanal varieties of Spanish ripened cheese, one made of ewe's raw milk and other of goat's raw milk, was presented. The effect of high hydrostatic pressure (HHP) treatments on both the levels of BA formed and on the main microbial groups present was also evaluated [27]. The aim of the present work has been to develop a fast, reliable and easy to perform screening method to evaluate the bacterial formation capacity of a wide range of BA, and evaluate the BA formation capability of the microbiota present in these two varieties of cheese to understand why HHP treatments reduce the formation of BA.

2. Materials and Methods

2.1. Cheese Manufacturing

Two types of artisan ripened cheeses elaborated in Spain were studied in this survey, both made of enzymatic curd and pressed paste. The first one was produced from goat's raw milk in the region of Catalonia, northeast of Spain, and the second was made from ewe's raw milk in Castilla y León, central Spain. The procedure of sampling as well as the HHP treatment applied have been described in a previous work [27]. Three independent batches of each type of cheese were produced following the usual manufacturing procedures used by the manufacturers. Cheese samples were separated in three batches: samples not HHP treated (Control samples); samples HHP treated before the 5th day of ripening (HHP1) and samples treated after 15 days of ripening (HHP15, only for ewe's milk cheeses). HHP treatments were performed at 400 MPa for 10 min at a temperature of 2 °C using an Alstom HHP equipment (Alstom, Nantes, France) with a 2 L pressure chamber.

2.2. Strain Isolation

Ten grams of each cheese sample were homogenized in 90 mL of sterile Buffered Peptone Water (Oxoid, Basingstoke, Hampshire, UK) with a BagMixer 400 paddle blender (Interscience, St Nom la Bretèche, France) and plated on M-17 agar (Oxoid) supplemented with a bacteriological grade lactose solution (5 g/L, Oxoid) and incubated at 30 °C, 48 h to isolate *Lactococcus* spp.; on de Man Rogosa Sharpe agar (MRS, Oxoid) incubated at 30 °C for 48 h to isolate Lactobacilli; on Kenner Fecal *Streptococcus* Agar (KF, Oxoid) supplemented with 2,3,5-triphenyltetrazolium chloride solution 1% (Oxoid) and incubated at 37 °C for 48 h to isolate Enterococci; Violet Red Bile Glucose Agar (VRBG, Oxoid) incubated at 37 °C for 24 h to isolate *Enterobactericeae* and Baird Parker Agar (BPA, BioMérieux, Marcy L'Etoile, France) incubated at 37 °C for 24–48 h to isolate *Staphylococcus* strains.

A total of 688 isolates were randomly picked out from the different selective media. The purification of each isolated was made by streaking single colonies on Petri plates with Tryptone Soy Agar (Oxoid) and incubating at 30 °C for 24–48 h. Two TY producing strains of *Lactobacillus brevis* and *Lactobacillus casei* and an HI producing strain of *Staphylococcus epidermidis*, isolated from previous surveys were used as positive controls [8,9]. These cultures were recovered in 10 mL of Tryptone Soy Broth (Oxoid) and incubated at 30 °C for 24 h. The purity of each culture was verified by subculturing the *Lactobacillus brevis* and *Lactobacillus casei* strains onto MRS agar (Oxoid), incubated at 30 °C for 24 h, and the *Staphylococcus epidermidis* strain on BPA (BioMérieux) incubated at 37 °C for 24 h. Before performing the decarboxylase assay, each strain was suspended in a tube with physiological solution of NaCl 0.85% (Panreac, Barcelona, Spain) until reaching a turbidity of about 0.5 in the McFarland scale.

2.3. Preparation of Decarboxylase Media

Table 1 shows the composition of the two synthetic media formulated to determine the ability to form the most toxic BA (HI and TY) and their enhancers (PU and CA): Low Nitrogen Broth (LND), prepared with the objective to decrease the incidence of false positive results of bacteria with a strong peptidase (or deaminase) activity; and the Low Glucose Broth (LGD) developed with the aim to decrease the incidence of false negative responses of bacteria with a great fermentative activity. Before performing the tests both base broth media were supplemented with the precursor amino acids (L-Lysine monohydrate (Merck, Darmstadt, Germany), L-Ornithine monohydrate (Sigma-Aldrich, Steinheim, Germany), L-Histidine monohydrochloride (Merck) and L-Tyrosine disodium salt (Sigma-Aldrich), individually, or adding a mixture of all them (described in the next section as total amino acid broth). The base broth without amino acids added was used as negative control. All media were adjusted to the pH values indicated in Table 1 and autoclaved at 120 °C during 5 min.

Table 1. Composition of broth media (g/100 mL)) used to evaluate decarboxylase ability of strains isolated from cheeses.

Reagent	Low Nitrogen Decarboxylase Broth (LND)	Low Glucose Decarboxylase Broth (LGD)	Adjusted pH
Tryptone	0.125	0.25	
Yeast extract	0.125	0.25	
NaCl	0.25	0.25	
CaCO$_3$	0.01	0.01	
Pyridoxal-5-phosphate	0.03	0.03	
Glucose	0.05	0.001	
Bromocresol purple	0.01	0.01	
All amino acids	1.0	1.0	5.5
L-Lysine	1.0	1.0	5.0
L-Ornithine	1.0	1.0	5.5
L-Histidine	1.0	1.0	5.7
L-Tyrosine	0.25	0.25	5.5

2.4. Assessment of Amino Acid Decarboxylase Activity

In order to detect the capacity of the isolated strains to form BA and to determine which of the two decarboxylase broths (LND and LGD) show the best results in each one, a screening test was performed on a 96-wells flat bottom Microtiter plate. Aliquots of 200 µL of total amino acid broth (TAB) and 20 µL saline solution were added into 6 wells of a 96 well (decarboxylase control assay: DCA); 200 µL of TAB and 20 µL of bacterial suspension were added into another 6 wells (positive decarboxylase assay: PDA); and 200 µL of broth base without amino acids with 20 µL of bacterial suspension were added into another 6 wells (negative decarboxylase assay: NDA). Microplates were incubated at 30 °C for 24 h. A positive result was considered in PDA wells when a purple color appeared due to an increase of alkalinity (Figure 1a). In LGD broth positive results were also considered when no color changes were observed in PDA wells and yellow color appeared in NDA wells, because a high acidification was produced due to the bacterial growth (Figure 1b). Negative results were considered when no color changes were observed in PDA wells (Figure 1a,b), or when a purple color appeared in NDA wells due to another alkaline compounds different than BA (Figure 1a).

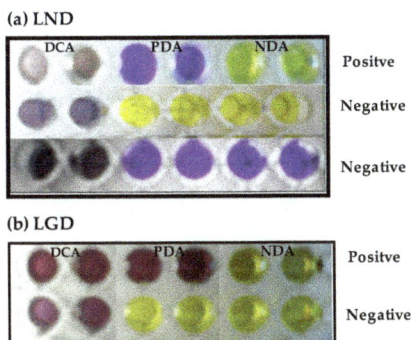

Figure 1. Example of negative and positive responses to amino acid decarboxylase activity in Low Nitrogen Decarboxylase (LND) and Low Glucose Decarboxylase (LGD) media. DCA: decarboxylase control assay; PDA: positive decarboxylase assay; NDA: negative decarboxylase assay.

2.5. Confirmation of Amino Acid Decarboxylase Activity by HPLC

Decarboxylase activity of strains was confirmed by the quantitative analysis of BA produced in the decarboxylase broths by means of reverse-phase High Performance Liquid Chromatography (HPLC), using an automated HPLC system (HPLC P680, Dionex, Sunnyvale, CA, USA) equipped with an Ultra Violet (UV) detector Dionex UVD170U (Thermo-Fisher Scientific, Waltham, MA, USA). Briefly: one mL of each bacterial suspension (0.5 McFarland) was inoculated into a tube containing 4 mL of the TAB version of LND or LGD broths (depending on the previous results for each strain). After 4 days of incubation at 30 °C, the media was centrifuged ($9000 \times g$, 10 min, 20 °C) and 3 mL of the supernatant was extracted with 2 mL of 0.4 M $HClO_4$ (Panreac). Determination of BA was carried out according to the RP-HPLC method described by Eerola et al. [28] and modified by Roig Sagués et al. [8] using dansyl chloride reagent (Sigma-Aldrich Chemical) to derivate the sample. The separation was performed on a Waters Spherisorb S5 ODS 2 45 × 150 mm column (Waters Corporation, Milford, MA, USA). All reagents were of analytical grade and all solvents involved in derivatization and in the separation process were of HPLC grade. The BA standards: putrescine (PU), cadaverine (CA), histamine (HI), tyrosine (TY), and the internal standard 1,7-diaminoheptane, were all purchased from Sigma-Aldrich Chemical.

2.6. Analytical Validation of the Qualitative Microplate Method of Amino Acid Decarboxylase Activity

The sensitivity, specificity, and the positive and negative predictive values were obtained to determine the diagnostic properties of the qualitative method [29–31] and were calculated by the following equations:

$$\text{Sensitivity} = \frac{TP}{TP + FN} \times 100$$

where TP is the truly positive amino acid decarboxylating isolates, correctly identified by the screening test and FN is the false negative responses obtained.

$$\text{Specifity} = \frac{TN}{TN + FP} \times 100$$

where TN is the truly negative (TN) amino acid decarboxylating isolates, correctly identified by the screening test and FP is the false positive responses obtained.

$$\text{PPV} = \frac{TP}{TP + FP} \times 100$$

where PPV is the positive predictive value and reflects the proportion of truly positive isolates confirmed by HPLC among all positive isolates evaluated by Microtiter plate screening.

$$NPV = \frac{TN}{TN + FN} \times 100$$

where negative predictive value (NPV) is reflects the proportion of truly negative isolates confirmed by HPLC among all negative isolates evaluated by Microtiter plate screening.

The Receiver Operating Characteristic (ROC) curves were assessed using the MedCalc statistical software, version 11.2.1 (MedCalc, Ostend, Belgium), to know the discriminative power of the qualitative method referred to the HPLC method with its 95% confidence interval. In a ROC curve the true positive rate (Sensitivity) is plotted in function of the false positive rate (100-Specificity). A test with perfect discrimination (no overlap in the two distributions) has a ROC curve that passes through the upper left corner (100% sensitivity, 100% specificity). Therefore, the closer the ROC curve is to the upper left corner, the higher the overall accuracy of the test [32]. The area under the ROC curve (AUC) is a measure of how well a parameter can distinguish between two groups (isolates with amino acid decarboxylase activity/isolates without this capacity). The better overall diagnostic performance of the test is when the AUC value is closer to 1 and the practical lower limit for the AUC of a diagnostic test is 0.5 [31,33]. A classification of diagnostic accuracy for the qualitative method is given according to AUC value: AUC 0.90–1.0 excellent, 0.80–0.90 good, 0.70–0.80 fair, 0.60–0.70 poor, 0.50–0.60 deficient and 0.50 null [34].

The point of intersection of the ROC curve with the diagonal line drawn from 100% sensibility to 100% 1-specificity was chosen as the best discriminator value. The optimal cut-off value showed the highest accuracy, the lowest false negative (FN) and the highest false positive (FP) results.

2.7. Identification of Strains with Decarboxylase Activity

Confirmed decarboxylase-positive strains were identified based on Gram stain and catalase and citochromooxidase activity [35]. Further identification to the species level was carried out by a variety of biochemical tests using API 20-E, API 20-Strep, API-Staph and API 50-CH strips (BioMérieux, Marcy l'Etoile, France).

3. Results and Discussion

3.1. Validation of the Qualitative Microplate Method of Amino Acid Decarboxylase Activity

ROC curve analysis was used to determine the discriminative power and the cuts-off of the amino acid decarboxylase screening method with both media (LND and LGD) to evaluate the specific amino acid decarboxylase activity (Table 2).

Tyrosine decarboxylase test showed an area under the ROC curve (AUC) around 0.98, with an optimal cut-off value at 25 mg/L and 20 mg/L of TY on LND and LGD broths, respectively. This means that the microplate screening method could discriminate the isolates with tyrosine decarboxylase activity the 98% of the time at optimal cut-off. The sensitivity and specificity values obtained with both broths were higher than 92%, reflecting that the number of false negative (FN) and false positive (FP) responses obtained by the qualitative method were generally low. However, the negative predictive value (NPV) was considered low (<66%). AUC for the lysine decarboxylase test displayed was greater than 0.930 with an optimal cut-off concentration of 15 mg/L and 10 mg/L for LND and LGD broths, respectively. In this case, the sensitivity and specificity values using LND broth were about 98% and 93%, respectively, while for LGD broth values were about 88% and 98%, respectively. The assay to detect ornithine decarboxylase with the LND broth showed the highest diagnostic values (over 98%) with the lowest cut-off concentration (10 mg/L) and an AUC higher than 0.995. On the other hand, for the same test using LGD broth an 84% of sensitivity and a 97.5% of specificity were reached at a cut-off value of 15 mg/L with an AUC of 0.907.

Histidine decarboxylase test showed the lowest sensitivity values (below 60%) using both broths with the highest optimal cut-off concentration set at 50 mg/L with specificity values up to 90%. Likewise, the AUC value was the lowest, possibly due to a 16.5% of FP and 24% of FN reactions observed at the optimal cut-off using LND broth, whereas a 9.9% and 25% of FP and FN were obtained, respectively, in LGD broth.

Table 2. Receiver Operating Characteristic (ROC) curve analysis of qualitative method to predict specific amino acid decarboxylase activity in isolates using Low Nitrogen Decarboxylase (LND) and Low Glucose Decarboxylase (LGD) broths.

Broth	Data of ROC Analysis	Ornithine	Lysine	Histidine	Tyrosine
LND	AUC	0.999 (0.974–1.00)	0.992 (0.961–1.00)	0.737 (0.659–0.806)	0.980 (0.943–0.996)
	Optimal cut-off (mg/L)	10	15	50	25
	Sensitivity at optimal cut-off (%)	98.53	98.68	65.38	92.25
	Specificity at optimal cut-off (%)	100	93.15	91.75	100
	PPV (%)	100	83.7	81	100
	NPV (%)	99.0	98.6	83.2	66.7
	FN at cut-off	3	6	11	1
	FP at cut-off	1	1	17	6
LGD	AUC	0.907 (0.849–0.948)	0.935 (0.883–0.969)	0.592 (0.509–0.672)	0.989 (0.956–0.999)
	Optimal cut-off (mg/L)	15	10	50	20
	Sensitivity at optimal cut-off (%)	84.37	88	30	93.06
	Specificity at optimal cut-off (%)	97.5	97.6	100	100
	PPV (%)	90	88	100	100
	NPV (%)	95.8	97.6	90.3	37.5
	FN at cut-off	5	5	2	0
	FP at cut-off	5	3	14	8

AUC: Area under ROC curve and 95% confidence interval; PPV: Positive predictive value; NPV: Negative predictive value; FN: Number of false negative: FP: Number of false positive.

In many occasions it has been reported that qualitative screening decarboxylase methods have some limitations in terms of sensitivity in detecting BA production. The presence of FP and FN reactions reported has not been insignificant. Hernández-Herrero et al. [9] observed that 96.5% of the suspected histamine formers detected by Niven decarboxylase media were finally considered as FP. Likewise, Roig-Sagués et al. [36] found that only a 15.8% of the total presumptively histamine-formers obtained in Joosten and Northolt media [37] were confirmed. Similar results were observed when tyramine decarboxylase capacity was tested in the same media, where only 8.4% of the suspected isolates with tyrosine decarboxylase activity were confirmed. The FP results were attributed to the production of other substances able to alkalinize the media [25]. Similarly, Moreno-Arribas et al. [38] used the Maijala modified decarboxylase media and noticed a high number of FP reactions to PU and agmatine production, but less than were found in the tyrosine decarboxylase activity test. On the contrary, de las Rivas et al. [15] did not find any correlation between the positive responses in the decarboxylase activity media and the BA detected by HPLC. They suggested that the screening Maijala modified decarboxylase media underestimates the number of BA-producing strains. On the contrary, Bover-Cid and Holzapfel [23], in their improved screening media tested on LAB, did not observe FP reactions and only 3 strains showed a negative response with the screening procedure. They justified these FN results due to the low amount of tyramine formed that did not neutralize the acid production of LAB. Although these authors proposed their improved decarboxylase medium as a rapid preliminary method to select strains with low decarboxylase activity, the optimal cut-off value was around 300 mg/L. Torracca et al. [17] reported, using the same decarboxylase medium described by Bover-Cid and Holzapfel [23], an optimal cut-off value of 631 mg/L and 810 mg/L for PUT and TY, respectively.

3.2. Amino Acid Decarboxylase Activity of the Control Strains

Table 3 shows the results after testing the control strains in the microplate screening method and the result of the confirmation by HPLC. Lactobacillus brevis and Lactobacillus casei showed tyrosine decarboxylase activity in LGD broth, and Staphylococcus epidermidis histidine decarboxylase capacity in LND broth. All strains were also able to produce low amounts of PU (around 1 mg/L) but were only detected by HPLC.

Table 3. Biogenic amine production by positive control bacteria strains in the amino acid decarboxylase microplate assay (DMA) and HPLC analysis (mg/L).

Strain	Broth	PU		CA		HI		TY	
		DMA	HPLC	DMA	HPLC	DMA	HPLC	DMA	HPLC
L. brevis	LGD	(−)	1.11	(−)	ND	ND	ND	(+)	109.8
L. casei	LGD	(−)	0.78	(−)	ND	ND	ND	(+)	77.14
S. epidermidis	LND	(−)	0.85	(−)	ND	(+)	46.48	(−)	ND

PU: putrescine; CA: cadaverine; HI: histamine; TY: tyramine; LND: Low Nitrogen Decarboxylase; LGD: Low Glucose Decarboxylase; DMA (detection on microplate assay): (+) Positive; (−) Negative; ND: not detected.

3.3. Biogenic Amine Production by Isolates from Goat's and Ewe's Milk Cheeses

3.3.1. Total Amino-Acid Decarboxylase Activity of the Isolated Strains

A total of 688 strains were obtained from the different culture media and a 43.02% of them gave a positive response in the microplate assay with TAB, being subsequently confirmed by HPLC. A 37.7% of the bacteria isolated from goat's milk cheeses and a 47% of the strains picked up from ewe's milk cheeses were BA-formers. The number of isolates obtained from VRBG and BPA media was much lower since these two groups of microorganisms are a minority among the microbiota of ripened cheeses, and their counts are usually low [27], but the percentage of decarboxylase positive results among these isolates was higher (87.5% and 92% on VRBG and BPA, respectively) than in KF, M-17 and MRS media (49.7, 36.8, and 35% respectively). Several studies found that the decarboxylase activity is more frequent in Enterobacteriaceae strains (from 80 to 95%) and in less extension among LAB strains (from 9.5 to 65%) [8,21,36,38,39]. Nevertheless, Enterobacteriaceae and Staphylococcus usually do not achieve high counts in ripened cheeses made under good hygienic practices and normally become undetectable after few days of ripening, reason why decarboxylase positive LAB is usually considered the main responsible for the formation of high concentrations of BA in cured cheeses [17,21].

3.3.2. Specific Amino Acid Decarboxylase Activity of the Isolated Strains

The assessment of the specific amino acid decarboxylase activity was done with the strains that gave positive responses in the screening assay with TAB. Up to 150 of these isolates were recovered and tested in LND and LGD media, respectively. In general, the capability to decarboxylate tyrosine was the most frequent activity detected (91.6% of the total isolates tested) in the specific amino acid decarboxylase screening assay, followed by the ability to decarboxylate lysine and ornithine (around 33.5%). In these cases, the 95%, 96% and 94% were confirmed by the HPLC analysis, respectively. However, histidine decarboxylase activity was detected in only a 24% of the isolates tested, a 76% of them confirmed by HPLC.

Table 4 shows the number of strains with HPLC-confirmed BA-producing capability obtained from goat´s and ewe's milk cheeses according to the culture media of origin. These results are shown as a whole without considering the HHP treatment to which cheese samples were subjected. Strains were grouped in four categories according to Aymerich et al. [38]: medium amine formers (25–50 mg/L), good amine formers (50–100 mg/L), strong amine formers (100–1000 mg/L) and prolific amine formers (>1000 mg/L). In general, strong amine formers were more frequent among strains with TY-producing capability, followed by those able to form CA and PU. Prolific amine production capacity was only observed in some diamine formers, while the formation of HI in amounts above 100 mg/L was a rare event (Table 4).

Table 4. Biogenic amine forming capacity of bacteria isolated from goat´s and ewe's milk cheeses according to the culture media. High hydrostatic pressure (HHP) treatments to which cheese samples were subjected are not considered.

Medium	BAP	PU				CA				HI				TY			
		±	+	++	+++	±	+	++	+++	±	+	++	+++	±	+	++	+++
VRBG	25	1	2	17	4	0	0	17	7	8	10	4	0	3	6	9	0
BPA	15	1	1	3	3	0	0	4	3	4	4	1	0	2	2	7	0
KF	89	5	3	22	4	13	0	22	9	19	21	4	0	4	8	71	0
M-17	98	7	2	5	0	7	1	2	1	14	6	2	0	5	6	69	0
MRS	72	2	2	2	3	0	0	2	3	7	0	0	0	5	8	50	0
Total	299	16	10	49	14	20	1	47	23	52	41	11	0	19	30	206	0

BAP: Number of positive BA producers; PU: putrescine; CA: cadaverine; Hi: histamine; TY: tyramine; Number of BA-forming isolates detected depending on their production (in mg L^{-1}): (±) 25–50, medium; (+) 50–100, good; (++) 100–1000, strong; and (+++), >1000, prolific.

The isolates obtained from VRBG medium showed the highest frequency of PU and CA forming activity. Lysine and ornithine decarboxylases are very common among enterobacteria, and their detection is commonly used for the biochemical identification of *Enterobacteriaceae* species. A 100% of the isolates picked up from goat's and ewe's milk cheeses showed strong activity (>100 mg/L) for CA, and 87.5% and 85.35% for PU, respectively. Strong TY forming capacity was detected in only one isolate obtained from goat's milk cheeses and in 8 from ewe's milk cheese. The isolates with histidine decarboxylase activity showed a weak production and only one isolate obtained from goat's milk cheese and three from ewe's milk cheese presented the ability to produce more than 100 mg/L. *Enterobacteriaceae* are known to decarboxylate several amino acids, specially arginine, lysine and ornithine [11,39–41] and histidine [17,25,42]. The number of isolates with amino acid decarboxylase activity obtained from BPA culture medium was low. Between 75–100% of these isolates displayed a strong PU, CA and TY forming ability. Whereas histamine accumulation was detected especially in a range of 25–100 mg/L in the 89% of the cases. Little information is available about the production of BA by *Staphylococcus* spp. in cheese. However, some species of this group have been related to a variable formation of TY, PU, CA and/or HI in meat and fish fermented products. Martin et al. [43] found in fermented sausages that TY was the main amine produced by this group and some strains also were able to produce PU, CA and HI. Hernández-Herrero et al. [9] reported that the main HI-formers detected in salted anchovies belonged to this genus and de las Rivas et al. [15] reported some strains isolated from "Chorizo", a Spanish ripened sausage, as TY-formers. Most of the isolates obtained from the KF medium were strong TY formers (76% from goat's milk cheese and 81% from ewe's milk cheese) and some of them were also able to produce above 100 mg/L of PU and CA. The ability to form TY was also frequent in the strains isolated from M-17 medium in both kind of cheeses (79% from goat's and 65% from ewe's milk cheeses, respectively) in amounts above 100 mg/L. In that case, the ability to form diamines was less frequent (around 5% of the isolates) in both type of cheeses. Similar results were found in the isolates obtained from MRS medium where around 69% of them were considered strong TY-producers, but no PU or CA formation was detected in any of the isolates obtained from goat's milk cheese and only a 10% of those obtained from ewe's milk cheese were able to from up to 100 mg/L.

3.3.3. Identification of BA-Producing Strains

The result of the identifications of the BA-forming strains isolated from cheese samples, as well as their BA forming capacity expressed as mg/L, are shown in Table 5 (goat's milk cheese) and Table 6 (ewe's milk cheese). It was not possible to establish a clear effect of the HHP treatments to which some of the cheese samples were submitted on the type of strain and its amine-forming capacity. Consequently, results are shown globally, without considering the type of treatment to which the samples were submitted. In both types of cheese, most of the strains with decarboxylase activity were Gram positive. In the case of the goat's milk cheese only 5 Gram negative strains showed decarboxylase activity, four of them identified as *Hafnia alvei*, all of them with a strong PU and CA forming capacity. One of these strains also showed a strong capacity to form TY, but much lower than diamines. Nevertheless, the maximum capacity to form these amines (and also HI) was shown by a strain that could not be precisely identified. In the case of the ewe's milk cheese, 11 Gram negative strains showed decarboxylase activity, most of them (5) also identified as Hafnia alvei and one of them showing the maximum capability to form CA, PU and HI. The maximum TY-forming capacity was shown by a strain of *Citrobacter freundii*. Strains of *Klebsiella oxytoca* and *Escherichia coli* with strong diamine production capacity were identified from ewe's cheese samples. However only one strain of each specie and other identified as *C. freundii* were able to produce CA and TY in a considerable amount (above 100 mg/L). *H. alvei*, *K. oxytoca* and *E. coli* have been previously associated with the formation of PU, CA and/or HI in foods [8,9,11,17,21,25,36,41]. Also, some strains of these species have been reported to possess the ability to decarboxylate tyrosine [21]. The formation of high amounts of PU and CA, as well as of TY, has been previously reported by *C. freundii* [21,40], indicating that this specie is more prolific forming PU than CA. Enterobacteria is usually a minor group among the microbiota present on fermented products. In the cheese object of this work, enterobacteria counts were usually below 3 Log_{10} after 60 days of ripening, but their counts at the beginning of the process were above 6 Log_{10} [27]. When unhygienic manufacturing practices allow for achieving high counts of enterobacteria at the beginning of the process, the fact that their counts would be later reduced during ripening does not necessarily imply the inhibition of their decarboxylase activity and consequently may contribute to BA formed in the final product [36].

Among strains identified as Gram positive, 50 that showed decarboxylase activity were obtained from the goat's milk cheese and 55 from the ewe's milk cheese, and in general this activity was much higher than the Gram negative strains. Among the positive decarboxylase bacteria isolated from the ewe's milk cheeses one strain of *Staphylococcus chromogenes* showed to be a prolific diamine former and strong TY former. Likewise, strains of *Staphylococcus xylosus* and *Staphylococcus aureus* with strong TY and HI production, respectively, were also found. On the other hand, the most frequent strains with high BA forming capacity obtained from goat's milk cheeses were identified as *Staphylococcus hominis*, that were capable to produce high levels of PU, CA and TY, and *Staphylococcus warneri*, which showed to be a strong diamine producer.

S. chromogenes was previously reported as a prolific PU former, good CA and strong TY and HI-forming bacteria in Spanish salted anchovies [44]. Masson et al. [45] detected a weak TY-production capacity in strains of *S. xylosus* isolated from fermented sausages and Martin et al. [43] found in slightly fermented sausages some strains of *S. xylosus* capable to produce strong and prolific amounts of TY, PU and/or HI, and in a lesser extent of CA. Silla-Santos [46] observed HI-production in the 76% of *S. xylosus* strains isolated from Spanish sausages. Strains of *S. warneri* have also been reported to possess tyrosine decarboxylase, but with great variability of production [15,43,45]. CA and PU formation, in medium-good and strong concentration, have also been described [39]. Drosinos et al. [47] isolated one strain of *S. hominis novobiosepticus* in traditional fermented sausages with lysine and tyrosine decarboxylase activity. As enterobacteria, *Staphylococcus* spp. are associated to the contamination of food during unhygienic handling, and consequently it is important to follow always good manufacturing practices to avoid the proliferation of this kind of bacteria in the product. Nevertheless, in the cheeses from where the studied strains were obtained *Staphylococcus* spp. counts were always low (below 3 Log_{10} at the beginning of the ripening), and their counts reduced during ripening until being undetectable after 60 days in most cases.

Enterococcus faecalis, *Enterococcus faecium* and *Enterococcus durans* were the most frequent amine producing bacteria identified from both types of cheese, with a varied production capacity of TY,

PU and CA (Tables 5 and 6). In addition, one strain of *E. faecium* and two of *E. durans* showed a strong HI formation capacity. Enteroccoci are commonly associated with unhygienic conditions during the production and processing of dairy, although they can play an important role for developing the aroma and flavor of certain type cheeses, especially traditional cheeses produced in the Mediterranean area [48]. Several authors have described *E. faecalis*, *E. faecium* and *E. durans* as the most frequently TY formers in food [12,17,21,23,25,49–52] and also some strains of *E. faecalis* and *E. faecium* have been registered as capable of producing amounts up to 100 mg/L of PU [12,17,21,53] and CA [17,21,50] and/or HI [17,21,54]. Tyrosine is a relevant amino acid for the formation of BA in cheese as it can be an inducer of PU production in *E. faecalis* and would be received by the enterococci cells as a signal to growth, what would lead in an increment in the number of BA-producing cells increasing the risk of accumulating TY and PU in cheese [55]. No references concerning histidine decarboxylase activity of *E. durans* have been found in the literature.

Several BA producing strains isolated from M-17 and MRS media of goat's milk cheeses were identified as *Lactococcus lactis* subsp. *lactis*, followed by Lactobacillus brevis, *Lactobacillus plantarum*, and *Leuconostoc* spp. All of them showed strong TY forming ability and *Leuconostoc* spp. also showed a strong-prolific PU formation. In the case of ewe's milk cheeses, *L. lactis* subsp. *lactis*, *L. lactis* subsp. *cremoris*, *Pediococcus pentosaceus*, *Lactobacillus paracasei* subsp. *paracasei*, *L. plantarum* and *Leuconostoc* spp. were often associated with a strong TY-forming capability. Moreover, two strains identified as *L. lactis* subsp. *lactis* showed strong PU and CA forming ability, respectively, while two strains of *P. pentosaseus* species were strong PU and prolific CA producers, respectively.

Table 5. Identification of strains obtained from goat's raw milk cheeses with decarboxylase activity and their BA production in decarboxylase broths (mg/mL). Results are shown without considering the HHP treatments to which some cheese samples were subjected.

Identification	N		PU		CA		HI		TY
Gram Negative									
Hafnia alvei	4	4	537.7–889.54	2	641–1001.30	4	30.43–95.68	4	22.44–151.54
Enterobacteriaceae	1	1	1037.52	1	1173.24	1	111.43	1	73.41
Gram Positive									
Staphylococcus cohni subsp. *cohni*	1	-		1	1.50	1	1.57	1	25.78
Staphylococcus warneri	2	2	69.29–753.55	2	240.94–694.22	1	88.63	1	65.50
Staphylococcus capitis	1	1	310.30	1	246.37	1	95.13	-	
Staphylococcus lentus	2	1	23.44	2	5.41–19	-		2	39.58–191.69
Staphylococcus hominis	2	1	890.96	1	998.20	1	91.63	2	102.42–245.22
Enterococcus faecalis	8	8	39.66–884.27	8	32.44–972	6	25.62–92.35	8	327.5–477.20
Enterococcus durans	2	-		-		-		2	337.40–357.44
Enterococcus avium	1	-		1	26.52	1	56.51	1	24.88
Enterococcus faecium	6	3	19.68–1113.80	3	1.32–1281.50	3	20.22–111.38	6	9.91–366.47
Lactococcus lactis subsp. *lactis*	10	1	7.25	1	7.50	2	21.24–33.38	10	198.77–450.77
Pediococcus pentosaceus	1	-		-		1	33.44	1	55.71
Lactobacillus brevis	5	-		-		1	22.51	5	212.58–519.52
Lactobacillus plantarum	3	1	22.41	-		1	24.5	3	307.62–528.45
Lactobacillus paracasei subps. *paracasei*	1	-		-		-		1	307.03
Leuconostoc spp.	3	2	252.88–749.05	2	14.47–28.83	2	22.74–25.75	3	330.01–417
Bacillus macerans	1	-		-		-		1	418.10
Bacillus licheniformis	1	-		-		-		1	403.07

N: number of strains identified; PU: putrescine; CA: cadaverine; HI: histamine; TY: tyramine.

Several studies have reported different species of LAB able to form BA, especially TY [8,11,16–18,21,23,25,50,54–57], but also HI [58,59]. Within the species of LAB that may occur in food some strains of *L. brevis* and *L. plantarum* were reported to possess the potential to form TY, PU and/or HI [15,18,21,23,25,60,61]. Strains of *P. pentosaceus* isolated from commercial starters [26] and ripened sausages [25] were reported to form TY. Although *L. lactis* subsp. *lactis*, *L. lactis* subsp. *cremoris* and *L. paracasei* subsp. *paracasei* are species usually used as starter cultures or probiotic strains and usually reported as non-decarboxylating strains [16–18,62], some strains of *L. lactis* and *L. paracasei* have been reported with the ability to form TY, HI, PU and/or CA in amounts up to 1000 mg/L [8,17,18,21,23,25,51,58] and *L. lactis* subsp. *cremoris* has been also described as a PU-producer [12]. *Leuconostoc* spp. has been frequently described among the microbiota present in several Spanish farm house cheeses [3,63,64], and has also been described to possess tyrosine, lysine and/or histidine decarboxylase activity [39]. Likewise, González de Llano et al. [63] reported the production of TY in a range of 100–1000 mg/L by strains of *Leuconostoc*. Pircher et al. [21] found that strains of this genus could form TY, PU, CA and /or HI in amounts up to 100 mg/L.

Strains belonging to the *Bacillus* genus isolated from ripened salted anchovies, cheese and raw sausages have previously been described as BA-formers [9,22,36]. One of them was *Bacillus macerans* isolated from Italian cheese, which was capable to from prolific amounts of HI. Formation of CA and PU was also observed in this strain [22].

Table 6. Identification of strains obtained from ewe's raw milk cheeses with decarboxylase activity and their BA production in decarboxylase broths (mg/mL). Results are shown without considering the HHP treatments to which some cheese samples were subjected.

Identification	N		PU		CA		HI		TY
Gram Negative									
Escherichia coli	2	2	746–857.48	2	762.35–983.93	2	41.37–74.12	2	21.7–281.48
Hafnia alvei	5	5	738.50–1049.51	5	787.20–1180.12	4	43.52–185.54	5	48.83–184.79
Klebsiella oxytoca	2	2	30.63–458.50	2	493.41–866.96	2	16.04–27.22	2	5.01–167.24
Citrobacter freundii	1	1	67.1	1	1095.3	1	45.62	1	372.38
Enterobacteriaceae	1	1	832.76	1	846.95	1	83.95	1	69.42
Gram Positive									
Staphylococcus xylosus	2	-		-		2	42.31–68.5	2	92.09–475.35
Staphylococcus chromogenes	1	1	1142.92	1	1760.06	1	31.12	1	441.40
Staphylococcus aureus	1	-		-		1	100.54	-	
Enterococcus faecalis	5	3	860.29–978.9	4	35.43–1394.87	5	20.7–92.98	5	338.01–461.50
Enterococcus durans	9	9	12.63–1160.44	8	13.32–1773.03	9	30.8–179.32	9	46.09–747.33
Enterococcus faecium	2	2	24.89–847.05	2	877.24–941.69	2	28.98–29.97	2	100.54–149.41
Enterococcus hirae	2	2	552.4–579.25	2	600.67–615.75	2	32.36–45.95	2	108.01–187.58
Enterococcus avium	2	-		1	15.17	1	2.18	2	19.93–434.25
Streptococcus salivarius	1	-		-		-		1	742.1
Lactococcus lactis subsp. *lactis*	9	3	34.39–795.26	2	14.95–844.13	2	16.38–30.72	9	211.93–566.2
Lactococcus lactis subsp. *cremoris*	4	3	4.78–37.63	3	5.23–22.94	2	11.15–19.03	4	229.47–406.87
Pediococcus pentosaceus	4	3	9.91–897.67	3	10.25–1018.4	3	27.06–88.78	4	34.58–411.66
Lactobacillus paracasei subsp. *paracasei*	6	2	17.42–45.05	1	17.18	1	51.93	6	365.79–575.73
Lactobacillus plantarum	4	1	8.58	1	17.58	1	26.26	4	45.1–353.31
Lactobacillus brevis	1	-		-		-		1	240.08
Lactobacillus pentosus	1	1	50.79	-		-		1	422.75
Leuconostoc spp.	5	2	11.13–1162.76	2	22.52–1781.26	3	29.63–33.82	5	392.01–626.42

N: number of strains identified; PU: putrescine; CA: cadaverine; HI: histamine; TY: tyramine.

3.4. Consequences of HHP Treatments on the Formation of BA

The effect of the HHP treatments on the microbial counts, the proteolytic activity and the formation of BA in the cheeses from where we obtained the studied strains are described in the previous work of Espinosa-Pesqueira et al. [27]. TY and PU were the main BA formed in control (untreated) cheeses, whereas in ewe's milk chesses the level of CA was also relevant. However, in cheeses that were pressurized on the 5th day of ripening (HHP1) the amounts TY formed at the end of the ripening (60th day) were about 93% and 88% lower than in control goat's and ewe's raw milk cheeses, respectively, and similar was the result for PU. The application of an HHP treatment on the 15th day of ripening (HHP15) showed to be less efficient reducing the formation of BA. HHP1 treatments caused a significant decrease on microbiological counts, specially of LAB, enteroccocci and enterobacteria, and also reduced the proteolytic activity, showing a reduction of about 34% and 49% of the free amino-acid content in goat's and ewe's cheeses, respectively. Although HHP15 samples also reduced the microbial counts, this did not affect the proteolysis, and consequently the release of amino acids. Several authors mentioned that the specificity of the amino acid decarboxylases is specially strain dependent [11,15,39,60] and a great variability in BA production by different groups and species of bacteria, either in type or amount, was found in this survey. LAB and enterococci were among the most efficient TY producers found in this work. LAB ruled the ripening process of cheeses from the beginning, especially *Lactococcus*, becoming the most important group of microorganisms in the ripened cheese.

The application of the HHP treatments in the early stages of ripening (HHP1) caused a significant reduction of LAB counts, although they recovered along the ripening process. It should be considered that the BA forming rate is greater during the first 15–30 days of ripening, and consequently, the reduction on the counts of the most prolific BA forming groups at these stages could have contributed to reduce the decarboxylase potential, as well as the availability of amino acids, reducing the final amounts of BA. Enterococci were also affected by HHP1 treatments and could not recover their initial counts during the rest of the ripening, affecting their contribution to the formation of BA.

The role of other minority bacterial groups, such as enterobacteria, on the BA formation is not so clear. In the cheeses object of this work their counts were above 6 Log_{10} at the beginning of the ripening, although practically disappeared after both HHP1 and HHP15 treatments samples. Different enterobacteria have been identified as prolific PU, CA and even TY formers, and consequently its elimination the first days of ripening could have reduced their contribution to the amino acid decarboxylase activity. When treatments are applied after 15 days of maturation, the decarboxylase capacity of these microorganisms in the early stages of maturation was still present and consequently their elimination from the 15th day of ripening reduced the consequences on the BA formation.

4. Conclusions

The microplate screening method allows for a rapid preliminary selection of strains with low decarboxylase activity, with a detection limit estimated around 50 mg/L. Moreover, the use of Microtiter plates allows for processing a large numbers of samples, reducing the volume of material and culture media needed. The data indicates that, in general, the specific amino acid decarboxylase assay with LND and LGD broths have satisfactory diagnostic parameters to discriminate bacterial isolates with ornithine, lysine and tyrosine decarboxylases. Moreover, the sensitivity and specificity values for ornithine, lysine and tyrosine decarboxylase test with both types of media were acceptable with low numbers of FP and FN responses. Generally, FN responses were due to weak BA producers. The detection of histidine decarboxylase activity in bacterial isolates using LND and LGD broths have a low sensitivity, but a high specificity value. About a 43% of the strains isolated from cheeses showed decarboxylase activity on one or more amino acids and most of them were later confirmed, especially of TY, PU and CA that were the most important present in cheeses after ripening. The application of the HHP treatments, especially in the early stages of ripening, caused a significant reduction among the most prolific BA forming groups including LAB. Considering that the BA forming rate is greater during the first 15–30 days of ripening, this effect on the microbiota reduces the decarboxylase potential, as well as the availability of amino acids, reducing the final amounts of BA formed. Most of the strains were LAB, including some species that are important for the development of the typical cheese characteristics, such as *L. lactis*, that can be used as starter culture, or *Lactobacillus casei/paracasei*, *Lactobacillus plantarum* and *Lactobacillus curvatus*, of which there is increasing interest to be employed in dairy products with 'protected geographic indication' [18]. Consequently, the formation of BA should always be considered among the selecting criteria for strains considered as suitable to be used as starter cultures.

Author Contributions: Conceived and designed the experiments: D.E.-P., M.M.H.-H. and A.X.R.-S.; performed the experiments: D.E.-P.; analysed the data; D.E.-P. and M.M.H.-H.; wrote the paper: D.E.-P. and A.X.R.-S.

Funding: This research received no external funding.

Conflicts of Interest: The authors declare no conflict of interest.

References

1. Karovicova, J.; Kohajdova, Z. Biogenic amines in food. *ChemInform* **2005**, *36*, 70–79. [CrossRef]
2. Collins, J.D.; Noerrung, B.; Budka, H.; Andreoletti, O.; Buncic, S.; Griffin, J.; Hald, T.; Havelaar, A.; Hope, J.; Klein, G.; et al. Scientific Opinion on risk based control of biogenic amine formation in fermented foods. *EFSA J.* **2011**, *9*, 2393. [CrossRef]
3. Costa, M.P.; Rodrigues, B.L.; Frasao, B.S.; Conte-junior, C.A. *Chemical Risk for Human Consumption*; Elsevier Inc.: Amsterdam, The Netherlands, 2018; ISBN 9780128114421.
4. Benkerroum, N. Biogenic amines in dairy products: Origin, incidence, and control means. *Compr. Rev. Food Sci. Food Saf.* **2016**, *15*, 801–826. [CrossRef]
5. Naila, A.; Flint, S.; Fletcher, G.; Bremer, P.; Meerdink, G. Control of biogenic amines in food-existing and emerging approaches. *J. Food Sci.* **2010**, *75*, R139–R150. [CrossRef] [PubMed]
6. Novella-Rodriguez, S.; Veciana-Nogues, M.T.; Izquierdo-Pulido, M.; Vidal-Carou, M.C. Distribution of biogenic amines and polyamines in cheese. *J. Food Sci.* **2003**, *68*, 750–756. [CrossRef]
7. Roig-Sagués, A.X.; Ruiz-Capillas, C.; Espinosa, D.; Hernández, M. The decarboxylating bacteria present in foodstuffs and the effect of emerging technologies on their formation. In *Biological Aspects of Biogenic Amines, Polyamines and Conjugates*; Dandrifosse, G., Ed.; Transworld Research Network: Trivandrum, India, 2009; pp. 201–230.
8. Roig-Sagués, A.; Molina, A.; Hernández-Herrero, M. Histamine and tyramine-forming microorganisms in Spanish traditional cheeses. *Eur. Food Res. Technol.* **2002**, *215*, 96–100. [CrossRef]
9. Hernández-Herrero, M.M.; Roig-Sagués, A.X.; Rodríguez-Jerez, J.J.; Mora-Ventura, M.T. Halotolerant and halophilic histamine-forming bacteria isolated during the ripening of salted anchovies (*Engraulis encrasicholus*). *J. Food Prot.* **1999**, *62*, 509–514. [CrossRef] [PubMed]
10. Kalac, P.; Abreu Gloria, M.B. Biogenic amine in cheeses, wines, beers and sauerkraut. In *Biological Aspects of Biogenic Amines, Poliamines and Conjugates*; Dandrifosse, G., Ed.; Transworld Research Network: Kerala, India, 2009; pp. 267–285, ISBN 9788178952499.
11. Bover-Cid, S.; Hugas, M.; Izquierdo-Pulido, M.; Vidal-Carou, M.C. Amino acid-decarboxylase activity of bacteria isolated from fermented pork sausages. *Int. J. Food Microbiol.* **2001**, *66*, 185–189. [CrossRef]
12. Ladero, V.; Fernández, M.; Calles-Enríquez, M.; Sánchez-Llana, E.; Cañedo, E.; Martín, M.C.; Alvarez, M.A. Is the production of the biogenic amines tyramine and putrescine a species-level trait in enterococci? *Food Microbiol.* **2012**, *30*, 132–138. [CrossRef] [PubMed]
13. Linares, D.M.; Martín, M.; Ladero, V.; Alvarez, M.A.; Fernández, M. Biogenic amines in dairy products. *Crit. Rev. Food Sci. Nutr.* **2011**, *51*, 691–703. [CrossRef] [PubMed]
14. Latorre-Moratalla, M.L.; Bover-Cid, S.; Veciana-Nogués, M.T.; Vidal-Carou, M.C. Control of biogenic amines in fermented sausages: Role of starter cultures. *Front. Microbiol.* **2012**, *3*, 169. [CrossRef] [PubMed]
15. De las Rivas, B.; Ruiz-Capillas, C.; Carrascosa, A.V.; Curiel, J.A.; Jiménez-Colmenero, F.; Muñoz, R. Biogenic amine production by Gram-positive bacteria isolated from Spanish dry-cured "chorizo" sausage treated with high pressure and kept in chilled storage. *Meat Sci.* **2008**, *80*, 272–277. [CrossRef] [PubMed]
16. Novella-Rodríguez, S.; Veciana-Nogués, M.T.; Roig-Sagués, A.X.; Trujillo-Mesa, A.J.; Vidal-Carou, M.C. Influence of starter and nonstarter on the formation of biogenic amine in goat cheese during ripening. *J. Dairy Sci.* **2002**, *85*, 2471–2478. [CrossRef]
17. Torracca, B.; Pedonese, F.; Turchi, B.; Fratini, F.; Nuvoloni, R. Qualitative and quantitative evaluation of biogenic amines in vitro production by bacteria isolated from ewes' milk cheeses. *Eur. Food Res. Technol.* **2018**, *244*, 721–728. [CrossRef]
18. Ladero, V.; Martín, M.C.; Redruello, B.; Mayo, B.; Flórez, A.B.; Fernández, M.; Alvarez, M.A. Genetic and functional analysis of biogenic amine production capacity among starter and non-starter lactic acid bacteria isolated from artisanal cheeses. *Eur. Food Res. Technol.* **2015**, *241*, 377–383. [CrossRef]
19. Sumner, S.S.; Speckhard, M.W.; Somers, E.B.; Taylor, S.L. Isolation of histamine-producing *Lactobacillus buchneri* from Swiss cheese implicated in a food poisoning outbreak. *Appl. Environ. Microbiol.* **1985**, *50*, 1094–1096. [PubMed]
20. Diaz, M.; del Rio, B.; Sanchez-Llana, E.; Ladero, V.; Redruello, B.; Fernández, M.; Martin, M.C.; Alvarez, M.A. *Lactobacillus parabuchneri* produces histamine in refrigerated cheese at a temperature-dependent rate. *Int. J. Food Sci. Technol.* **2018**, *53*, 2342–2348. [CrossRef]

21. Pircher, A.; Bauer, F.; Paulsen, P. Formation of cadaverine, histamine, putrescine and tyramine by bacteria isolated from meat, fermented sausages and cheeses. *Eur. Food Res. Technol.* **2007**, *226*, 225–231. [CrossRef]

22. Rodriguez-Jerez, J.J.; Giaccone, V.; Colavita, G.; Parisi, E. *Bacillus macerans*—A new potent histamine producing micro-organism isolated from Italian cheese. *Food Microbiol.* **1994**, *11*, 409–415. [CrossRef]

23. Bover-Cid, S.; Holzapfel, W.H. Improved screening procedure for biogenic amine production by lactic acid bacteria. *Int. J. Food Microbiol.* **1999**, *53*, 33–41. [CrossRef]

24. Marcobal, Á.; de las Rivas, B.; Moreno-Arribas, M.V.; Muñoz, R. Multiplex PCR method for the simultaneous detection of histamine-, tyramine-, and putrescine-producing lactic acid bacteria in foods. *J. Food Prot.* **2005**, *68*, 874–878. [CrossRef] [PubMed]

25. Roig-Sagués, A.X.; Hernàndez-Herrero, M.M.; López-Sabater, E.I.; Rodríguez-Jerez, J.J.; Mora-Ventura, M.T. Evaluation of three decarboxylating agar media to detect histamine and tyramine-producing bacteria in ripened sausages. *Lett. Appl. Microbiol.* **1997**, *25*, 309–312. [CrossRef] [PubMed]

26. Maijala, R.L. Formation of histamine and tyramine by some lactic acid bacteria in MRS-broth and modified decarboxylation agar. *Lett. Appl. Microbiol.* **1993**, *17*, 40–43. [CrossRef]

27. Espinosa-Pesqueira, D.; Hernández-Herrero, M.; Roig-Sagués, A.; Espinosa-Pesqueira, D.; Hernández-Herrero, M.M.; Roig-Sagués, A.X. High Hydrostatic Pressure as a tool to reduce formation of biogenic amines in artisanal Spanish cheeses. *Foods* **2018**, *7*, 137. [CrossRef] [PubMed]

28. Eerola, S.; Hinkkanen, R.; Lindfors, E.; Hirvi, T. Liquid chromatographic determination of biogenic amines in dry sausages. *J. AOAC Int.* **1993**, *76*, 575–577. [PubMed]

29. Altman, D.G.; Bland, J.M. Statistics Notes: Diagnostic tests 1: Sensitivity and specificity. *BMJ* **1994**, *308*, 1552. [CrossRef] [PubMed]

30. Altman, D.G.; Bland, J.M. Statistics Notes: Diagnostic tests 2: Predictive values. *BMJ* **1994**, *309*, 102. [CrossRef] [PubMed]

31. Altman, D.G.; Bland, J.M. Statistics Notes: Diagnostic tests 3: Receiver operating characteristic plots. *BMJ* **1994**, *309*, 188. [CrossRef] [PubMed]

32. Zweig, M.H.; Campbell, G. Receiver-operating characteristic (ROC) plots: A fundamental evaluation tool in clinical medicine. *Clin. Chem.* **1993**, *39*, 561–577. [PubMed]

33. Park, S.H.; Goo, J.M.; Jo, C.-H. Receiver Operating Characteristic (ROC) Curve: Practical review for radiologists. *Korean J. Radiol.* **2004**, *5*, 11–18. [CrossRef] [PubMed]

34. Hanley, J.A.; McNeil, B.J. The meaning and use of the area under a receiver operating characteristic (ROC) curve. *Radiology* **1982**, *143*, 29–36. [CrossRef] [PubMed]

35. Harrigan, W.F. *Laboratory Methods in Food Microbiology*; Academic Press: Cambridge, MA, USA, 1998; ISBN 0123260434.

36. Roig-Sagues, A.X.; Hernandez-Herrero, M.; Lopez-Sabater, E.I.; Rodriguez-Jerez, J.J.; Mora-Ventura, M.T. Histidine decarboxylase activity of bacteria isolated from raw and ripened salchichón, a Spanish cured sausage. *J. Food Prot.* **1996**, *59*, 516–520. [CrossRef]

37. Joosten, H.M.; Northolt, M.D. Detection, growth, and amine-producing capacity of lactobacilli in cheese. *Appl. Environ. Microbiol.* **1989**, *55*, 2356–2359. [PubMed]

38. Moreno-Arribas, M.V.; Polo, M.C.; Jorganes, F.; Muñoz, R. Screening of biogenic amine production by lactic acid bacteria isolated from grape must and wine. *Int. J. Food Microbiol.* **2003**, *84*, 117–123. [CrossRef]

39. Marino, M.; Maifreni, M.; Bartolomeoli, I.; Rondinini, G. Evaluation of amino acid-decarboxylative microbiota throughout the ripening of an Italian PDO cheese produced using different manufacturing practices. *J. Appl. Microbiol.* **2008**, *105*, 540–549. [CrossRef] [PubMed]

40. Suzzi, G.; Gardini, F. Biogenic amines in dry fermented sausages: A review. *Int. J. Food Microbiol.* **2003**, *88*, 41–54. [CrossRef]

41. Marino, M.; Maifreni, M.; Moret, S.; Rondinini, G. The capacity of *Enterobacteriaceae* species to produce biogenic amines in cheese. *Lett. Appl. Microbiol.* **2000**, *31*, 169–173. [CrossRef] [PubMed]

42. Halász, A.; Baráth, Á.; Simon-Sarkadi, L.; Holzapfel, W. Biogenic amines and their production by microorganisms in food. *Trends Food Sci. Technol.* **1994**, *5*, 42–49. [CrossRef]

43. Martín, B.; Garriga, M.; Hugas, M.; Bover-Cid, S.; Veciana-Nogués, M.T.; Aymerich, T. Molecular, technological and safety characterization of Gram-positive catalase-positive cocci from slightly fermented sausages. *Int. J. Food Microbiol.* **2006**, *107*, 148–158. [CrossRef] [PubMed]

44. Pons-Sánchez-Cascado, S.; Vidal-Carou, M.C.; Mariné-Font, A.; Veciana-Nogués, M.T. Influence of the Freshness grade of raw fish on the formation of volatile and biogenic amines during the manufacture and storage of vinegar-marinated anchovies. *J. Agric. Food Chem.* **2005**, *53*, 8586–8592. [CrossRef] [PubMed]
45. Masson, F.; Talon, R.; Montel, M.C. Histamine and tyramine production by bacteria from meat products. *Int. J. Food Microbiol.* **1996**, *32*, 199–207. [CrossRef]
46. Santos, M.H. Amino acid decarboxylase capability of microorganisms isolated in Spanish fermented meat products. *Int. J. Food Microbiol.* **1998**, *39*, 227–230. [CrossRef]
47. Drosinos, E.H.; Paramithiotis, S.; Kolovos, G.; Tsikouras, I.; Metaxopoulos, I. Phenotypic and technological diversity of lactic acid bacteria and staphylococci isolated from traditionally fermented sausages in Southern Greece. *Food Microbiol.* **2007**, *24*, 260–270. [CrossRef] [PubMed]
48. Foulquié Moreno, M.R.; Sarantinopoulos, P.; Tsakalidou, E.; De Vuyst, L. The role and application of enterococci in food and health. *Int. J. Food Microbiol.* **2006**, *106*, 1–24. [CrossRef] [PubMed]
49. Leuschner, R.G.K.; Kurihara, R.; Hammes, W.P. Formation of biogenic amines by proteolytic enterococci during cheese ripening. *J. Sci. Food Agric.* **1999**, *79*, 1141–1144. [CrossRef]
50. Galgano, F.; Suzzi, G.; Favati, F.; Caruso, M.; Martuscelli, M.; Gardini, F.; Salzano, G. Biogenic amines during ripening in 'Semicotto Caprino' cheese: Role of enterococci. *Int. J. Food Sci. Technol.* **2001**, *36*, 153–160. [CrossRef]
51. Landete, J.M.; Pardo, I.; Ferrer, S. Tyramine and phenylethylamine production among lactic acid bacteria isolated from wine. *Int. J. Food Microbiol.* **2007**, *115*, 364–368. [CrossRef] [PubMed]
52. Burdychova, R.; Komprda, T. Biogenic amine-forming microbial communities in cheese. *FEMS Microbiol. Lett.* **2007**, *276*, 149–155. [CrossRef] [PubMed]
53. Martuscelli, M.; Gardini, F.; Torriani, S.; Mastrocola, D.; Serio, A.; Chaves-López, C.; Schirone, M.; Suzzi, G. Production of biogenic amines during the ripening of Pecorino Abruzzese cheese. *Int. Dairy J.* **2005**, *15*, 571–578. [CrossRef]
54. Tham, W.; Karp, G.; Danielsson-Tham, M.L. Histamine formation by enterococci in goat cheese. *Int. J. Food Microbiol.* **1990**, *11*, 225–229. [CrossRef]
55. Perez, M.; Ladero, V.; del Rio, B.; Redruello, B.; de Jong, A.; Kuipers, O.; Kok, J.; Martin, M.C.; Fernandez, M.; Alvarez, M.A. The relationship among tyrosine decarboxylase and agmatine deiminase pathways in *Enterococcus faecalis*. *Front. Microbiol.* **2017**, *8*, 2107. [CrossRef] [PubMed]
56. Novella-Rodriguez, S.; Veciana-Nogues, M.T.; Trujillo-Mesa, A.J.; Vidal-Carou, M.C. Profile of biogenic amines in goat cheese made from pasteurized and pressurized milks. *J. Food Sci.* **2002**, *67*, 2940–2944. [CrossRef]
57. Fernández-García, E.; Tomillo, J.; Núñez, M. Effect of added proteinases and level of starter culture on the formation of biogenic amines in raw milk Manchego cheese. *Int. J. Food Microbiol.* **1999**, *52*, 189–196. [CrossRef]
58. Sumner, S.S.; Taylor, S.L. Detection method for histamine-producing, dairy-related bacteria using diamine oxidase and leucocrystal violet. *J. Food Prot.* **1989**, *52*, 105–108. [CrossRef]
59. Ascone, P.; Maurer, J.; Haldemann, J.; Irmler, S.; Berthoud, H.; Portmann, R.; Fröhlich-Wyder, M.-T.; Wechsler, D. Prevalence and diversity of histamine-forming *Lactobacillus parabuchneri* strains in raw milk and cheese—A case study. *Int. Dairy J.* **2017**, *70*, 26–33. [CrossRef]
60. Kung, H.F.; Tsai, Y.H.; Hwang, C.C.; Lee, Y.H.; Hwang, J.H.; Wei, C.I.; Hwang, D.F. Hygienic quality and incidence of histamine-forming *Lactobacillus* species in natural and processed cheese in Taiwan. *J. Food Drug Anal.* **2005**, *13*, 51–56.
61. Leuschner, R.G.; Heidel, M.; Hammes, W.P. Histamine and tyramine degradation by food fermenting microorganisms. *Int. J. Food Microbiol.* **1998**, *39*, 1–10. [CrossRef]
62. Straub, B.W.; Kicherer, M.; Schilcher, S.M.; Hammes, W.P. The formation of biogenic amines by fermentation organisms. *Z. Lebensm. Unters. Forsch.* **1995**, *201*, 79–82. [CrossRef] [PubMed]
63. González de Llano, D.; Ramos, M.; Rodriguez, A.; Montilla, A.; Juarez, M. Microbiological and physicochemical characteristics of Gamonedo blue cheese during ripening. *Int. Dairy J.* **1992**, *2*, 121–135. [CrossRef]
64. Fontecha, J.; Peláez, C.; Juárez, M.; Requena, T.; Gómez, C.; Ramos, M. Biochemical and microbiological characteristics of artisanal hard goat's cheese. *J. Dairy Sci.* **1990**, *73*, 1150–1157. [CrossRef]

foods

Article

The Occurrence of Biogenic Amines and Determination of Biogenic Amine-Producing Lactic Acid Bacteria in *Kkakdugi* and *Chonggak* Kimchi

Young Hun Jin, Jae Hoan Lee, Young Kyung Park, Jun-Hee Lee and Jae-Hyung Mah *

Department of Food and Biotechnology, Korea University, 2511 Sejong-ro, Sejong 30019, Korea;
younghoonjin3090@korea.ac.kr (Y.H.J.); jae-lee@korea.ac.kr (J.H.L.); eskimo@korea.ac.kr (Y.K.P.);
bory92@korea.ac.kr (J.-H.L.)
* Correspondence: nextbio@korea.ac.kr; Tel.: +82-44-860-1431

Received: 3 February 2019; Accepted: 12 February 2019; Published: 14 February 2019

Abstract: In this study, biogenic amine content in two types of fermented radish kimchi (*Kkakdugi* and *Chonggak* kimchi) was determined by high performance liquid chromatography (HPLC). While most samples had low levels of biogenic amines, some samples contained histamine content over the toxicity limit. Additionally, significant amounts of total biogenic amines were detected in certain samples due to high levels of putrefactive amines. As one of the significant factors influencing biogenic amine content in both radish kimchi, *Myeolchi-aekjoet* appeared to be important source of histamine. Besides, tyramine-producing strains of lactic acid bacteria existed in both radish kimchi. Through 16s rRNA sequencing analysis, the dominant species of tyramine-producing strains was identified as *Lactobacillus brevis*, which suggests that the species is responsible for tyramine formation in both radish kimchi. During fermentation, a higher tyramine accumulation was observed in both radish kimchi when *L. brevis* strains were used as inocula. The addition of *Myeolchi-aekjeot* affected the initial concentrations of histamine and cadaverine in both radish kimchi. Therefore, this study suggests that reducing the ratio of *Myeolchi-aekjeot* to other ingredients (and/or using *Myeolchi-aekjeot* with low biogenic amine content) and using starter cultures with ability to degrade and/or inability to produce biogenic amines would be effective in reducing biogenic amine content in *Kkakdugi* and *Chonggak* kimchi.

Keywords: kimchi; *Kkakdugi*; *Chonggak* kimchi; radish kimchi; biogenic amines; tyramine; lactic acid bacteria; *Lactobacillus brevis*

1. Introduction

Biogenic amines (BA) have been considered to be toxic compounds in foods. Several authors have proposed the maximum tolerable limits of some toxicologically important BA in foods as follows: histamine, 100 mg/kg; tyramine, 100–800 mg/kg; β-phenylethylamine, 30 mg/kg; total BA, 1000 mg/kg [1,2]. In addition, polyamines such as putrescine and cadaverine have been known to potentiate the toxicity of BA, especially histamine and tyramine, in foods, although they are less toxic [1]. Consumption of foods containing excessive BA may cause symptoms such as migraines, sweating, nausea, hypotension, and hypertension, unless human intestinal amine oxidases—such as monoamine oxidase (MAO), diamine oxidase (DAO), and polyamine oxidase (PAO)—quickly metabolize and detoxify BA [3]. Thus, it is important to know that, although relatively low levels of BA naturally exist in common foods, microbial decarboxylation of amino acids may sometimes lead to a significant increment of BA in fermented or contaminated foods [2]. In lactic acid fermented foods such as cheese and fermented sausage, some species of lactic acid bacteria (LAB) have been considered as producers of BA, particularly tyramine [4]. On the other hand, several reports have indicated that use of LAB starter cultures unable to produce BA may reduce BA accumulation during fermentation and storage [5,6].

Kimchi is a generic term of Korean traditional lactic fermented vegetables. According to Codex standard [7], for preparation of kimchi, salted Chinese cabbage (as a main ingredient) is mixed with seasoning paste consisting of red pepper powder, radish, garlic, green onion, and ginger, and then fermented properly, however, which, in reality, refers to *Baechu* kimchi. Alongside the Chinese cabbage, various vegetables such as radish, ponytail radish, cucumber, and green onion are also used as main ingredients of kimchi depending on kimchi varieties in Korea [8]. Among numerous kimchi varieties prepared with different vegetables, *Baechu* kimchi, *Kkakdugi* (diced radish kimchi), and *Chonggak* kimchi (ponytail radish kimchi) are the most popular varieties of kimchi in Korea [9]. In the meantime, for improving sensory quality of kimchi, various types of salted and fermented seafood (*Jeotgal*) and sauces thereof (*Aekjeot*) are usually used for kimchi preparation in Korea [10]. Particularly, *Myeolchi-jeotgal* (salted and fermented anchovy), *Saeu-jeotgal* (salted and fermented shrimp), *Myeolchi-aekjeot* (a sauce prepared from *Myeolchi-jeotgal*) are commonly used *Jeotgal* and *Aekjeot* [11]. As *Jeotgal* and *Aekjeot* contain high levels of proteins and amino acids, when kimchi is prepared with them, BA accumulation may occur during kimchi fermentation [12]. Hence, several authors have intensively investigated BA content and BA-producing LAB in *Baechu* kimchi [13–15]. On the other hand, there is a lack of study on BA content and BA-producing LAB in *Kkakdugi* and *Chonggak* kimchi, although the two types of radish kimchi are as popular as *Baechu* kimchi in Korea.

In this study, therefore, BA content in *Kkakdugi* and *Chonggak* kimchi was determined to evaluate BA-related risks. Several possible contributing factors to BA content, including physicochemical properties and microbial BA production, were also investigated in the study. Finally, fermentation of both radish kimchi was carried out to determine the most important bacterial species contributing to BA formation in the radish kimchi, employing LAB strains with distinguishable BA-producing activities as fermenting microorganisms. This is the first study describing that *Lactobacillus brevis* is the species responsible for tyramine formation in kimchi variety throughout fermentation period.

2. Materials and Methods

2.1. Sampling

Two types of radish kimchi (*Kkakdugi* and *Chonggak* kimchi) samples of five popular kimchi manufacturers made within 30 days were obtained from the retail markets. After arrival, samples were stored at 4 °C or immediately analyzed for BA content, physicochemical parameters, and microbial measurement.

2.2. Physicochemical Measurements

pH, acidity, salinity, and water activity of *Kkakdugi* and *Chonggak* kimchi samples were determined. The pH of the samples was determined by Orion 3-star Benchtop pH meter (Thermo Scientific, Waltham, MA, USA). Acidity and salinity were measured according to the AOAC method [16]. The water activity was determined by water activity meter (AquaLab Pre; Meter Group, Inc., Pullman, WA, USA).

2.3. Microbial Measurement, Isolation, and Identification of Strains

Lactic acid bacterial counts and total aerobic bacterial counts were determined on de Man, Rogosa, and Sharpe (MRS, Laboratorios Conda Co., Madrid, Spain) agar and Plate Count Agar (PCA, Difco, Becton Dickinson, Sparks, MD, USA). According to manufacturer's instructions, MRS agar was incubated at 37 °C for 48–72 h, and PCA at 37 °C for 24 h. After incubation, enumeration was carried out on plates with 30–300 colonies.

LAB strains were isolated on MRS agar. Individual colonies on MRS agar were randomly selected and streaked on the same media. The single colonies were transferred to MRS broth at 37 °C for 48–72 h. Then, the cultured broth was stored in the presence of 20% glycerol (*v/v*) at −80 °C. In *Kkakdugi* and

Chonggak kimchi samples, 130 and 120 LAB strains were isolated, respectively. The strains were identified by 16s rRNA gene sequence analysis with the universal bacterial primer pair (518F and 805R, Solgent Co., Daejeon, Korea).

2.4. BA Extraction from Samples and Bacterial Cultures for HPLC Analysis

BA extraction from *Kkakdugi* and *Chonggak* kimchi samples was conducted by the methods developed by Eerola et al. [17], with minor modification. The sample broth (5 g) was mixed with 20 mL of perchloric acid (0.4 M). The mixture was incubated at 4 °C for 2 h and centrifuged at 3000× g at 4 °C for 10 min. After collecting the supernatant, the pellet was extracted again with equal volumes of perchloric acid under the same conditions. The total volume of supernatant was adjusted to 50 mL with perchloric acid. The extract was filtered using Whatman paper no. 1 and stored before analysis.

BA extraction from bacterial cultures was carried out based on the procedures described by Ben-Gigirey et al. [18,19], with minor modification. A loopful of a strain was inoculated in 5 mL of BA production assay medium. The compositions of BA production assay medium are as follows: MRS broth with 0.5% of L-ornithine monohydrochloride, L-lysine monohydrochloride, L-histidine monohydrochloride monohydrate, and L-tyrosine disodium salt hydrate (all Sigma-Aldrich Chemical Co., St. Louis, MO, USA); 0.0005% of pyridoxal-HCl (Sigma-Aldrich); pH of the broth was adjusted to 5.8 by adding hydrochloride solution (2 M). After incubating the strain at 37 °C for 48 h, 100 μL of the culture was inoculated into the same broth and incubated under the same conditions. Subsequently, after being mixed with 0.4 M perchloric acid at a volume ratio of 1:9, the mixture was incubated at 4 °C for 2 h and stored before analysis.

2.5. Preparation of Standard Solutions for HPLC Analysis

Tryptamine, β-phenylethylamine hydrochloride, putrescine dihydrochloride, cadaverine dihydrochloride, histamine dihydrochloride, tyramine hydrochloride, spermidine trihydrochloride, and spermine tetrahydrochloride (all Sigma-Aldrich) were used for standard solutions, and 1,7-diaminoheptane (Sigma-Aldrich) was applied for an internal standard. The concentrations of all standard solutions were adjusted to 0, 10, 50, 100, and 1000 ppm.

2.6. Derivatization of Extracts and Standards

The procedures of derivatization of BA in the extract were carried out by the method developed by Eerola et al. [17]. Briefly, 200 μL of 2 M sodium hydroxide and 300 μL of saturated sodium bicarbonate were added to 1 mL of the extract/standard solutions. Then, 2 mL of 1% dansyl chloride solution (dissolved in acetone) was mixed with the solution and then incubated for 45 min at 40 °C in dark room. The incubated solution was mixed with 100 μL of 25% ammonium hydroxide and reacted for 30 min at room temperature. The volume of the sample solution was adjusted to 5 mL by adding acetonitrile. The sample solution was centrifuged at 3000× g for 5 min, and the supernatant was filtered by using a 0.2 μm-pore-size filter (Millipore Co., Bedford, MA, USA).

2.7. HPLC Analysis

HPLC analysis was carried out according to the procedure developed by Eerola et al. [17] and modified by Ben-Gigirey et al. [18]. YL9100 HPLC system equipped with YL9120 UV–vis detector (all Younglin, Anyang, Korea) was employed and the data were analyzed with Autochro-3000 data system (Younglin). For the gradient HPLC method, 0.1 M ammonium acetate (solvent A; Sigma-Aldrich) and HPLC-grade acetonitrile (solvent B; SK chemicals, Ulsan, Korea) were used as the mobile phases. The chromatographic separation was carried out using Nova-Pak C18 column (4 μm, 4.6 × 150 mm; Waters, Milford, MA, USA) held in 40 °C at a flow rate of 1 mL/min. The gradient elution mode was as follows; 50:50 (A:B) to 10:90 for 19 min, 50:50 at 20 min, isocratic with 50:50 before next analysis. The analysis was conducted at 254 nm, and 10 μL of the sample solution was injected.

The detection limits were within the range of 0.01 to 0.10 mg/kg for food matrices [20]. The validation parameters, including detection limits, of the analytical procedure used in the study were reported in our earlier study [20]. Figure S1 illustrates the procedure, from extraction to HPLC analysis, for BA analysis.

2.8. Fermentation of Two Types of Radish Kimchi: Kkakdugi and Chonggak Kimchi

For preparation of *Kkakdugi* and *Chonggak* kimchi, diced white radish ($2 \times 2 \times 2 \text{ cm}^3$) or halved ponytail radish were soaked in 10% w/v salt brine for 30 min, respectively. Then, each salted radish was rinsed with tap water three times and drained for 3 h. *Kkakdugi* and *Chonggak* kimchi samples were prepared in triplicate, as shown in Table 1, according to the standard recipes developed by the National Institute of Agricultural Sciences [21]. The salinity of all samples was adjusted to 2.5%. The *Kkakdugi* and *Chonggak* kimchi samples were divided into five experimental groups, respectively, based on the presence or absence of *Myeolchi-aekjeot* and *Saeu-jeotgal* and LAB inoculum. The experimental groups designed for the present study were B group ("Blank" samples prepared with neither *Myeolchi-aekjeot* and *Saeu-jeotgal* nor inoculum), C group ("Control" samples prepared with *Myeolchi-aekjeot* and *Saeu-jeotgal*, but without inoculum), PC group ("Positive Control" samples prepared with *Myeolchi-aekjeot* and *Saeu-jeotgal*, and inoculated with *L. brevis* JCM 1170 as a reference strain), LB group ("L. brevis" samples prepared with *Myeolchi-aekjeot* and *Saeu-jeotgal*, and inoculated with tyramine-producing *L. brevis* strains, i.e., KD3M5 strain for *Kkakdugi* and CG2M15 strain for *Chonggak* kimchi, respectively), and LP group ("L. plantarum" samples prepared with *Myeolchi-aekjeot* and *Saeu-jeotgal*, and inoculated with *L. plantarum* strains, i.e., KD3M15 strain for *Kkakdugi* and CG3M21 strain for *Chonggak* kimchi, respectively). The samples belonging to respective experimental groups were fermented at 25 °C for three days. Changes on the physicochemical and microbial properties, and BA content were measured in triplicate during fermentation.

Table 1. Ingredients used for preparation of *Kkakdugi* and *Chonggak* kimchi.

Ingredients (g)	Salted Radish	Red Pepper Powder	Garlic	Ginger	Sesame Seed	Sugar	Glutinous Rice Paste	Myeolchi-aekjeot	Saeu-jeotgal
Kkakdugi	100	3	3	1.5	1	2	5	2	2
Chonggak kimchi	100	3.5	3	1.5	0.5	1.5	4	2	2

2.9. Statistical Analyses

Statistical analyses were performed with Minitab statistical software version 12.11 (Minitab Inc. State College, PA, USA). The data were presented as means ± standard deviations of the three independent replicates. The mean values were compared by one-way analysis of variance (ANOVA) with Tukey's honest significant difference (HSD) test and a probability (*p*) values of less than 0.05 were considered statistically significant.

3. Results and Discussion

3.1. Determination of BA Content in Radish Kimchi: Kkakdugi and Chonggak Kimchi

As shown in Table 2, BA content in *Kkakdugi* and *Chonggak* kimchi samples produced by popular manufacturers in Korea was determined, and human health risk of BA in both radish kimchi was estimated based on the suggestions of both Ten Brink et al. [1] and Silla Santos [2]. In all the samples of *Kkakdugi* and *Chonggak* kimchi, low levels of tyramine (<100 mg/kg), tryptamine, β-phenylethylamine, spermidine, and spermine (<30 mg/kg) were detected, which are within safe levels for human consumption. However, one *Kkakdugi* sample (KD2) had 127.78 ± 26.78 mg/kg of histamine, which is over the toxicity limit (100 mg/kg) suggested by Ten Brink et al. [1]. Another *Kkakdugi* sample (KD5) contained putrescine and cadaverine at concentrations of 982.32 ± 19.42 mg/kg and 124.60 ± 108.78 mg/kg, respectively, consequently exceeding the 1000 mg/kg limit for total BA which is considered to provoke toxicity [2]. In *Chonggak* kimchi samples, 131.20 ± 7.90 mg/kg of histamine was detected in one sample (CG5), which also contained 853.7 ± 36.80 mg/kg of putrescine and 112.10 ± 3.60 mg/kg of cadaverine. The amounts of histamine and total BA in the sample were found to exceed toxicity limits. Meanwhile, the BA content detected in both types of radish kimchi samples varied widely in the present study, which is similar to respective BA levels in *Baechu* kimchi reported previously [13,22]. On the other hand, Mah et al. [12] reported lower concentrations of putrescine, cadaverine, histamine, tyramine, spermidine, and spermine in both *Kkakdugi* and *Chonggak* kimchi than those detected in the same kinds of kimchi used in this study. This may be due to the differences in manufacturing methods, main ingredients, and storage conditions between kimchi samples used in the present and previous studies [9]. In the meantime, Mah et al. [12] also reported that the amounts of tyramine and other BA increased during the ripening of *Baechu* kimchi. Therefore, although tyramine was detected at low levels in all the samples of *Kkakdugi* and *Chonggak* kimchi in the present study, the significance and risk of tyramine formation in both types of radish kimchi should not be overlooked.

Table 2. BA content in two types of radish kimchi samples: *Kkakdugi* and *Chonggak* kimchi.

Samples [2]	BA Content (mg/kg) [1]							
	Trp	Phe	Put	Cad	His	Tyr	Spd	Spm
KD1	ND [3]	ND	10.85 ± 1.17 [4]	2.57 ± 0.62	18.75 ± 1.16	2.97 ± 0.33	12.27 ± 0.98	0.56 ± 0.96
KD2	ND	1.93 ± 1.69	563.59 ± 45.64	ND	127.78 ± 26.78	14.73 ± 1.96	12.66 ± 2.75	ND
KD3	ND	ND	19.00 ± 2.00	6.10 ± 0.40	24.50 ± 4.00	10.80 ± 0.40	ND	ND
KD4	ND	0.86 ± 1.49	97.45 ± 77.05	3.15 ± 5.46	40.82 ± 29.05	21.67 ± 17.81	5.30 ± 4.85	3.10 ± 2.82
KD5	ND	15.24 ± 1.87	982.32 ± 19.42	124.60 ± 108.78	67.84 ± 17.46	76.95 ± 4.25	16.76 ± 0.87	1.48 ± 0.08
Average	ND	3.61 ± 6.55	334.64 ± 427.97	27.28 ± 54.44	55.94 ± 44.45	25.42 ± 29.59	9.40 ± 6.68	1.03 ± 1.31
CG1	ND	ND	8.97 ± 2.02	2.38 ± 2.12	38.61 ± 6.03	4.85 ± 4.60	9.22 ± 2.16	20.74 ± 3.47
CG2	ND	ND	3.89 ± 1.68	2.00 ± 0.77	8.24 ± 2.09	0.79 ± 0.69	8.27 ± 2.90	2.12 ± 0.53
CG3	12.30 ± 6.30	ND	175.10 ± 7.30	55.40 ± 2.80	46.30 ± 6.70	18.70 ± 2.40	7.70 ± 5.50	ND
CG4	9.10 ± 7.10	1.10 ± 1.00	303.70 ± 20.20	148.50 ± 9.00	69.30 ± 20.90	11.10 ± 2.20	6.10 ± 3.70	8.30 ± 5.60
CG5	23.70 ± 6.10	2.80 ± 1.20	853.70 ± 36.80	112.10 ± 3.60	131.20 ± 7.90	7.00 ± 2.20	14.00 ± 5.30	ND
Average	9.02 ± 9.86	0.78 ± 1.23	269.07 ± 349.93	64.08 ± 65.51	58.73 ± 46.02	8.49 ± 6.80	9.06 ± 2.99	6.23 ± 8.79

[1] Trp: tryptamine, Phe: β-phenylethylamine, Put: putrescine, Cad: cadaverine, His: histamine, Tyr: tyramine, Spd: spermidine, Spm: spermine; [2] KD: *Kkakdugi* (diced radish kimchi), CG: *Chonggak* kimchi (ponytail radish kimchi); [3] ND: not detected (<0.1 mg/kg); [4] mean ± standard deviation.

According to Tsai et al. [13], a high level of histamine in kimchi may result from the addition of salted and fermented fish products. *Myeolchi-aekjeot* is the most widely used salted and fermented fish product for the preparation of kimchi variety, and approximately 2–4% of *Kkakdugi* (on the basis of weight percent) and 2–5% of *Chonggak* kimchi, respectively, are commonly added to main ingredients during kimchi preparation [21,23–26]. *Saeu-jeotgal* is also added, alone or together with *Myeolchi-aekjeot*, to main ingredients of kimchi, but Mah et al. [12] reported that *Myeolchi-aekjeot* contains a significantly higher level of histamine (up to 1154.7 mg/kg) than *Saeu-jeotgal*. In this study, all radish kimchi samples were prepared with both *Myeolchi-aekjeot* and *Saeu-jeotgal* as ingredients. Altogether, the excessive level of histamine in several radish kimchi samples could be due to the amount of added *Myeolchi-aekjeot* with high histamine content. Unfortunately, the food labels of the samples used in this study just provided the list of ingredients.

An overdose of histamine may provoke undesirable symptoms such as a migraine, sweating, and hypotension [3]. In addition, high levels of putrescine and cadaverine can potentiate histamine toxicity by inhibiting intestinal diamine oxidase and histamine-N-methyltransferase [27] and potentially react with nitrites to form carcinogenic N-nitrosamines [28]. Taking this into account, although most *Kkakdugi* and *Chonggak* kimchi samples seem to be safe for consumption, the fact that several samples contained relatively high levels of putrescine and cadaverine in the present study indicates that it is necessary to monitor and reduce BA content, particularly histamine, putrescine, and cadaverine.

3.2. Physicochemical and Microbial Properties of Radish Kimchi: Kkakdugi and Chonggak Kimchi

To predict possible reasons as to why some samples of two types of radish kimchi contained higher levels of BA, pH, acidity, salinity, water activity (a_w), and lactic acid bacterial and total aerobic bacterial counts of *Kkakdugi* and *Chonggak* kimchi samples were determined. In *Kkakdugi* samples, the values of the parameters were as follows: pH, 4.16 ± 0.17 (minimum to maximum range of 3.94–4.41); acidity (%), 0.86 ± 0.31 (0.51–1.27); salinity (%), 3.36 ± 1.21 (1.40–4.50); a_w, 0.983 ± 0.003 (0.977–0.988); lactic acid bacterial counts, 8.52 ± 0.61 Log CFU/mL (7.88–9.38 Log CFU/mL); total aerobic bacterial counts, 8.37 ± 0.96 Log CFU/mL (6.83–9.32 Log CFU/mL). In case of *Chonggak* kimchi samples, the measured values were as follows: pH, 4.96 ± 1.17 (3.98–6.36); acidity (%), 0.71 ± 0.43 (0.19–1.10); salinity (%), 3.83 ± 1.67 (2.15–6.48); a_w, 0.984 ± 0.004 (0.979–0.991); lactic acid bacterial counts, 7.83 ± 0.48 Log CFU/mL (7.42–8.60 Log CFU/mL); total aerobic bacterial counts, 8.18 ± 1.07 Log CFU/mL (6.88–9.48 Log CFU/mL). The values are in accordance with those of previous reports [13,29]. Linear regression analysis was performed to determine the contributors influencing BA content. Results revealed weak correlations between physiochemical parameters, as well as microbial properties, and BA content (data not shown). Nonetheless, several reports have shown that physicochemical and microbial properties may affect BA content in fermented foods [2,30,31]. Altogether, the results indicate that, besides physicochemical and microbial properties, there are complex factors affecting BA content in both radish kimchi, for instance, kinds of salted and fermented fish products used for kimchi preparation as described above.

3.3. BA Production by LAB Strains Isolated from Radish Kimchi: Kkakdugi and Chonggak Kimchi

BA production by LAB strains isolated from *Kkakdugi* and *Chonggak* kimchi samples was examined to determine BA-producing LAB species in two types of radish kimchi. All the strains showed low production (below the detection limit) of tryptamine, β-phenylethylamine, putrescine, cadaverine, histamine, spermidine, and spermine. However, 39 strains (30%) of 130 LAB isolated from *Kkakdugi* samples produced higher levels of tyramine (287.23–386.17 µg/mL) than other strains (below the detection limit). Among the 120 LAB strains isolated from *Chonggak* kimchi, 16 strains (13%) also showed a stronger tyramine production capability (260.93–339.56 µg/mL), while other strains revealed lower capability (below the detection limit). In addition, the tyramine-producing LAB strains, which were isolated from either *Kkakdugi* or *Chonggak* kimchi samples, revealed a similar ability to produce tyramine, as described right above. Meanwhile, despite the low level of tyramine detected in all the samples of *Kkakdugi* and *Chonggak* kimchi, the fact that parts of LAB strains isolated from both radish kimchi samples were highly capable of producing tyramine supports that tyramine increment may occur during the ripening of the kimchi [12].

To further determine microorganisms responsible for BA formation in radish kimchi at species level, the strains were divided into two groups: (i) 55 tyramine-producing LAB strains (39 strains from *Kkakdugi*; 16 strains from *Chonggak* kimchi) and (ii) 195 LAB strains unable to produce BA. In the two groups, several strains were randomly selected and subsequently identified based on 16s rRNA sequencing analysis. Then, the selected strains able to produce tyramine were all identified as *L. brevis*, which indicates that the species is probably responsible for tyramine formation in both types of radish kimchi. On the other hand, the selected strains unable to produce BA were identified as *Leuconostoc* (*Leu.*) *mesenteroides*, *Weissella cibaria*, *W. paramesenteroides*, *L. pentosus*, and *L. plantarum*. The results are in agreement with previous reports in which *Leuconostoc*, *Weissella*, and *Lactobacillus* spp. were suggested to be responsible for kimchi fermentation [8,32]. Meanwhile, tyramine production by *L. brevis* in various fermented foods, including wine and fermented sausage, as well as *Baechu* kimchi, has been previously reported [14,33,34]. In the reports, tyramine production by *L. brevis* isolated from wine ranged from 441.6 to 1070.0 µg/mL, which is higher than that of the present study. On the contrary, *L. brevis* isolated from fermented sausage and *Baechu* kimchi produced tyramine at the range from 138.51 to 169.47 µg/mL and from 282 to 388 µg/mL, respectively, which are similar or lower than that of this study. In addition, several authors also isolated tyramine-producing *Leu. mesenteroides*, *W. cibaria*, and *W. paramesenteroides* from *Baechu* kimchi [14,15] and *L. plantarum* from wine [35]. Interestingly, as described right above, there are somewhat disparate results between the present and previous studies, which indicates that the strains belonging to the same species may possess different ability to produce tyramine especially depending upon the kinds of foods. Thus, microbial BA production in radish kimchi is likely determined at strain level, probably adapting to the respective food ecosystems, as suggested by previous reports [36,37]. Another implication is that the strains unable to produce BA isolated in the current study have potential as starter cultures for kimchi fermentation. Further investigations are needed to use them as starter cultures, which may involve tests to examine if the strains fulfill the criteria of starter culture, including the technical properties of strains, food safety requirements, and quality expectations [38].

3.4. Changes in Tyramine and Other BA Content during Fermentation of Radish Kimchi: Kkakdugi and Chonggak Kimchi

Fermentation of *Kkakdugi* and *Chonggak* kimchi was performed to investigate the influences of *Myeolchi-aekjeot* (together with *Saeu-jeotgal*) and LAB strains (particularly *L. brevis*) on BA content (especially tyramine) of both radish kimchi. Five groups of *Kkakdugi* and *Chonggak* kimchi samples were prepared based on the presence or absence of *Myeolchi-aekjeot* and types of LAB inocula. *L. brevis* strains of KD3M5 and CG2M15 with the highest tyramine production activity among the identified tyramine-producing LAB strains were used to see if the species is practically responsible for tyramine formation during fermentation of *Kkakdugi* and *Chonggak* kimchi. On the other hand, *L. plantarum* strains of KD3M15 and CG3M21 unable to produce BA were used for two reasons. (i) *L. plantarum*, like *L. brevis*, is predominant species in kimchi [39]. (ii) Differently from *L. brevis*, *L. plantarum* has been found to be negative for tyramine production in the present and previous studies [33,40,41].

As shown in Figures 1 and 2, changes in physicochemical and microbial properties of *Kkakdugi* and *Chonggak* kimchi during the fermentation for 3 days were similar with those of several previous reports [25,29,42]. In detail, the pH of all radish kimchi groups decreased during day 1 of fermentation, and stayed constantly thereafter. On the contrary, the counts of total aerobic bacteria and lactic acid bacteria, and the acidity of all radish kimchi groups increased during day 1 and day 2, respectively, and remained constantly thereafter, which indicates that an appropriate fermentation process of *Kkakdugi* and *Chonggak* kimchi took place. It is mention worthy that the initial pH of C, PC, LB, and LP groups of both radish kimchi was slightly higher than that of B group, which might be because the neutral pH of *Saeu-jeotgal* affected the pH values of the former groups [43]. Nonetheless, the initial acidity of all groups, belonging to either *Kkakdugi* or *Chonggak* kimchi, was similar to each other. The salinity of all radish kimchi groups decreased slightly during fermentation. According to Shin, Ann, and Kim [44], osmosis between radish and broth (containing seasoning paste) occurs during fermentation, which results in a steady reduction of salinity. Regardless of the drop in salinity, water activity of all radish kimchi groups was constant during fermentation. In addition, the initial counts of total aerobic bacteria and lactic acid bacteria of PC, LB, and LP groups inoculated with any of LAB strains were higher than those of B and C groups to be fermented naturally without any inocula, as expected.

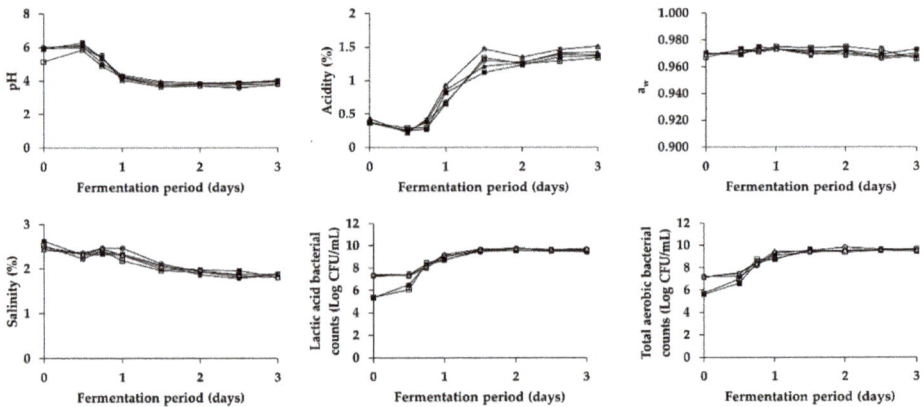

Figure 1. Changes in physicochemical and microbial properties of *Kkakdugi* during fermentation. □: B (no addition of *Myeolchi-aekjeot* and *Saeu-jeotgal*, no inoculum), ■: C (addition of *Myeolchi-aekjeot* and *Saeu-jeotgal*, no inoculum), ▲: PC (addition of *Myeolchi-aekjeot* and *Saeu-jeotgal*, *L. brevis* JCM 1170), △: LB (addition of *Myeolchi-aekjeot* and *Saeu-jeotgal*, *L. brevis* KD3M5), ○: LP (addition of *Myeolchi-aekjeot* and *Saeu-jeotgal*, *L. plantarum* KD3M15).

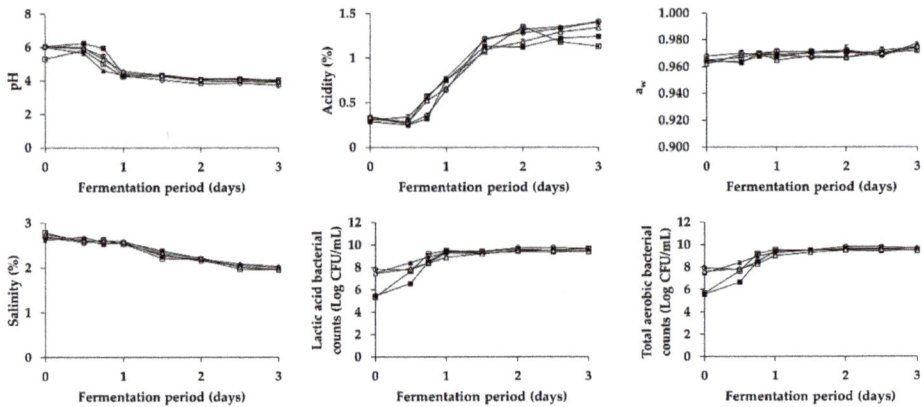

Figure 2. Changes in physicochemical and microbial properties of *Chonggak* kimchi during fermentation. □: B (no addition of *Myeolchi-aekjeot* and *Saeu-jeotgal*, no inoculum), ■: C (addition of *Myeolchi-aekjeot* and *Saeu-jeotgal*, no inoculum), ▲: PC (addition of *Myeolchi-aekjeot* and *Saeu-jeotgal*, *L. brevis* JCM 1170), △: LB (addition of *Myeolchi-aekjeot* and *Saeu-jeotgal*, *L. brevis* CG2M15), ○: LP (addition of *Myeolchi-aekjeot* and *Saeu-jeotgal*, *L. plantarum* CG3M21).

Changes in BA content (except for tryptamine and β-phenylethylamine not detected) during fermentation of *Kkakdugi* and *Chonggak* kimchi were shown in Figures 3 and 4, respectively. There appeared an increment of tyramine content in most groups (except for LP group) of both radish kimchi over the fermentation period, probably resulting from tyramine production by either inoculated or indigenous *L. brevis* strains (refer to Section 3.3). Also, the increment of tyramine content in PC and LB groups was higher than that in B and C groups of both radish kimchi (except for day 3 of *Chonggak* kimchi fermentation). This might be due to higher lactic acid bacterial counts of PC and LB groups, resulting from the inoculation of tyramine-producing *L. brevis* strains, than those of B and C groups of both radish kimchi. In the meantime, tyramine content in B and C groups of *Chonggak* kimchi steadily increased during fermentation, while that in the same groups of *Kkakdugi* increased slightly (but at a low level compared to *Chonggak* kimchi), both of which are likely associated with tyramine production by indigenous LAB strains (probably *L. brevis*). The observations are consistent with previous reports described right below. In short, Choi et al. [45] reported a dramatic increase of tyramine during natural fermentation of *Baechu* kimchi, whereas Kim et al. [46] reported that *Baechu* kimchi had a constantly low level of tyramine during natural fermentation. It is also noteworthy that, in the case of *Chonggak* kimchi, tyramine content in PC and LB groups dramatically increased during day 1 of fermentation, which was higher (and also showed a faster increment) than that in the same groups of *Kkakdugi*. The results, together with the comparison of tyramine content in B and C groups between two types of radish kimchi described above, can be explained by two speculations. The first is the difference in the ability of *L. brevis* strains to produce tyramine. The second is the distinguishable adaptation of the strains to different food ecosystems, i.e., differences in the main ingredients and/or ratio of ingredients in seasoning paste between two types of radish kimchi. Since KD3M5 strain served as an inoculum for *Kkakdugi* revealed a stronger ability to produce tyramine (377.35 ± 4.36 μg/mL) than CG2M15 strain for *Chonggak* kimchi (328.48 ± 2.61 μg/mL) when compared in vitro (refer to Section 3.3), the second speculation seems to be more probable than the first one. In addition, it is well known that bacteria produce BA to neutralize acidic environments as part of homeostatic regulation [47]. In this study, however, both radish kimchi samples of PC and LB groups showed similar patterns of acidity changes, so that the homeostatic regulation was excluded from possible reasons. Either way, there seem to be much complicated cross effects by the combinations of factors influencing the intensity of BA production by LAB during fermentation of kimchi variety. Interestingly, LP group of both radish kimchi had significantly lower levels of tyramine than the other groups. Thus, it seems that *L. plantarum* strains unable to produce BA in vitro not only have incapability of producing BA during fermentation, but also may inhibit tyramine production by indigenous LAB strains. This indicates the applicability of this species as a starter culture for reducing BA in kimchi variety.

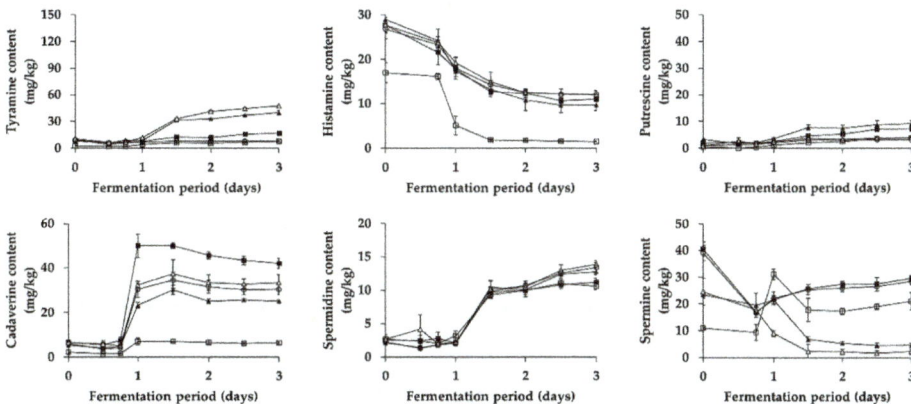

Figure 3. Changes in BA content in *Kkakdugi* during fermentation. □: B (no addition of *Myeolchi-aekjeot* and *Saeu-jeotgal*, no inoculum), ■: C (addition of *Myeolchi-aekjeot* and *Saeu-jeotgal*, no inoculum), ▲: PC (addition of *Myeolchi-aekjeot* and *Saeu-jeotgal*, *L. brevis* JCM 1170), △: LB (addition of *Myeolchi-aekjeot* and *Saeu-jeotgal*, *L. brevis* KD3M5), ○: LP (addition of *Myeolchi-aekjeot* and *Saeu-jeotgal*, *L. plantarum* KD3M15).

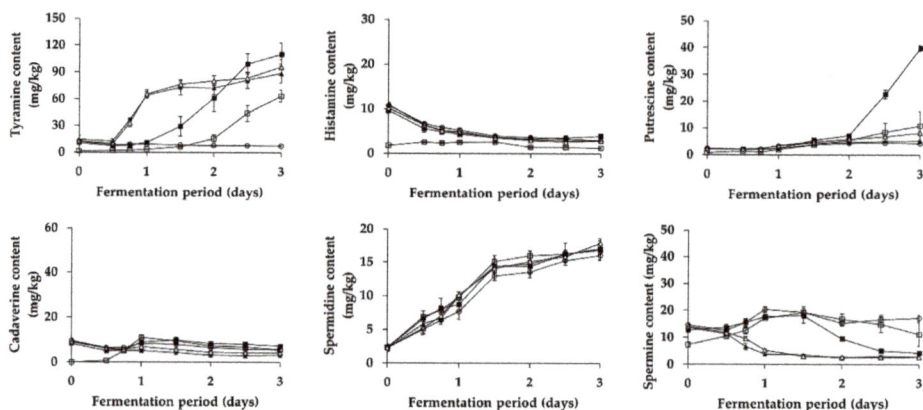

Figure 4. Changes in BA content in *Chonggak* kimchi during fermentation. □: B (no addition of *Myeolchi-aekjeot* and *Saeu-jeotgal*, no inoculum), ■: C (addition of *Myeolchi-aekjeot* and *Saeu-jeotgal*, no inoculum), ▲: PC (addition of *Myeolchi-aekjeot* and *Saeu-jeotgal*, *L. brevis* JCM 1170), △: LB (addition of *Myeolchi-aekjeot* and *Saeu-jeotgal*, *L. brevis* CG2M15), ○: LP (addition of *Myeolchi-aekjeot* and *Saeu-jeotgal*, *L. plantarum* CG3M21).

Differently from tyramine, histamine content in all groups of both radish kimchi gradually decreased during fermentation. This result might be because there were some indigenous LAB strains with histamine-degrading activity. Similarly, Kim et al. [48] reported a significant reduction of histamine content in *Baechu* kimchi inoculated with type strains of different LAB species including *L. sakei*, *L. plantarum*, *Leu. carnosum*, and *Leu. mesenteroides*, when compared with non-inoculated kimchi, suggesting that some LAB stains in kimchi are capable of degrading histamine. Meanwhile, the experimental groups of *Kkakdugi* and *Chonggak* kimchi prepared with *Myeolchi-aekjeot* (C, PC, LB, and LP groups) contained a significantly higher level of histamine than B group, which is in accordance with the suggestion of previous studies [12,22]. In the studies, the authors assumed that histamine level in *Baechu* kimchi could be affected by histamine in *Myeolchi-aekjeot*. Taking this into account, histamine content of *Kkakdugi* and *Chonggak* kimchi in the present study seems to come from *Myeolchi-aekjeot* rather than microbial histamine production during fermentation.

Putrescine and spermidine content steadily increased in all groups of *Kkakdugi* and *Chonggak* kimchi during fermentation, which is in agreement with previous reports [12,46]. There was a small and insignificant difference in putrescine and spermidine content among the groups of both radish kimchi during fermentation, which indicates that LAB strains—including *L. brevis* and *L. plantarum*—produced the polyamines during fermentation. Meanwhile, the initial concentrations putrescine and spermidine in *Kkakdugi* and *Chonggak* kimchi might be come from main ingredients, i.e., white radish and ponytail radish, respectively. In addition, a sharp increment of putrescine was observed during day 3 of fermentation, in the case of C group of *Chonggak* kimchi. To ignore the possibility of outliers, the fermentation experiment was repeatedly performed; however, the same results were observed, and the reason for such observation was not clear.

Somewhat differently from above, cadaverine content in all groups of *Kkakdugi* and *Chonggak* kimchi showed an increment during day 1 of fermentation and slight decline thereafter, although the increased cadaverine amount was mostly higher in *Kkakdugi* than in *Chonggak* kimchi. The difference in the intensity of cadaverine formation between two types of radish kimchi seems to be attributed to the complex combinations of factors described above to explain difference in the kinetics of tyramine formation between two radish kimchi. Interestingly, the initial cadaverine content in C, PC, LB, and LP groups of both radish kimchi was higher than that in B group, which might be come from *Myeolchi-aekjeot* rather than *Saeu-jeotgal*. The speculation is supported by a study by Cho et al. [22]

who reported a significantly higher level of cadaverine in *Myeolchi-aekjeot* (up to 263.6 mg/kg) than that in *Saeu-jeotgal* (up to 7.0 mg/kg). For both radish kimchi, C group contained the highest level of cadaverine, as compared to the other groups, over the fermentation period. This may be explained by a presumption that while cadaverine-producing bacteria derived from *Myeolchi-aekjeot* are probably responsible for cadaverine formation during fermentation of both radish kimchi, LAB strains (*L. brevis* and *L. plantarum*) served as inocula are probably capable of degrading cadaverine. Supporting this presumption, Mah et al. [49] reported that *Bacillus* strains isolated from *Myeolchi-jeotgal* were highly capable of producing cadaverine. Capozzi et al. [50] also reported that *L. plantarum* strains isolated from wine were capable of degrading cadaverine. At present, however, investigations on cadaverine-degrading activity of *L. brevis* are rarely found in literature.

As for change in spermine content, there appeared difference among groups of *Kkakdugi* and *Chonggak* kimchi. In PC and LB groups of both radish kimchi, a gradual decrease of spermine content was observed over the fermentation period, and the content was relatively lower than that in the other groups of both types of radish kimchi. This implies that *L. brevis* could be able to degrade spermine, although relevant reports are scarce to date. It is worth nothing that in B, C, and LP groups, spermine content decreased for day 1 of fermentation and slightly increased thereafter in *Kkakdugi*, whereas that in *Chonggak* kimchi increased for day 1 and slightly decreased thereafter. The different kinetics of spermine formation seems to result from the complex combinations of factors mentioned above. Therefore, it would be interesting in a future study to identify the factors (and combinations thereof) associated with BA formation or degradation by LAB strains during fermentation of *Kkakdugi* and *Chonggak* kimchi. The factors may involve time-related successional changes and/or interactions of microorganisms during fermentation as well as ingredients of foods and metabolic activities of strains [51]. In addition, recent studies suggested that results of in vitro BA production by food fermenting microorganisms were in disagreement with those of BA formation during fermentation of the corresponding foods [52,53]. In the present study, however, *L. brevis* was considered to be responsible for tyramine formation not only in vitro but also during practical fermentation of *Kkakdugi* and *Chonggak* kimchi.

4. Conclusions

The present study indicated that the amounts of BA in most samples of *Kkakdugi* and *Chonggak* kimchi were considered safe for consumption, but some samples contained histamine and total BA at concentrations over toxicity limits ($\geq$100 mg/kg and $\geq$1000 mg/kg, respectively). It was also found that, while *Myeolchi-aekjeot* seems to be an important source of histamine in both types of radish kimchi, *L. brevis* strains isolated from *Kkakdugi* and *Chonggak* kimchi are highly capable of producing tyramine in assay media. On the other hand, the physicochemical and microbial properties of both radish kimchi revealed weak correlations with BA content in the respective kimchi types in the present study. Through the practical fermentation of *Kkakdugi* and *Chonggak* kimchi, it turned out that *L. brevis* is responsible for tyramine formation, and *Myeolchi-aekjeot* influences histamine and cadaverine content in both radish kimchi. Consequently, this study suggests strategies for reducing BA in radish kimchi: the alteration of the ratio of ingredients used for kimchi preparation, particularly reducing ratio of *Myeolchi-aekjeot* to others, and use of starter cultures other than tyramine-producing *L. brevis* strains, especially BA-degrading LAB starter cultures. Studies on other contributing factors influencing the intensity of BA production by LAB are also required to understand complex kinetics of BA formation in the kimchi.

Supplementary Materials: The following is available online at http://www.mdpi.com/2304-8158/8/2/73/s1, Figure S1: Scheme of procedure for BA analysis.

Author Contributions: Conceptualization: J.-H.M.; Investigation: Y.H.J., J.H.L., Y.K.P., and J.-H.L.; Writing—original draft: Y.H.J. and J.-H.M.; Writing—review and editing: Y.H.J. and J.-H.M.; Supervision: J.-H.M.

Funding: This work was supported by the National Research Foundation of Korea (NRF) grant funded by the Korea government (MSIT) (no. 2016R1A2B4012161).

Acknowledgments: The authors thank Junsu Lee of Department of Food and Biotechnology at Korea University for technical assistance.

Conflicts of Interest: The authors declare no conflict of interest.

References

1. Ten Brink, B.; Damink, C.; Joosten, H.M.L.J.; Huis In't Veld, J.H.J. Occurrence and formation of biologically active amines in foods. *Int. J. Food Microbiol.* **1990**, *11*, 73–84. [CrossRef]
2. Silla Santos, M.H. Biogenic amines: Their importance in foods. *Int. J. Food Microbiol.* **1996**, *29*, 213–231. [CrossRef]
3. Ladero, V.; Calles-Enríquez, M.; Fernández, M.; Alvarez, M.A. Toxicological effects of dietary biogenic amines. *Curr. Nutr. Food Sci.* **2010**, *6*, 145–156. [CrossRef]
4. Marcobal, A.; De Las Rivas, B.; Landete, J.M.; Tabera, L.; Muñoz, R. Tyramine and phenylethylamine biosynthesis by food bacteria. *Crit. Rev. Food Sci. Nutr.* **2012**, *52*, 448–467. [CrossRef]
5. Bover-Cid, S.; Hugas, M.; Izquierdo-Pulido, M.; Vidal-Carou, M.C. Reduction of biogenic amine formation using a negative amino acid-decarboxylase starter culture for fermentation of *Fuet* sausages. *J. Food Prot.* **2000**, *63*, 237–243. [CrossRef]
6. Bover-Cid, S.; Hugas, M.; Izquierdo-Pulido, M.; Vidal-Carou, M.C. Effect of the interaction between a low tyramine-producing *Lactobacillus* and proteolytic *staphylococci* on biogenic amine production during ripening and storage of dry sausages. *Int. J. Food Microbiol.* **2001**, *65*, 113–123. [CrossRef]
7. Codex Alimentarius Commission. *Codex Standard for kimchi, Codex Stan 223-2001*; Food and Agriculture Organization of the United Nations: Rome, Italy, 2001.
8. Park, E.-J.; Chun, J.; Cha, C.-J.; Park, W.-S.; Jeon, C.O.; Bae, J.-W. Bacterial community analysis during fermentation of ten representative kinds of kimchi with barcoded pyrosequencing. *Food Microbiol.* **2012**, *30*, 197–204. [CrossRef]
9. Cheigh, H.-S.; Park, K.-Y. Biochemical, microbiological, and nutritional aspects of kimchi (Korean fermented vegetable products). *Crit. Rev. Food Sci. Nutr.* **1994**, *34*, 175–203. [CrossRef]
10. Jang, K.-S.; Kim, M.-J.; Oh, Y.-A.; Kim, I.-D.; No, H.-K.; Kim, S.-D. Effects of various sub-ingredients on sensory quality of Korean cabbage kimchi. *J. Korean Soc. Food Nutr.* **1991**, *20*, 233–240.
11. Park, D.-C.; Park, J.-H.; Gu, Y.-S.; Han, J.-H.; Byun, D.-S.; Kim, E.-M.; Kim, Y.-M.; Kim, S.-B. Effects of salted-fermented fish products and their alternatives on angiotensin converting enzyme inhibitory activity of *Kimchi* during fermentation. *Korean J. Food Sci. Technol.* **2000**, *32*, 920–927.
12. Mah, J.-H.; Kim, Y.J.; No, H.-K.; Hwang, H.-J. Determination of biogenic amines in *kimchi*, Korean traditional fermented vegetable products. *Food Sci. Biotechnol.* **2004**, *13*, 826–829.
13. Tsai, Y.-H.; Kung, H.-F.; Lin, Q.-L.; Hwang, J.-H.; Cheng, S.-H.; Wei, C.-I.; Hwang, D.-F. Occurrence of histamine and histamine-forming bacteria in kimchi products in Taiwan. *Food Chem.* **2005**, *90*, 635–641. [CrossRef]
14. Kim, M.-J.; Kim, K.-S. Tyramine production among lactic acid bacteria and other species isolated from kimchi. *LWT-Food Sci. Technol.* **2014**, *56*, 406–413. [CrossRef]
15. Jeong, D.-W.; Lee, J.-H. Antibiotic resistance, hemolysis and biogenic amine production assessments of *Leuconostoc* and *Weissella* isolates for kimchi starter development. *LWT-Food Sci. Technol.* **2015**, *64*, 1078–1084. [CrossRef]
16. AOAC. *Official Methods of Analysis of AOAC International*, 17th ed.; AOAC International: Gaithersburg, MD, USA, 2000.
17. Eerola, S.; Hinkkanen, R.; Lindfors, E.; Hirvi, T. Liquid chromatographic determination of biogenic amines in dry sausages. *J. AOAC Int.* **1993**, *76*, 575–577. [PubMed]
18. Ben-Gigirey, B.; De Sousa, J.M.V.B.; Villa, T.G.; Barros-Velazquez, J. Changes in biogenic amines and microbiological analysis in albacore (*Thunnus alalunga*) muscle during frozen storage. *J. Food Prot.* **1998**, *61*, 608–615. [CrossRef]
19. Ben-Gigirey, B.; De Sousa, J.M.V.B.; Villa, T.G.; Barros-Velazquez, J. Histamine and cadaverine production by bacteria isolated from fresh and frozen albacore (*Thunnus alalunga*). *J. Food Prot.* **1999**, *62*, 933–939. [CrossRef]

20. Yoon, H.; Park, J.H.; Choi, A.; Hwang, H.-J.; Mah, J.-H. Validation of an HPLC analytical method for determination of biogenic amines in agricultural products and monitoring of biogenic amines in Korean fermented agricultural products. *Toxicol. Res.* **2015**, *31*, 299–305. [CrossRef]

21. National Institute of Agricultural Sciences. Available online: http://koreanfood.rda.go.kr/kfi/kimchi/kimchi_01 (accessed on 2 February 2019).

22. Cho, T.-Y.; Han, G.-H.; Bahn, K.-N.; Son, Y.-W.; Jang, M.-R.; Lee, C.-H.; Kim, S.-H.; Kim, D.-B.; Kim, S.-B. Evaluation of biogenic amines in Korean commercial fermented foods. *Korean J. Food Sci. Technol.* **2006**, *38*, 730–737.

23. Kim, M.R.; Oh, Y.; Oh, S. Physicochemical and sensory properties of Kagdugi prepared with fermentation northern sand sauce during fermentation. *Korean J. Soc. Food Sci.* **2000**, *16*, 602–608.

24. Park, S.-O.; Kim, W.-K.; Park, D.-J.; Lee, S.-J. Effect of blanching time on the quality characteristics of elderly-friendly *kkakdugi*. *Food Sci. Biotechnol.* **2017**, *26*, 419–425. [CrossRef] [PubMed]

25. Kang, J.-H.; Kang, S.-H.; Ahn, E.-S.; Chung, H.-J. Quality properties of *Chonggak* kimchi fermented at different combination of temperature and time. *J. Korean Soc. Food Cult.* **2003**, *18*, 551–561.

26. Kim, Y.-J.; Jin, Y.-Y.; Song, K.-B. Study of quality change in Chonggak-kimchi during storage, for development of a freshness indicator. *Korean J. Food Preserv.* **2008**, *15*, 491–496.

27. Stratton, J.E.; Hutkins, R.W.; Taylor, S.L. Biogenic amines in cheese and other fermented foods: A review. *J. Food Prot.* **1991**, *54*, 460–470. [CrossRef]

28. Warthesen, J.J.; Scanlan, R.A.; Bills, D.D.; Libbey, L.M. Formation of heterocyclic *N*-nitrosamines from the reaction of nitrite and selected primary diamines and amino acids. *J. Agric. Food Chem.* **1975**, *23*, 898–902. [CrossRef] [PubMed]

29. Mheen, T.-I.; Kwon, T.-W. Effect of temperature and salt concentration on *Kimchi* fermentation. *Korean J. Food Sci. Technol.* **1984**, *16*, 443–450.

30. Lu, S.; Xu, X.; Shu, R.; Zhou, G.; Meng, Y.; Sun, Y.; Chen, Y.; Wang, P. Characterization of biogenic amines and factors influencing their formation in traditional Chinese sausages. *J. Food Sci.* **2010**, *75*, M366–M372. [CrossRef]

31. Özdestan, Ö.; Üren, A. Biogenic amine content of tarhana: A traditional fermented food. *Int. J. Food Prop.* **2013**, *16*, 416–428. [CrossRef]

32. Kim, M.; Chun, J. Bacterial community structure in kimchi, a Korean fermented vegetable food, as revealed by 16S rRNA gene analysis. *Int. J. Food Microbiol.* **2005**, *103*, 91–96. [CrossRef]

33. Landete, J.M.; Ferrer, S.; Pardo, I. Biogenic amine production by lactic acid bacteria, acetic bacteria and yeast isolated from wine. *Food Control* **2007**, *18*, 1569–1574. [CrossRef]

34. Latorre-Moratalla, M.L.; Bover-Cid, S.; Talon, R.; Garriga, M.; Zanardi, E.; Ianieri, A.; Fraqueza, M.J.; Elias, M.; Drosinos, E.H. Vidal-Carou, M.C. Strategies to reduce biogenic amine accumulation in traditional sausage manufacturing. *LWT-Food Sci. Technol.* **2010**, *43*, 20–25. [CrossRef]

35. Arena, M.E.; Fiocco, D.; Manca de Nadra, M.C.; Pardo, I.; Spano, G. Characterization of a *Lactobacillus plantarum* strain able to produce tyramine and partial cloning of a putative tyrosine decarboxylase gene. *Curr. Microbiol.* **2007**, *55*, 205–210. [CrossRef] [PubMed]

36. Bover-Cid, S.; Holzapfel, W.H. Improved screening procedure for biogenic amine production by lactic acid bacteria. *Int. J. Food Microbiol.* **1999**, *53*, 33–41. [CrossRef]

37. Landete, J.M.; Ferrer, S.; Polo, L.; Pardo, I. Biogenic amines in wines from three Spanish regions. *J. Agric. Food Chem.* **2005**, *53*, 1119–1124. [CrossRef] [PubMed]

38. Holzapfel, W.H. Appropriate starter culture technologies for small-scale fermentation in developing countries. *Int. J. Food Microbiol.* **2002**, *75*, 197–212. [CrossRef]

39. Lee, C.-H. Lactic acid fermented foods and their benefits in Asia. *Food Control* **1997**, *8*, 259–269. [CrossRef]

40. Moreno-Arribas, M.V.; Polo, M.C.; Jorganes, F.; Muñoz, R. Screening of biogenic amine production by lactic acid bacteria isolated from grape must and wine. *Int. J. Food Microbiol.* **2003**, *84*, 117–123. [CrossRef]

41. Park, S.; Ji, Y.; Park, H.; Lee, K.; Park, H.; Beck, B.R.; Shin, H.; Holzapfel, W.H. Evaluation of functional properties of lactobacilli isolated from Korean white kimchi. *Food Control* **2016**, *69*, 5–12. [CrossRef]

42. Kim, S.-D.; Jang, M.-S. Effects of fermentation temperature on the sensory, physicochemical and microbiological properties of *Kakdugi*. *J. Kor. Soc. Food Sci. Nutr.* **1997**, *26*, 800–806.

43. Um, M.-N.; Lee, C.-H. Isolation and identification of *Staphylococcus* sp. from Korean fermented fish products. *J. Microbiol. Biotechnol.* **1996**, *6*, 340–346.

44. Shin, Y.-H.; Ann, G.-J.; Kim, J.-E. The changes of hardness and microstructure of Dongchimi according to different kinds of water. *Korean J. Food Cookery Sci.* **2004**, *20*, 86–94.

45. Choi, Y.-J.; Jang, M.-S.; Lee, M.-A. Physicochemical changes in kimchi containing skate (*Raja kenojei*) pretreated with organic acids during fermentation. *Food Sci. Biotechnol.* **2016**, *25*, 1369–1377. [CrossRef] [PubMed]

46. Kim, S.-H.; Kang, K.H.; Kim, S.H.; Lee, S.; Lee, S.-H.; Ha, E.-S.; Sung, N.-J.; Kim, J.G.; Chung, M.J. Lactic acid bacteria directly degrade N-nitrosodimethylamine and increase the nitrite-scavenging ability in kimchi. *Food Control* **2017**, *71*, 101–109. [CrossRef]

47. Arena, M.E.; Manca de Nadra, M.C. Biogenic amine production by *Lactobacillus*. *J. Appl. Microbiol.* **2001**, *90*, 158–162. [CrossRef]

48. Kim, S.-H.; Kim, S.H.; Kang, K.H.; Lee, S.; Kim, S.J.; Kim, J.G.; Chung, M.J. Kimchi probiotic bacteria contribute to reduced amounts of N-nitrosodimethylamine in lactic acid bacteria-fortified kimchi. *LWT-Food Sci. Technol.* **2017**, *84*, 196–203. [CrossRef]

49. Mah, J.-H.; Ahn, J.-B.; Park, J.-H.; Sung, H.-C.; Hwang, H.-J. Characterization of biogenic amine-producing microorganisms isolated from Myeolchi-Jeot, Korean salted and fermented anchovy. *J. Microbiol. Biotechnol.* **2003**, *13*, 692–699.

50. Capozzi, V.; Russo, P.; Ladero, V.; Fernández, M.; Fiocco, D.; Alvarez, M.A.; Grieco, F.; Spano, G. Biogenic amines degradation by *Lactobacillus plantarum*: Toward a potential application in wine. *Front. Microbiol.* **2012**, *3*, 122. [CrossRef]

51. Yılmaz, C.; Gökmen, V. Formation of tyramine in yoghurt during fermentation—Interaction between yoghurt starter bacteria and *Lactobacillus plantarum*. *Food Res. Int.* **2017**, *97*, 288–295. [CrossRef]

52. Nie, X.; Zhang, Q.; Lin, S. Biogenic amine accumulation in silver carp sausage inoculated with *Lactobacillus plantarum* plus *Saccharomyces cerevisiae*. *Food Chem.* **2014**, *153*, 432–436. [CrossRef]

53. Jeon, A.R.; Lee, J.H.; Mah, J.-H. Biogenic amine formation and bacterial contribution in *Cheonggukjang*, a Korean traditional fermented soybean food. *LWT-Food Sci. Technol.* **2018**, *92*, 282–289. [CrossRef]

Review

Biogenic Amines in Plant-Origin Foods: Are they Frequently Underestimated in Low-Histamine Diets?

Sònia Sánchez-Pérez [1,2,3], Oriol Comas-Basté [1,2,3], Judit Rabell-González [1,2,3],
M. Teresa Veciana-Nogués [1,2,3], M. Luz Latorre-Moratalla [1,2,3] and M. Carmen Vidal-Carou [1,2,3,*]

[1] Departament de Nutrició, Ciències de l'Alimentació i Gastronomia, Facultat de Farmàcia i Ciències de
    l'Alimentació, Universitat de Barcelona (UB), Av. Prat de la Riba 171, 08921 Santa Coloma de Gramenet,
    Spain; soniasanchezperez@ub.edu (S.S.-P.); oriolcomas@ub.edu (O.C.-B.);
    juditrabellgonzalez@hotmail.com (J.R.-G.); veciana@ub.edu (M.T.V.-N.); mariluzlatorre@ub.edu (M.L.L.-M.)
[2] Institut de Recerca en Nutrició i Seguretat Alimentària (INSA·UB), Universitat de Barcelona (UB), Av. Prat
    de la Riba 171, 08921 Santa Coloma de Gramenet, Spain
[3] Xarxa de Referència en Tecnologia dels Aliments de la Generalitat de Catalunya (XaRTA), C/ Baldiri Reixac
    4, 08028 Barcelona, Spain
*   Correspondence: mcvidal@ub.edu; Tel.: +34-934-033-786

Received: 8 November 2018; Accepted: 12 December 2018; Published: 14 December 2018

Abstract: Low-histamine diets are currently used to reduce symptoms of histamine intolerance, a disorder in histamine homeostasis that increases plasma levels, mainly due to reduced diamine-oxidase (DAO) activity. These diets exclude foods, many of them of plant origin, which patients associate with the onset of the symptomatology. This study aimed to review the existing data on histamine and other biogenic amine contents in nonfermented plant-origin foods, as well as on their origin and evolution during the storage or culinary process. The only plant-origin products with significant levels of histamine were eggplant, spinach, tomato, and avocado, each showing a great variability in content. Putrescine has been found in practically all plant-origin foods, probably due to its physiological origin. The high contents of putrescine in certain products could also be related to the triggering of the symptomatology by enzymatic competition with histamine. Additionally, high spermidine contents found in some foods should also be taken into account in these diets, because it can also be metabolized by DAO, albeit with a lower affinity. It is recommended to consume plant-origin foods that are boiled or are of maximum freshness to reduce biogenic amine intake.

Keywords: histamine; putrescine; tyramine; cadaverine; biogenic amines; histamine intolerance; low-histamine diet; plant-origin foods; culinary process; storage conditions

1. Introduction

In recent years, various diets have been proposed for the treatment of histamine intolerance [1–8]. These diets, known as low- or free-histamine diets, usually exclude foods that patients associate with the onset of intolerance symptoms. Such foods tend to be rich in histamine, but some, surprisingly, are not usually regarded as sources of this amine.

As described in the literature and scientific reports issued by the European Food Safety Authority (EFSA) and a joint Food and Agriculture Organization of the United Nations (FAO)/World Health Organization (WHO) committee, histamine intolerance (also called food histaminosis or food histamine sensitivity) is a disorder associated with increased plasma histamine levels and is recognized as clinically different from the more established histamine intoxication [9,10]. Although in both cases, histamine is the causative agent, the etiology of the disorders differs. Intoxication appears after the consumption of foods with unusually high histamine concentrations, while intolerance is due to a deficiency in histamine metabolism, so that symptoms may be triggered even by the intake of low amounts [1,9–11].

Diamine oxidase (DAO) is the main enzyme responsible for the metabolism of histamine and other amines at the intestinal level, and impaired DAO activity is one of the main causes of histamine intolerance [1,12,13]. This enzymatic deficit may have its origins in genetic mutations. Different polymorphisms of a single nucleotide in the gene that encodes this enzyme (*AOC1* on chromosome 7) have been associated with lower DAO activity [14–16]. The deficit may also be due to acquired causes such as inflammatory bowel diseases that block the secretion of DAO [1,3,12], or to the inhibitory action of drugs, some of them with a very widespread use (e.g., acetylcysteine, clavulanic acid, metoclopramide, verapamil) [1,17]. Another enzyme involved in histamine metabolization is monoamine oxidase (MAO) [13]. Therefore, MAO inhibitor drugs, such as selegiline or rasagiline, could also favor the plasmatic accumulation of histamine and the onset of symptoms of histamine intolerance. In addition, the presence of other biogenic amines, mainly putrescine and cadaverine, may compromise the intestinal degradation of histamine by enzymatic competition with DAO [9].

The symptoms of histamine intolerance are numerous and highly variable, due to the effects and functions of histamine in multiple organs and systems of the body. They include gastrointestinal (abdominal pain, diarrhea, vomiting), dermatological (urticaria, dermatitis, or pruritus), respiratory (rhinitis, nasal congestion, and asthma), cardiovascular (hypotonia and arrhythmias), and neurological (headaches) symptoms, and it is common for more than one disorder to occur simultaneously [1,11,12]. Several clinical studies have shown that patients with a potential diagnosis of histamine intolerance or with a diagnosis of migraine, intestinal, or dermatological diseases (atopic dermatitis, eczema, or chronic urticaria) have a higher prevalence of DAO deficits compared to the control population [3,6,18–28].

In order to carry out a correct dietary treatment of histamine intolerance, it is necessary to know what foods may contain this amine and what factors influence its accumulation. Likewise, it is also important to consider the occurrence of other amines that are also metabolized by the DAO enzyme. In contrast to plant-origin foods, there is more available information on the contents of histamine and other amines in fish and fish derivatives and all types of fermented products (cheeses, sausages, sauerkraut, wines, beer), in which their presence is attributed to the aminogenic activity of spoilage microorganisms and also to fermentative microorganisms [9,10,29]. Therefore, the freshness of the food and the hygienic conditions of the raw materials and manufacturing processes, as well as the adequate selection of starter cultures without decarboxylase activity, are of vital importance to avoid or reduce the formation of these compounds [9,29–31].

Due to the information available on the contents of biogenic amines in nonfermented plant-origin foods being scarce, the aim of this study was to review the existing data on the contents of histamine and other biogenic amines in these types of products, as well as their origin and evolution during storage or cooking.

2. Methods

A selective search of scientific literature dealing with biogenic amine contents in nonfermented plant-origin foods, including vegetables, fruits, and cereals, was performed. The bibliographic search was carried out in the PubMed and Web of Science databases using the following keywords: "histamine", "biogenic amines", "tyramine", "putrescine", "cadaverine", "plant-origin food", "food samples", "storage", "cooking", "fruit", "vegetable", "legume", "cereal", "spinach", "eggplant", "tomato", "citrus", "modified atmosphere packaging", and "microbial decarboxylase activity". Original analytical studies, reviews, and table compilations of content in food were included. Articles published before 1990 were excluded from this review.

Apart from data obtained from the literature, data on the biogenic amine content of plant-origin foods from our own database of Spanish market products were also used. Specifically, histamine, tyramine, putrescine, and cadaverine contents of 25 types of vegetables, 19 fruits, and 8 cereals were included.

3. Content of Biogenic Amines in Plant-Origin Foods

In this section, the contents of biogenic amines (histamine, tyramine, putrescine, and cadaverine) in different plant-origin foods are reviewed, using our own database and data from studies published by other authors. A total of 20 studies reporting data on biogenic amine contents in such foods were found. Most provided data on putrescine contents (normally together with the polyamines spermine and spermidine, not dealt with in this section), and only a few included other amines, such as histamine, tyramine, and cadaverine.

3.1. Vegetables and Legumes

Table 1 shows the contents of biogenic amines in different types of vegetables and legumes (nonfermented).

The only products found to contain significant levels of histamine were eggplant, spinach, and tomato, each showing a great variability in content, both in samples from the same study and among different studies. Histamine values ranged from 4.2 to 100.6 mg/kg in eggplant, from 9.5 to 69.7 mg/kg in spinach, and from not detected to 17.1 mg/kg in tomato. In the case of asparagus, pumpkin, and chard, histamine was found in only a few samples and at very low levels (<2 mg/kg).

Histamine occurs naturally in certain foods [29,32], which explains why it was recorded in practically all samples of spinach, eggplant, and tomato. The variability observed may have been due to botanical variety, as reported by Kumar et al. [33] for eggplant. However, as occurs in foods of animal origin, the presence of high contents of histamine and other amines in plant-origin products could also be associated with microbial activity [29,32,34]. Lavizzari et al. [32] attributed the high contents of histamine in spinach to the activity of contaminating bacteria during storage, belonging mainly to the groups Enterobacteriaceae and Pseudomonadaceae. There is currently a need for more research to understand in more detail the origin of histamine in plant foods such as spinach, eggplant, and tomatoes.

Tyramine has been found in more foods than histamine, although in lower concentrations, in no case exceeding 10 mg/kg. It should be noted that histamine-containing foods also contained tyramine (eggplants, tomatoes, spinach, chard, and asparagus). Although there is very little information about the origin of tyramine in nonfermented vegetables, its presence seems to be associated with microbial aminogenic activity. The ability to form tyramine has been reported for bacteria of the genus *Enterococcus* isolated from plants and fruits, mainly *E. faecium*, *E. mundtii*, and *E. casseliflavus* [35].

Putrescine has been detected in all the studied vegetables and legumes, although its content varied greatly among foods and sometimes also within the same product. In most vegetables and legumes, the average values ranged from 1 to 25 mg/kg. However, some samples of green pepper, eggplant, sweet corn, green and purple beans, spinach, tomato ketchup, soybeans, and peas had strikingly high putrescine contents, in some cases exceeding 200 mg/kg (Table 1). The putrescine found in food can have a dual origin. In plant-origin foods, low contents of this amine generally have a physiological source, as it performs different functions in plants, as do the polyamines spermidine and spermine, ranging from the activation of organogenesis to protection against stress [34,36,37]. On the other hand, the presence of putrescine is also associated with the decarboxylase activity of different groups of spoilage bacteria, mainly Enterobacteriaceae and *Clostridium* spp. [36]. According to Kalač et al. [38], the high amounts of putrescine found in frozen peas are due to bacterial activity in the period between harvesting and freezing or during thawing. However, high putrescine contents cannot always be attributed to bacterial decarboxylase activity. Toro-Funes et al. [39] have suggested that the considerable levels of putrescine found in soybean sprouts arise from the germination process, as this amine is a plant growth factor. In general, based on the available information, and due to the great variability in the reported contents, it is difficult to establish to what degree the presence of putrescine in plant-origin products can be considered physiological or the result of bacterial activity.

Cadaverine, like tyramine, has been described in few vegetables and legumes and in relatively low concentrations, with average values that in no case exceeded 8 mg/kg. The values reported by Nishimura et al. [40] in onion (29 mg/kg) and tofu (18 mg/kg) were an exception.

3.2. Fruits and Nuts

Table 2 shows the content of biogenic amines in different types of fresh fruits, fruit juices, and nuts. There were fewer publications reporting amine data for this type of food than for vegetables and legumes. In general, the contents were low, putrescine being in many cases the only amine found (in addition to the polyamines spermidine and spermine).

Avocado and kiwi, and grapefruit, orange, and pineapple juices, are the only products in this category for which the presence of histamine has been reported, but not in all studies. The 23 mg/kg of histamine in avocado reported by Jarisch et al. [12] stands out, although no relevant information about its possible origin was provided. A study conducted by Preti et al. [41] concluded that the presence of histamine in grapefruit, orange, and pineapple juices is due to a lack of hygienic quality during processing or storage, since this amine is not found in the original fresh fruit.

Similarly, very few fruits contained tyramine, and levels have always been low (Table 2). Avocado and plum stand out for their content of this amine, although in no case has it exceeded 7 mg/kg.

Putrescine has been found in practically all the fruits and nuts, with the highest levels in orange, orange juice, mandarin, grapefruit, grapefruit juice, banana, passion fruit, and pistachio. The range of contents of this amine in citrus fruits and their juices has been very broad, varying from not detected to as high as 200 mg/kg. Suggested explanations for this variability have included different origins, cultivation, and transport and storage conditions [41–44]. As reported by Gonzalez-Aguilar et al. [45], the contents of putrescine in mandarin (flavedo) can be increased by a drop in temperature before harvesting and by damage of mechanical origin. Its presence in most of the samples, unaccompanied by high levels of other amines (related to bacterial activity), seemed to indicate that, with some exceptions, putrescine in fruits has a physiological origin. To confirm this, it would be necessary to carry out more studies analyzing the fruit at the moment of collection. The only fruits reported as having no putrescine were avocado and plum, although interestingly, these did contain histamine and tyramine.

The only fruits with a notable content of cadaverine were bananas and sunflower seeds, for which Nishimura et al. [40] reported average levels of 11 and 22 mg/kg, respectively, although these data were from the analysis of only two samples.

3.3. Cereals and Derivatives

Table 3 shows the contents of biogenic amines in cereals and some derivatives such as breakfast cereals, pasta, and bread. The quantitative information available on amines in cereals is very limited. In principle, these foods do not contain amines other than putrescine, which has a physiological origin [36]. The only standout source of putrescine is wheat germ, which, like soya bean sprouts, has a high rate of cell division, in which putrescine and polyamines play a significant role [36].

Table 1. Biogenic amine contents (mg/kg fresh weight) found in vegetables and legumes. Data are presented as average (standard deviation) and range (minimum–maximum).

Food Categories	n	Occurrence of Biogenic Amines (mg/kg)								Reference
		Histamine		Tyramine		Putrescine		Cadaverine		
		Mean (SD)	Range	Mean (SD)	Range	Mean (SD)	Range	Mean (SD)	Range	
Vegetables and Vegetable Products										
Asparagus (wheat)	5	0.34 (0.62)	nd–1.42	0.69 (0.88)	nd–2.1	13.08 (2.81)	8.58–16.27	0.12 (0.19)	nd–0.43	#
Beans (green)	12	nd	-	2.46 (3.24)	nd–9.86	10.30 (8.61)	2.98–28.81	nd	-	#
Beans (purple)	-	nd	-	nd	-	34.9 (6.2)	-	nd	-	[46]
Beans (yellow)	-	nd	-	nd	-	77.8 (7.7)	-	nd	-	[46]
Broccoli	4	-	-	-	-	14.8 (1.4)	-	-	-	[47]
	5					9	7–10.5			[42]
	10					6.4 (2.9)	3.4–10.8			[48]
	5					5.7	-			[49]
	2					1.9	1.3–2.6			[40]
	3						8.37–20.53	0.1		[47]
Cabbage	-			<0.5		16				[34]
	2					14.18		6.03		[40]
Cabbage (white)	10					6.6				[48]
	8					2.7				[49]
Cauliflower	3	nd		1.05 (0.91)	nd–1.71	3 (1.49)	0.7–6.2	0.16 (0.27)	nd–0.48	[50]
	3					4.9	1.61–4.58			[47]
	7					5.3 (2.1)	3.1–4.5			[42]
	5					3.7	2.2–7.6			[40]
	2					9	3.3–8.9			[34]
Carrot	13	nd		<0.5		2.27 (2.20)	0.35–8.92	<0.5		#
	3					2.8	1.2–1.8	nd		[50]
	4					3.5	2–3.9			[47]
	2					1.5 (0.7)	0.7–2.7			[51]
	6					0.7				[42]
	10					12.1 (3.9)	7–18			[48]
	4					14.8	8–24.7			[52]
	5			1		5				[49]
	-						5.73–14.10			[34]
	2						3.7–7.7			[40]
Celeriac	2					6.1		2.55		[47]
Chard	3	0.79 (0.41)	nd–1.33	1.90 (0.98)	0.74–3.48	6.38 (3.22)	2.4–11.94	0.13 (0.11)	nd–0.24	#
Courgette	8	nd		nd		7.94 (4.12)	2.74–24.81	0.32 (0.7)	nd–2.07	#
	41			2		4				[34]
Cucumber	10	nd		0.61 (0.89)	nd–2.33	5.42 (3.13)	1.32–10.62	nd		[50]
	3					3.2				[42]
	5					6.9 (1.4)	5.5–8.7			[48]
	10					8.7				[48]

Table 1. *Cont.*

Food Categories	n	Histamine		Tyramine		Putrescine		Cadaverine		Reference
		Mean (SD)	Range	Mean (SD)	Range	Mean (SD)	Range	Mean (SD)	Range	
				Vegetables and Vegetable Products						
	5	-	-	-	-	13.1	2.1–28.8	-	-	[49]
	1	-	-	1	-	25	-	-	-	[34]
	2	-	-	-	-	9.87	-	2.96	-	[40]
Eggplant	23	39.42 (30.66)	4.17–100.6	0.60 (0.90)	nd–2.27	34.30 (6.98)	24.10–48.63	nd	-	#
	2	-	-	-	-	18.35	-	-	-	[12]
	4	nd	-	nd	-	2.85 (0.75)	2.20–3.90	5	-	[40]
Lettuce	3	-	-	-	-	3.3–4.8	3.3–4.8	nd	-	#
	3	-	-	-	-	5.6 (1.3)	4.5–7.3	-	-	[50]
	10	-	-	-	-	7.9	-	-	-	[42]
	7	-	-	-	-	20.7	-	-	-	[48]
	2	-	-	-	-	6.87	10.2–42.3	3.27	-	[49]
Mushroom	11	nd	-	nd	-	1.29 (1.20)	0.02–3.65	0.10 (0.4)	nd–1.59	[40]
	10	-	-	-	-	11.7	-	-	-	[48]
	213	-	-	-	-	-	nd–156	-	-	[53]
Onion	4	nd	-	nd	-	nd	-	nd	-	#
	3	-	-	-	-	0.5	5.5–7.2	-	-	[50]
	10	-	-	-	-	0.6	0.2–1	-	-	[48]
	6	-	-	3	-	2	-	-	-	[49]
	-	-	-	-	-	3.96	-	-	-	[34]
Pepper (green)	2	-	-	-	-	90.04 (41.65)	117–148.9	29.32	-	[40]
	9	nd	-	nd	-	-	-	0.05 (0.14)	nd–0.41	#
	2	-	-	-	-	70 (31)	104–237	5.62	-	[40]
	5	-	-	-	-	13.2–96.9	13.2–96.9	-	-	[52]
Pepper (red)	8	nd	-	nd	-	2.42 (2.21)	0.59–5.35	nd	-	#
Potato	10	nd	-	0.58 (0.64)	nd–2.2	4.14 (3.06)	1.05–11.68	0.22 (0.54)	nd–1.75	#
	3	-	-	-	-	9.7	-	-	-	[50]
	3	-	-	-	-	17.6	-	-	-	[51]
	6	-	-	-	-	9.7 (2.1)	5.8–12.8	-	-	[42]
	2	-	-	-	-	-	0.1–22.4	-	-	[40]
	10	-	-	-	-	2.8	-	-	-	[48]
	6	-	-	-	-	7.2	1.1–10.5	-	-	[49]
	-	-	-	5	-	8	-	<0.5	-	[34]
Pumpkin	12	0.28 (0.54)	nd–1.90	nd	-	9.87 (6.19)	2.95–24.23	0.54 (0.76)	nd–2.15	[49]
Spinach	18	31.77 (17.02)	9.46–69.71	2.05 (0.83)	0.785–4.28	4.48 (2.46)	0.14–9.19	nd	-	#
	5	-	-	6	-	4.8	1.8–13.5	1	-	[49]
	-	16	30–60	-	-	6.0	-	-	-	[34]
	-	-	-	-	-	-	-	-	-	[12]
	-	37.5	-	-	-	-	-	-	-	[5]
	2	-	-	-	-	4.41	-	8.48	-	[40]
	-	61 (1.5)	-	nd	-	7.8 (0.1)	-	nd	-	[54]

Table 1. Cont.

Food Categories	n	Occurrence of Biogenic Amines (mg/kg)								Reference
		Histamine		Tyramine		Putrescine		Cadaverine		
		Mean (SD)	Range	Mean (SD)	Range	Mean (SD)	Range	Mean (SD)	Range	
Vegetables and Vegetable Products										
Sweet corn	32		-	-	-	12.9	nd–119	0.18 (0.25)	nd–0.46	[52]
Tomato	5	nd	-	-	-	38.44 (9.50)	30.5–119.2	0.50 (0.48)	nd–2.33	#
	53	2.51 (4.08)	nd–17.07	0.49 (0.92)	nd–6.38	16.48 (6.93)	6.29–35.55	-	-	#
	3	-	-	-	-		9.3–122	-	-	[50]
	2	-	-	-	-	10.6	-	-	-	[51]
	5	-	-	-	-	10	5.3–20.7	-	-	[49]
	-	6	-	<0.5	-	23	-	<0.5	-	[34]
	2	22	-	-	-	23.96	-	1.63	-	[5]
Tomato (concentrated)	19	-	-	-	-	25.9 (8.2)	7.9–41.1	-	-	[40]
Tomato (crushed)	3	1.22 (1.69)	0.24–3.17	0.14 (0.02)	0.12–0.16	9.66 (8.78)	4.5–19.80	0.18 (0.06)	0.12–0.23	#
	-	2	-	2	-	20	nd–4	-	-	[34]
Tomato (ketchup)	3	0.37 (0.64)	nd–1.11	nd	-	1.07 (0.08)	1–1.15	nd	-	#
	24	22	-	-	-	52.5 (54.1)	nd–165	-	-	[38]
										[12]
Legumes and Derivatives										
Beans (white)	6	nd	-	nd	-	0.66 (0.64)	0.35–1.96	nd	-	#
Beans (red kidney)	3	-	-	2	-	3	-	-	-	[34]
	5	-	-	-	-	-	0.3–0.4	-	-	[50]
		-	-	3	-	1	nd–4	-	-	[34]
Chickpeas	4	nd	-	nd	-	3.63 (2.49)	0.90–6.39	nd	-	#
	-	-	-	<0.50	-	2	-	<0.50	-	[34]
Lentils	7	nd	-	nd	-	8.19 (8.36)	1.96–21.81	nd	-	#
	5	-	-	-	-	3	nd–20.2	-	-	[34]
Peanuts	7	nd	-	nd	-	0.87 (1.01)	nd–2.56	nd	-	#
Peas	9	nd	-	nd	-	34.28 (13.50)	8.74–54.44	nd	-	[34]
	10	-	-	-	-	17.3	-	-	-	#
Peas (frozen)	6	-	-	-	-	32.3	5.5–51.1	-	-	[48]
Soybean, dried	14	-	-	-	-	46.3 (27)	11.7–107	-	-	[49]
	3	-	-	-	-	17	1.6–6.5	-	-	[38]
	1	-	-	-	-	41	-	-	-	[50]
	2	-	-	-	-		-	-	-	[47]
	13	-	-	-	-	30.9 (15.5)	3.7–16.8	-	-	[51]
	4	-	-	-	-		16.3–57	-	-	[55]
	2	-	-	-	-		35.2–57.2	-	-	[52]
Soybean milk	5	-	-	-	-	17.1	6.4–24.2	-	-	[40]
	3	nd	-	nd	-	1.02 (0.73)	0.39–1.81	0.28 (0.24)	nd–0.42	[39]
	2	-	-	-	-	2.11	-	13.9	-	[40]

Table 1. *Cont.*

Food Categories	n	Occurrence of Biogenic Amines (mg/kg)								Reference
		Histamine		Tyramine		Putrescine		Cadaverine		
		Mean (SD)	Range	Mean (SD)	Range	Mean (SD)	Range	Mean (SD)	Range	
		Vegetables and Vegetable Products								
Soybean sprouts	3	-	-	-	-	44.71 (3.21)	41.13–47.43	0.21 (0.18)	nd–0.33	[39]
Tofu	6	nd	-	nd	-	0.76 (0.55)	nd–1.49	0.67 (0.49)	nd–1.42	#
	4	-	-	-	-	nd	-	-	-	[49]
	19	-	-	-	-	2.6 (1.4)	nd–5	-	-	[56]
	2	-	-	-	-	1.76	-	18.4	-	[40]

Here, *n*: number of samples; SD: standard deviation; nd: not detected; -: values not reported by the study; #: data on the biogenic amine content from our own database of Spanish market products.

Table 2. Biogenic amine contents (mg/kg fresh weight) found in fruits and nuts. Data are presented as average (standard deviation) and range (minimum–maximum).

Food Categories	n	Occurrence of Biogenic Amines (mg/kg)								Reference
		Histamine		Tyramine		Putrescine		Cadaverine		
		Mean (SD)	Range	Mean (SD)	Range	Mean (SD)	Range	Mean (SD)	Range	
		Fruit and Fruit Products								
Apple	3	-	-	-	-	-	0.4–1.7	-	-	[50]
	2	-	-	-	-	nd	-	-	-	[51]
	2	-	-	-	-	1.5	-	nd	-	[40]
Apple juice	10	nd	-	0.67 (0.50)	nd–1.6	1.02 (0.35)	0.59–1.68	2.30 (1.53)	0.55–4.27	[41]
Avocado	5	nd	-	1.81 (2.06)	0.58–5.44	nd	-	nd	-	#
	2	23	-	-	-	-	-	-	-	[12]
Banana	8	nd	-	0.53 (0.79)	nd–1.85	nd	-	nd	-	[40]
	2	-	-	-	-	37.94 (8.32)	25.50–49.49	nd	-	#
Cherry	5	nd	-	nd	-	-	15.86–41.05	10.83	-	[40]
	2	-	-	-	-	3.08 (0.51)	2.35–3.46	nd	-	#
Grape	2	-	-	-	-	4.67	-	nd	-	[40]
Grapefruit	2	nd	-	nd	-	9.34	-	5.93	-	[40]
	2	-	-	-	-	55.55 (12.8)	46.52–64.57	nd	-	#
	3	-	-	-	-	51.1	-	nd	-	[40]
Grapefruit juice	10	0.31 (0.58)	nd–1.74	nd	-	98.6	-	-	-	[50]
	21	-	-	-	-	10.08 (4.11)	7.17–20.8	1 (0.64)	0.38–2.28	[41]
Guava	13	nd	-	nd	-	1	0.4–1.8	nd	-	[57]
Kiwi	2	-	-	-	-	2.49 (3.96)	0.5–15.57	nd	-	#
	-	-	-	-	-	1.06	-	nd	-	[40]
Lemon	3	1.9 (0.1)	-	nd	-	3.1 (0.1)	nd–3.67	nd	-	[54]
Mandarin	21	nd	-	nd	-	2.33 (2.02)	12.29–173.8	nd	-	#
	10	nd	-	0.94 (1.31)	nd–5.76	90.16 (36.6)	67.3–200	-	-	[42]
						122 (44.2)				

Table 2. *Cont.*

Food Categories	n	Occurrence of Biogenic Amines (mg/kg)								Reference
		Histamine		Tyramine		Putrescine		Cadaverine		
		Mean (SD)	Range	Mean (SD)	Range	Mean (SD)	Range	Mean (SD)	Range	
Fruit and Fruit Products										
Mango	21	-	-	-	-	0.9	nd–2.7	-	-	[57]
Orange	12	nd	-	nd	-	91.24 (41.7)	11.34–151.1	nd	-	#
	3	-	-	-	-	117	95.1–140	-	-	[50]
	2	-	-	-	-	137 (11.3)	119–153	-	-	[51]
	5	-	-	-	-	-	54.62–119.82	2.04	-	[42]
	2	-	-	-	-	-	-	nd	-	[40]
Orange juice	3	nd	-	nd	-	45.51 (10.5)	37.35–57.3	-	-	#
	3	-	-	-	-	85 (11.4)	76.6–100	nd	-	[42]
Papaya	11	0.46 (0.41)	nd–1.32	nd	-	45.51 (8.35)	34.70–60.97	-	-	[41]
	21	-	-	-	-	11	5.3–19.3	nd	-	[57]
Passion fruit	2	-	-	-	-	4.67	-	-	-	[57]
Pear	3	-	-	-	-	17.9	6.5–40.5	nd	-	[50]
	2	-	-	-	-	1.5	23.6–24.2	0.41	-	[40]
Peach	2	nd	-	nd	-	1.92 (0.14)	1.82–2.02	nd	-	#
	2	-	-	-	-	0.35	-	<0.10	-	[40]
Pineapple	6	nd	-	nd	-	4.20 (2.17)	1.39–7.96	nd	-	#
	2	-	-	-	-	4.05	-	3.07	-	[40]
	21	-	-	-	-	1.1	nd–2.5	-	-	[57]
Pineapple juice	12	2.44 (1.59)	nd–4.61	0.87 (0.86)	nd–1.93	1.79 (0.16)	1.53–1.98	1.21 (1.22)	nd–3.14	[41]
Plum	2	nd	-	4.02 (4.32)	0.96–7.07	nd	-	nd	-	#
Strawberry	9	nd	-	nd	-	3.77 (1.52)	2.04–6.42	nd	-	#
	2	-	-	-	-	0.97	-	4.29	-	[40]
Nuts										
Almonds	7	nd	-	nd	-	2.47 (1.24)	nd–4.36	nd	-	#
	2	-	-	-	-	4.32	-	5.57	-	[40]
Chestnuts	2	nd	-	nd	-	4.53 (3.40)	2.12–6.93	nd	-	#
	2	-	-	-	-	5.2	-	1.33	-	[40]
Hazelnuts	9	nd	-	0.49 (0.85)	nd–2.63	1.18 (1.09)	nd–3.19	nd	-	#
Nuts	6	nd	-	nd	-	5.64 (4.17)	2.82–13.79	nd	-	#
Pistachios	7	nd	-	nd	-	14.84 (14.0)	4.31–39.51	1.65 (4.37)	nd–11.58	#
	2	-	-	-	-	43	-	3.27	-	[40]
Sunflower seeds	2	nd	-	nd	-	0.50 (0.19)	0.36–0.63	nd	-	#
	2	-	-	-	-	3	-	22.58	-	[40]

Here, *n*: number of samples; SD: standard deviation; nd: not detected; -: values not reported by the study; #: data on the biogenic amine content from our own database of Spanish market products.

Table 3. Biogenic amine contents (mg/kg fresh weight) found in cereals and cereal-based products. Data are presented as average (standard deviation) and range (minimum–maximum).

Food Categories	n	Occurrence of Biogenic Amines (mg/kg)									Reference
		Histamine		Tyramine		Putrescine		Cadaverine			
		Mean (SD)	Range	Mean (SD)	Range	Mean (SD)	Range	Mean (SD)	Range		
Barley	2	nd	-	nd	-	2.19 (1.55)	1.09–3.28	nd	-		#
Bread, white	2	nd	-	nd	-	nd	-	nd	-		#
	3	-	-	-	-	-	1.5–1.8	-	-		[50]
	10	-	-	-	-	1.1	-	-	-		[48]
	2	-	-	-	-	1.32	-	2.35	-		[40]
Bread, wholemeal	6	nd	-	nd	-	1.96 (1.45)	nd–4.32	nd	-		#
Cereal (corn, chocolate)	8	nd	-	nd	-	0.32 (0.44)	nd–0.93	nd	-		#
	10	-	-	-	-	-	2–2.2	-	-		[50]
Oats	2	nd	-	nd	-	0.67 (0.32)	0.44–0.89	nd	-		#
Pasta (wheat)	7	nd	-	nd	-	1.56 (1.65)	0.81–4.52	nd	-		#
Rice	2	nd	-	nd	-	2.4 (0.03)	2.38–2.42	nd	-		#
	2	-	-	-	-	<0.9	-	-	-		[51]
	2	-	-	-	-	1.2	-	-	-		[40]
	6	-	-	-	-	0.2	0.2–0.3	-	-		[49]
	10	-	-	-	-	0.2	-	-	-		[48]
Wheat germ	2	nd	-	nd	-	31.64 (0.35)	31.39–31.9	0.63 (0.08)	0.57–0.69		#
	2	-	-	-	-	62.1	-	-	-		[40]

Here, *n*: number of samples; SD: standard deviation; nd: not detected; -: values not reported by the study; #: data on the biogenic amine content from our own database of Spanish market products.

Putrescine contents in wholemeal bread were slightly higher than in bread made with refined flour. In white bread, low contents of cadaverine have also been reported, although only in one study, and from the analysis of two samples.

4. Evolution of Amine Contents during Storage and Cooking

The variability of amine contents observed among samples of the same product can be attributed mainly to conditions of production, transport, and storage [42].

The storage temperature is one of the most important factors in the formation of biogenic amines [11,29]. Refrigeration delays or reduces the aminogenic potential of microorganisms, although the formation of amines at refrigeration temperatures (4–10 °C) has been reported. The influence of the conservation temperature has been widely studied in foods such as meats, fish, and fermented products [29,30,58], but scarcely in plant-origin foods.

A study conducted by Simon-Sarkadi et al. [59] showed a clear increase in putrescine in different types of green leafy vegetables (lettuce, endives, Chinese cabbage, and radicchio) during six days of storage at 5 °C. The authors concluded that there was a positive correlation between putrescine contents and the hygienic state of these foods (total microorganism counts). Tyramine contents also showed a tendency to increase slightly. Histamine was present only in Chinese cabbage and in very low concentrations, remaining stable throughout the study period. In contrast, when Moret et al. [34] studied the effect of storage temperature on the amine content in various vegetables (parsley, zucchini, broccoli, and cucumber), no significant changes in histamine, tyramine, putrescine, and cadaverine were observed after three weeks of refrigeration.

Lavizzari et al. [32] also reported an increase in histamine in different spinach samples over 12–15 days of storage at 6 °C, noting that the relatively high pH of this vegetable favored the growth of Gram-negative bacteria, which could have been responsible for the formation of this amine during storage. The contents of tyramine and putrescine did not undergo significant changes under these storage conditions. It should be noted that in two of the five trials carried out in this study, histamine levels decreased in the last days of storage. The authors suggested that this histamine degradation could have been due to the action of bacteria with DAO activity, as well as the effect of the pH, which reached values above 8 [32]. Another study also recently reported the complete degradation of histamine in a spinach sample (61 mg/kg) after three weeks of storage at 4 °C [54].

Modified atmosphere packaging, together with low storage temperatures, is commonly used to extend the life of fresh vegetables and fruits. This type of packaging can influence the capacity of microorganisms to form amines [30,58,60]. Esti et al. [43] monitored the contents of amines during the ripening of cherries and apricots packaged in modified atmospheres and stored at 0 °C, and found that after 20 days of storage the contents of amines (mainly putrescine) had decreased by 20% compared to the initial value. Although the authors did not provide an explanation for this reduction, it could have been due to putrescine serving as a substrate for polyamine formation [36].

Another factor that can affect the content of biogenic amines in foods of plant origin, especially vegetables, is the culinary process. Again, the results reported in the literature were variable, depending on the type of cooking and the amine in question.

Latorre-Moratalla et al. [61] evaluated the effect of cooking spinach in water, with or without salt. The cooking process reduced the histamine content in all the samples by an average of 83% with respect to the raw product (after a correction for the dilution effect of the cooking). Analysis confirmed a transfer of histamine to the cooking water, which was not enhanced by the addition of salt. Likewise, Kumar et al. [33] observed the loss of 11–14% histamine in eggplants boiled at 100 °C for 10 min. Veciana et al. [62] also concluded that the putrescine content in certain vegetables (spinach, cauliflower, Swiss chard, potato, and green beans) is reduced by transfer to the cooking water. However, this heat treatment had no effect on the putrescine content in other vegetables such as pepper, pea, and asparagus. Eliassen et al. [42] also found no significant differences in putrescine levels among different types of raw and boiled vegetables (carrot, broccoli, cauliflower, and potato), although they acknowledged that the low number of samples analyzed (two per food) was a limitation when trying to reach a conclusion.

Conversely, three recent studies have shown an increase in amine levels after a cooking process. According to Lo Scalzo et al. [63], boiling and grilling enhanced the putrescine content in a specific variety of eggplant by 55% and 32%, respectively. In the other two varieties of eggplant tested, the cooking had no effect. Similarly, Preti et al. [46] reported a significant increase in putrescine in green beans after boiling, whereas steaming did not modify the contents. According to the work performed by Chung et al. [64], frying brought about a 2.5- and 4-fold increase in histamine in carrots and seaweed, respectively. The authors attributed this increase to the loss of water caused by the high heat treatment. The same process had no effect on spinach and onions. However, it should be noted that in this study, the contents of histamine in all foods were well below 1 mg/kg, both before and after frying.

Amines are thermostable compounds, so in principle changes in contents can only be due to their transfer to the cooking water or by dilution or concentration effects of the culinary process, in which the food gains or loses water.

5. Plant-Origin Foods in Low-Histamine Diets

At present, the main strategy to prevent the onset of histamine intolerance symptoms is to follow a low-histamine diet. Its efficacy has been demonstrated in different clinical studies, which have always described an improvement or remission of gastrointestinal, dermatological, and neurological symptoms [3,6,18–20,22,24,27,65–67] if the diet was followed.

Current low-histamine diets exclude foods that patients associate with the onset of symptoms [1–8], such as blue fish and their preserves, and all kinds of fermented products (cheeses, sausages, wine, beer, sauerkraut, and fermented soy derivatives), all of which are susceptible to having high contents of histamine and other amines. A high number of nonfermented plant-origin foods are also excluded: The average contents of biogenic amines and polyamines in these foods are shown in Table 4. As can be seen, with the exception of spinach, eggplant, tomatoes, and avocado, for which high amounts of histamine have been described, the rest contained very little or no histamine, so a priori should not be responsible for triggering symptoms. However, some of them had relatively high contents of other biogenic amines and polyamines.

Table 4. Content of histamine and other biogenic amines (mg/kg fresh weight) in plant-origin foods excluded from different low-histamine diets [1–8]. Data obtained from own database and from different scientific studies [5,12,34,38–42,47–55,57].

Food Items	Histamine	Putrescine	Cadaverine	Tyramine	Spermidine	Spermine
Spinach [a]	9–70	nd–119	nd–9	1–10	14–53	nd–9
Eggplant [a]	4–101	24–49	nd–5	nd–2	2–12	nd–6
Tomato [a]	nd–17	5–122	nd–2	nd–6	2–16	nd–2
Ketchup [a]	nd–22	nd–165	nd	nd	nd–33	nd–12
Avocado [a]	nd–23	nd	nd	0.5–5	nd–7	2–8
Citrus (fresh and juices) [b]	nd–2	7–200	nd–2	nd–5	nd–12	nd–5
Mushroom [b]	nd	nd–156	nd	nd	9–155	nd–13
Banana [b]	nd	15–50	nd–10	nd–2	8–16	nd–3
Soybean or soybean sprouts [b]	nd	2–57	nd–0.3	nd	33–389	7–114
Nuts [b]	nd	nd–40	nd–23	nd–3	6–40	2–33
Pears [b]	-	2–25	nd–0.4	-	30–76	8–49
Lentils [b]	nd	nd–21	nd	nd	15–107	5–18
Chickpeas [b]	nd	1–6	nd–0.5	nd–0.5	15–85	4–32
Peanuts [c]	nd	nd–3	nd	nd	23–48	5–13
Kiwi [c]	nd–2	nd–15	nd	nd	3–6	nd–2
Papaya [c]	-	5–20	nd	-	4–8	nd–2
Strawberry [c]	nd	2–6	nd–4	nd	5–10	nd–2
Pineapple [c]	nd	nd–8	nd–3	nd	nd–3	nd–1
Plum [c]	nd	nd	nd	1–7	2–3	nd–4

Here, nd: not detected; -: values not reported by the studies; [a] plant-origin foods with histamine; [b] plant-origin foods without histamine but with high contents of other amines; [c] plant-origin foods with low levels of all amines.

Putrescine, cadaverine, and tyramine are all substrates of the DAO enzyme, so if present in high amounts they may increase the adverse effects of histamine by competing as rival substrates or for binding sites in the intestinal mucosa [1,9,68,69]. The high putrescine contents found in citrus fruits, mushrooms, soybeans, bananas, and nuts could thus explain why patients associate their consumption with the onset of histamine intolerance symptoms. However, it should be noted that some foods with similar or even much higher putrescine contents, such as green pepper, peas, or corn, are permitted in low-histamine diets (Table 1).

The polyamines spermidine and spermine can also be metabolized by DAO, albeit with a lower affinity [68,69], and therefore their presence should also be taken into account in this type of diet (Table 4). Thus, the exclusion of foods such as soybeans, mushrooms, lentils, chickpeas, peanuts, nuts, and pears may be justified by their high polyamine content.

Finally, the levels of biogenic amines and polyamines found in kiwi, papaya, strawberry, pineapple, and plum are too low to justify their exclusion. Some authors consider these foods, along with others such as milk, shellfish, and eggs, as endogenous histamine releasers, although by mechanisms still not well understood [1,11,70].

6. Conclusions

Biogenic amine data in nonfermented plant-origin foods from the different reviewed studies showed a great variability both within the same food item and among them. Putrescine was the most frequent biogenic amine found in fresh vegetables, legumes, fruits, and cereals, and only a limited number of products contained relevant levels of histamine (eggplant, spinach, tomato, and avocado). Tyramine and cadaverine were usually more scarcely found in plant-origin foods. Generally, low levels of histamine and putrescine may have a physiological origin. However, undesirable microbial enzymatic activity during production or storage may lead to the accumulation of high levels of these amines.

No single trend has emerged in the evolution of amine contents during refrigerated storage, which might be at least partly due to the different experimental designs of the studies. In some cases, refrigeration seems to have prevented the formation of certain amines, but this remains a hypothesis, as no study performed a comparative analysis of samples stored under refrigeration and at room temperature. The increase in the biogenic amine content during refrigerated storage reported by other authors may be attributed to bacterial activity. Additionally, some studies have observed an influence of culinary process on the biogenic amine content, mainly derived from the transfer of these compounds to the boiling water or by dilution or concentration effects of the applied treatment.

The exclusion of a high number of plant-origin foods from low-histamine diets cannot be accounted for by their histamine contents, but is more likely due to high levels of putrescine or spermidine. The plant-origin foods consumed by people with histamine intolerance should be of maximum freshness, since histamine and other amines may continue to form during refrigerated storage. The cooking of vegetables in water (boiling) is another relevant strategy for this population, since it can reduce the contents of histamine and other amines in the food.

Author Contributions: Conceptualization, M.T.V.-N., M.L.L.-M., and M.C.V.-C.; Investigation, S.S.-P., O.C.-B., and J.R.-G.; Writing—Original draft preparation, S.S.-P., O.C.-B., and M.L.L.-M.; Writing—Review and editing, M.T.V.-N., M.L.L.-M., and M.C.V.-C.; Supervision, M.C.V.-C.

Funding: This research received no external funding.

Acknowledgments: Sònia Sánchez-Pérez is a recipient of a doctoral fellowship from the University of Barcelona (APIF2018).

Conflicts of Interest: The authors declare no conflicts of interest.

References

1. Maintz, L.; Novak, N. Histamine and histamine intolerance. *Am. J. Clin. Nutr.* **2007**, *85*, 1185–1196. [CrossRef] [PubMed]
2. Böhn, L.; Störsrud, S.; Törnblom, H.; Bengtsson, U.; Simrén, M. Self-reported food-related gastrointestinal symptoms in IBS are common and associated with more severe symptoms and reduced quality of life. *Am. J. Gastroenterol.* **2013**, *108*, 634–641. [CrossRef] [PubMed]
3. Rosell-Camps, A.; Zibetti, S.; Pérez-Esteban, G.; Vila-Vidal, M.; Ferrés Ramis, L.; García-Teresa-García, E. Intolerancia a la histamina como causa de síntomas digestivos crónicos en pacientes pediátricos. *Rev. Esp. Enferm. Dig.* **2013**, *105*, 201–207. [CrossRef] [PubMed]
4. Veciana-Nogués, M.T.; Vidal-Carou, M.C. Dieta baja en histamina. In *Nutrición y dietética clínica*, 3rd ed.; Salas-Salvadó, J., Bonada-Sanjaume, A., Trallero-Casaña, R., Saló-Solà, M., Burgos-Peláez, R., Eds.; Ediciones Elsevier España: Barcelona, Spain, 2014; pp. 443–448.
5. Lefèvre, S.; Astier, C.; Kanny, G. Histamine intolerance or false food allergy with histamine mechanism. *Rev. Fr. Allergol.* **2016**, *57*, 24–34. [CrossRef]
6. Wagner, N.; Dirk, D.; Peveling-Oberhag, A.; Reese, I.; Rady-Pizarro, U.; Mitzel, H.; Staubach, P.A. Popular myth—Low-histamine diet improves chronic spontaneous urticaria—Fact or fiction? *J. Eur. Acad. Dermatol. Venereol.* **2017**, *31*, 650–655. [CrossRef] [PubMed]
7. Ede, G. Histamine intolerance: Why freshness matters? *J. Evol. Health* **2017**, *2*, 11. [CrossRef]
8. Swiss Interest Group Histamine Intolerance (SIGHI)—Leaflet Histamine Elimination Diet. Available online: http://www.histaminintoleranz.ch/downloads/SIGHI-Leaflet_HistamineEliminationDiet.pdf (accessed on 25 October 2018).
9. EFSA Panel on Biological Hazards (BIOHAZ). Scientific opinion on risk based control of biogenic amines formation in fermented foods. *EFSA J.* **2011**, *9*, 2393. [CrossRef]
10. World Health Organization and Food and Agriculture Organization of the United Nations. *Joint FAO/WHO Expert Meeting on the Public Health Risks of Histamine and other Biogenic Amines from Fish and Fishery Products: Meeting Report*; World Health Organization: Geneva, Switzerland, 2013.
11. Kovacova-Hanuskova, E.; Buday, T.; Gavliakova, S.; Plevkova, J. Histamine, histamine intoxication and intolerance. *Allergol. Immunopathol.* **2015**, *43*, 498–506. [CrossRef]
12. Jarisch, R.; Wantke, F.; Raithel, M.; Hemmer, W. Histamine and biogenic amines. In *Histamine Intolerance. Histamine and Seasickness*; Jarisch, R., Ed.; Springer: Stuttgart, Germany, 2014; pp. 3–44.
13. Comas-Basté, O.; Latorre-Moratalla, M.L.; Bernacchia, R.; Veciana-Nogués, M.T.; Vidal-Carou, M.C. New approach for the diagnosis of histamine intolerance based on the determination of histamine and methylhistamine in urine. *J. Pharm. Biomed. Anal.* **2017**, *14*, 379–385. [CrossRef]
14. Maintz, L.; Yu, C.F.; Rodríguez, E.; Baurecht, H.; Bieber, T.; Illig, T.; Weidinger, S.; Novak, N. Association of single nucleotide polymorphisms in the diamine oxidase gene with diamine oxidase serum activities. *Allergy* **2011**, *66*, 893–902. [CrossRef]
15. García-Martín, E.; Martínez, C.; Serrador, M.; Alonso-Navarro, H.; Ayuso, P.; Navacerrada, F.; Agúndez, J.A.G.; Jiménez-Jiménez, F.J. Diamine oxidase rs10156191 and rs2052129 variants are associated with the risk for migraine. *Headache* **2015**, *55*, 276–286. [CrossRef] [PubMed]
16. Meza-Velázquez, R.; López-Márquez, F.; Espinosa-Padilla, S.; Rivera-Guillen, M.; Ávila-Hernández, J.; Rosales-González, M. Association of diamine oxidase and histamine N-methyltransferase polymorphisms with presence of migraine in a group of Mexican mothers of children with allergies. *Neulología* **2017**, *32*, 500–507. [CrossRef]
17. Sattler, J.; Häfner, D.; Klotter, H.J.; Lorenz, W.; Wagner, P.K. Food-induced histaminosis as an epidemiological problem: Plasma histamine elevation and haemodynamic alterations after oral histamine administration and blockade of diamine oxidase (DAO). *Agent Action* **1988**, *23*, 361–365. [CrossRef]
18. Steinbrecher, I.; Jarisch, R. Histamine and headache. *Allergologie* **2005**, *28*, 85–91. [CrossRef]
19. Maintz, L.; Benfadal, S.; Allam, J.P.; Hagemann, T.; Fimmers, R.; Novak, N. Evidence for a reduced histamine degradation capacity in a subgroup of patients with atopic eczema. *J. Allergy Clin. Immunol.* **2006**, *117*, 1106–1112. [CrossRef] [PubMed]

20. Worm, M.; Fielder, E.; Döle, A.S.; Schink, T.; Hemmer, W.; Jarisch, R.; Suberbier, T. Exogenous histamine aggravates eczema in a subgroup of patients with atopic dermatitis. *Acta Derm. Venereol.* **2009**, *89*, 52–56. [CrossRef]

21. Honzawa, Y.; Nakase, H.; Matsuura, M.; Chiba, T. Clinical significance of serum diamine oxidase activity in inflammatory bowel disease: Importance of evaluation of small intestinal permeability. *Inflamm. Bowel Dis.* **2011**, *17*, E23–E25. [CrossRef] [PubMed]

22. Mušič, E.; Korošec, P.; Šilar, M.; Adamič, K.; Košnik, M.; Rijavec, M. Serum diamine oxidase activity as a diagnostic test for histamine intolerance. *Wien. Klin. Wochenschr.* **2013**, *12*, 239–243. [CrossRef]

23. Manzotti, G.; Breda, D.; Gioacchino, M.; Burastero, S.E. Serum diamine oxidase activity in patients with histamine intolerance. *Int. J. Immunopath. Pharmacol.* **2016**, *29*, 105–111. [CrossRef]

24. Hoffmann, M.; Gruber, E.; Deutschmann, A.; Jahnel, J.; Hauer, A. Histamine intolerance in children with chronic abdominal pain. *Arch. Dis. Child.* **2016**, *98*, 832–833. [CrossRef]

25. Pinzer, T.C.; Tietz, E.; Waldmann, E.; Schink, M.; Neurath, M.F.; Zopf, Y. Circadian profiling reveals higher histamine plasma levels and lower diamine oxidase serum activities in 24% of patients with suspected histamine intolerance compared to food allergy and controls. *Allergy* **2018**, *73*, 949–957. [CrossRef] [PubMed]

26. Kacik, J.; Wróblewska, B.; Lewicki, S.; Zdanowski, R.; Kalicki, B. Serum diamine oxidase in pseudoallergy in the pediatric population. *Adv. Exp. Med. Biol.* **2018**, *1039*, 35–44. [CrossRef] [PubMed]

27. Son, J.H.; Chung, B.Y.; Kim, H.O.; Park, C.W. A histamine-free diet is helpful for treatment of adult patients with chronic spontaneous urticaria. *Ann. Dermatol.* **2018**, *30*, 164–172. [CrossRef] [PubMed]

28. Izquierdo-Casas, J.; Comas-Basté, O.; Latorre-Moratalla, M.L.; Lorente-Gascón, M.; Duelo, A.; Vidal-Carou, M.C.; Soler-Singla, L. Low serum diamine oxidase (DAO) activity levels in patients with migraine. *J. Physiol. Biochem.* **2018**, *74*, 93–99. [CrossRef] [PubMed]

29. Bover-Cid, S.; Latorre-Moratalla, M.L.; Veciana-Nogués, M.T.; Vidal-Carou, M.C. Processing contaminants: Biogenic amines. In *Encyclopedia of Food Safety*; Motarjemi, Y., Moy, G.G., Todd, E.C.D., Eds.; Elsevier Inc.: Burlington, MA, USA, 2014; Volume 2, pp. 381–391.

30. Gardini, F.; Özogul, Y.; Suzzi, G.; Tabanelli, G.; Özogul, F. Technological factors affecting biogenic amine content in foods: A review. *Front. Microbiol.* **2016**, *7*, 1218. [CrossRef] [PubMed]

31. Vidal-Carou, M.C.; Veciana-Nogués, M.T.; Latorre-Moratalla, M.L.; Bover-Cid, S. Biogenic amines: Risks and control. In *Handbook of Fermented Meat and Poultry*, 2nd ed.; Toldrá, F., Hui, Y.H., Astiasarán, I., Sebranek, J.G., Talon, R., Eds.; Wiley-Blackwell: Hoboken, NJ, USA, 2014; pp. 413–428.

32. Lavizzari, T.; Veciana-Nogués, M.T.; Weingart, O.; Bover-Cid, S.; Mariné-Font, A.; Vidal-Carou, M.C. Occurrence of biogenic amines and polyamines in spinach and changes during storage under refrigeration. *J. Agric. Food Chem.* **2007**, *55*, 9514–9519. [CrossRef] [PubMed]

33. Kumar, M.N.K.; Bbu, B.N.H.; Venkayesh, Y.P. Higher histamine sensitivity in non-atopic subjects by skin prick test may result in misdiagnosis of eggplant allergy. *Immunol. Investig.* **2009**, *38*, 93–103. [CrossRef]

34. Moret, S.; Smela, D.; Populin, T.; Conte, L. A survey on free biogenic amine content of fresh and preserved vegetables. *Food Chem.* **2005**, *89*, 355–361. [CrossRef]

35. Trivedi, K.; Borkovcová, I.; Karpíšková, R. Tyramine production by enterococci from various foodstuffs: A threat to the consumers. *Czech J. Food Sci.* **2009**, *27*, S357–S360. [CrossRef]

36. Kalač, P. Health effects and occurrence of dietary polyamines: A review for the period 2005–mid 2013. *Food Chem.* **2014**, *161*, 27–39. [CrossRef]

37. Bouchereau, A.; Aziz, A.; Larher, F.; Martin-Tanguy, J. Polyamines and environmental challenges: Recent development. *Plant Sci.* **1999**, *140*, 103–125. [CrossRef]

38. Kalač, P.; Svecová, S.; Pelikánová, T. Levels of biogenic amines in typical vegetable products. *Food Chem.* **2002**, *77*, 349–351. [CrossRef]

39. Toro-Funes, N.; Bosch-Fuste, J.; Latorre-Moratalla, M.L.; Veciana-Nogués, M.T.; Vidal-Carou, M.C. Biologically active amines in fermented and non-fermented commercial soybean products from the Spanish market. *Food Chem.* **2015**, *173*, 1119–1124. [CrossRef] [PubMed]

40. Nishimura, K.; Shiina, R.; Kashiwagi, K.; Igarashi, K. Decrease in polyamines with aging and their ingestion from food and drink. *J. Biochem.* **2006**, *139*, 81–90. [CrossRef] [PubMed]

41. Preti, R.; Bernacchia, R.; Vinci, G. Chemometric evaluation of biogenic amines in commercial fruit juices. *Eur. Food Res. Technol.* **2016**, *242*, 2031–2039. [CrossRef]

42. Eliassen, K.A.; Reistad, R.; Risøen, U.; Rønning, H.F. Dietary polyamines. *Food Chem.* **2002**, *78*, 273–280. [CrossRef]

43. Esti, M.; Volpe, G.; Masignan, D.; Compagnone, E.; La Notte, E.; Palleschi, G. Determination of amines in fresh and modified atmosphere packaged fruits using electrochemical biosensors. *J. Agric. Food Chem.* **1998**, *46*, 4233–4237. [CrossRef]

44. Kalač, P.; Křížek, M.; Pelikánová, T.; Langová, M.; Veškrna, O. Contents of polyamines in selected foods. *Food Chem.* **2005**, *90*, 561–564. [CrossRef]

45. González-Aguilar, G.A.; Zacarias, L.; Perez-Amador, M.A.; Carbonell, J.; Lafuente, M.T. Polyamine content and chilling susceptibility are affected by seasonal changes in temperature and by conditioning temperature in cold-stored "Fortune" mandarin fruit. *Physiol. Plant* **2000**, *108*, 140–146. [CrossRef]

46. Preti, R.; Rapa, M.; Vinci, G. Effect of Steaming and boiling on the antioxidant properties and biogenic amines content in Green Bean (*Phaseolus vulgaris*) varieties of different colours. *J. Food Quality* **2017**. [CrossRef]

47. Ziegler, W.; Hahn, M.; Wallnöfer, P.R. Changes in biogenic amine contents during processing of several plant foods. *Deut. Lebensm. Rundsch.* **1994**, *90*, 108–112.

48. Cipolla, B.G.; Havouis, R.; Moulinoux, J.P. Polyamine contents in current foods: A basis for polyamine reduced diet and a study of its long-term observance and tolerance in prostate carcinoma patients. *Amino Acids* **2007**, *33*, 203–212. [CrossRef] [PubMed]

49. Nishibori, N.; Fujihara, S.; Akatuki, T. Amounts of polyamines in foods in Japan and intake by Japanese. *Food Chem.* **2007**, *100*, 491–497. [CrossRef]

50. Bardócz, S.; Grant, G.; Brown, D.S.; Ralph, A.; Pusztai, A. Polyamines in food—Implications for growth and health. *J. Nutr. Biochem.* **1993**, *4*, 66–71. [CrossRef]

51. Okamoto, A.; Sugi, E.; Koizumi, Y.; Yanadiga, F.; Udaka, S. Polyamine content of ordinary foodstuffs and various fermented foods. *Biosci. Biotechnol. Biochem.* **1997**, *61*, 1582–1584. [CrossRef] [PubMed]

52. Kalač, P.; Krausová, P. A review of dietary polyamines: Formation, implications for growth and health and occurrence in foods. *Food Chem.* **2005**, *90*, 219–230. [CrossRef]

53. Dadáková, E.; Pelikánová, T.; Kalač, P. Content of biogenic amines and polyamines in some species of European wild-growing edible mushrooms. *Eur. Food Res. Technol.* **2009**, *230*, 163–171. [CrossRef]

54. Dionex. Determination of biogenic amines in fruit, vegetables, and chocolate using ion chromatography with suppressed, conductivity and integrated pulsed amperometric detections. *Appl. Update* **2016**, *162*, 1–8.

55. Glória, M.B.A.; Tavares-Neto, J.; Labanca, R.A.; Carvalho, M.S. Influence of cultivar and germination on bioactive amines in soybeans (*Glycine max* L. Merril). *J. Agric. Food Chem.* **2005**, *53*, 7480–7485. [CrossRef]

56. Byun, B.Y.; Bai, X.; Mah, J.H. Occurrence of biogenic amines in Doubanjiang and Tofu. *Food Sci. Biotech.* **2013**, *22*, 55–62. [CrossRef]

57. Santiago-Silva, P.; Labanca, R.; Gloria, B. Functional potential of tropical fruits with respect to free bioactive amines. *Food Res. Int.* **2011**, *44*, 1264–1268. [CrossRef]

58. Naila, A.; Flint, S.; Fletcher, G.; Bremer, P.; Meerdink, G. Control of biogenic amines in food-existing and emerging approaches. *J. Food Sci.* **2010**, *75*, 139–150. [CrossRef] [PubMed]

59. Simon-Sarkadi, L.; Holazapfel, W.H.; Halasz, A. Biogenic amine content and microbial contamination of leafy vegetables during storage at 5C. *J. Food Biochem.* **1994**, *17*, 407–418. [CrossRef]

60. Chong, C.Y.; Abu Bakar, F.; Russly, A.R.; Jamilah, B.; Mahyudin, N.A. The effects of food processing on biogenic amines formation. *Int. Food Res. J.* **2011**, *18*, 867–876.

61. Latorre-Moratalla, M.L.; Comas-Basté, O.; Veciana-Nogués, M.T.; Vidal-Carou, M.C. La cocción reduce el contenido de histamina de las espinacas. In *11a Reunión anual de la Sociedad Española de Seguridad Alimentaria*; Sociedad Española de Seguridad Alimentaria: Pamplona, Spain, 2015.

62. Veciana-Nogués, M.T.; Latorre-Moratalla, M.L.; Toro-Funes, N.; Bosh-Fusté, J.; Vidal-Carou, M.C. Efecto de la cocción con y sin sal en el contenido de poliaminas de las verduras. *Nutr. Hosp.* **2014**, *30*, 53.

63. Lo Scalzo, R.; Fibiani, M.; Francese, G.; D'Alessandro, A.; Rotino, G.L.; Conte, P.; Mennella, G. Cooking influence on physico-chemical fruit characteristics of eggplant (*Solanum melongena* L.). *Food Chem.* **2016**, *194*, 835–842. [CrossRef] [PubMed]

64. Chung, B.Y.; Park, S.Y.; Byun, Y.S.; Son, J.H.; Choi, Y.W.; Cho, Y.S.; Kim, H.O.; Park, C.W. Effect of different cooking methods on histamine levels in selected foods. *Ann. Dermatol.* **2017**, *29*, 706–714. [CrossRef]

65. Wantke, F.; Gotz, M.; Jarisch, R. Histamine-free diet: Treatment of choice for histamine-induced food intolerance and supporting treatment for chronic headaches. *Clin. Exp. Allergy* **1993**, *23*, 982–985. [CrossRef]

66. Guida, B.; De Martino, C.; De Martino, S.; Tritto, G.; Patella, V.; Trio, R.; D'Agostino, C.; Pecoraro, P.; D'Agostino, L. Histamine plasma levels and elimination diet in chronic idiopathic urticaria. *Eur. J. Clin. Nutr.* **2000**, *54*, 155–158. [CrossRef]

67. Siebenhaar, L.; Melde, A.; Magerl, T.; Zuberier, T.; Church, M.K.; Maurer, M. Histamine intolerance in patients with chronic spontaneous urticaria. *J. Eur. Acad. Dermatol. Venereol.* **2016**, *30*, 1774–1777. [CrossRef]

68. Schwelberger, H.G.; Bodner, E. Purification and characterization of diamine oxidase from porcine kidney and intestine. *Biochim. Biophys. Acta* **1997**, *1340*, 152–164. [CrossRef]

69. Finney, J.; Moon, H.J.; Ronnebaum, T.; Lantz, M.; Mure, M. Human copper-dependent amine oxidases. *Arch Biochem. Biophys.* **2014**, *546*, 19–32. [CrossRef] [PubMed]

70. Vlieg-Boerstra, B.J.; Van der Heide, S.; Oude, J.N.G.; Kluin-Nelemans, J.C.; Dubois, A.E. Mastocytosis and adverse reactions to biogènic amines and histamine-releasing foods. What is the evidence? *Neth. J. Med.* **2005**, *63*, 244–249. [PubMed]

MDPI

St. Alban-Anlage 66

4052 Basel

Switzerland

Tel. +41 61 683 77 34

Fax +41 61 302 89 18

www.mdpi.com

Foods Editorial Office

E-mail: foods@mdpi.com

www.mdpi.com/journal/foods

www.ingramcontent.com/pod-product-compliance
Lightning Source LLC
Chambersburg PA
CBHW051849210326
41597CB00033B/5836